Tony L. Schmitz · Kevin S. Smith

Machining Dynamics

Frequency Response to Improved Productivity

Springer

Tony L. Schmitz
University of Florida
Mechanical and Aerospace Engineering
Gainesville, FL 32611-6250
tschmitz@ufl.edu

Kevin S. Smith
University of North Carolina
Mechanical Engineering and Engineering Science Dept.
Charlotte, NC 28223-0001
kssmith@uncc.edu

Additional material to this book can be downloaded from http://extras.springer.com.

ISBN: 978-1-4614-9938-1 ISBN 978-0-387-09645-2 (eBook)
DOI 10.1007/978-0-387-09645-2

Softcover re-print of the Hardcover 1st edition 2009

Printed on acid-free paper

springer.com

To our wives, Christine and Jan

Machining Dynamics

Preface

Through this book we demonstrate the importance of considering the role of process dynamics in machining performance. We based the book on graduate courses in mechanical vibrations and manufacturing that we have previously offered, but also included aspects of our research programs in machining dynamics and precision engineering. We developed the text to be applied in a traditional 15 week course format with an intended audience of upper division and graduate level engineering students, as well as the practicing engineer.

We organized the book into seven chapters. The chapter topics are summarized here.

- Chapter 1 – We provide a of list potential obstacles to machining productivity and highlight the focus areas for this text. We direct the reader to Fig. 1.1.1 for a graphical identification of these areas.
- Chapter 2 - We first review the fundamentals of single and two degree of freedom free and forced vibrations. We then continue with a description of the frequency response function, including experimental techniques.
- Chapter 3 – The purpose of this chapter is to describe regenerative chatter in turning and introduce the stability lobe diagram. We detail both analytical and time-domain simulations to determine stable and unstable cutting conditions.
- Chapter 4 – In this chapter we focus on milling and describe the corresponding analytical and time-domain simulations for stability prediction.
- Chapter 5 – Our goal for this chapter is to investigate the influence of forced vibrations during stable milling on part geometric accuracy. Both analytical and time-domain approaches are provided.
- Chapter 6 - We analyze the frequency content of stable and unstable milling signals; update the time-domain simulations developed in Chapters 4 and 5 to include runout of the cutter teeth and variable teeth spacing; discuss stability of low radial immersion milling; and describe the uncertainty evaluation for stability boundaries.
- Chapter 7 - In this chapter, we apply receptance coupling to prediction of the tool point frequency response function. We also review Euler-Bernoulli

beam theory and provide expressions for beam frequency response functions under various boundary conditions.

A key aspect of this book is the inclusion of functional MATLAB® code (in the form of m-files) on the companion CD. We use this code to support numerical examples throughout the text and demonstrate the analytical and time-domain algorithms we describe within the individual chapters. We organized the code by chapter on the CD and named the files according to the example that they support. To demonstrate the naming convention, Example 3.5.2 has three supporting programs: p_3_5_2_1.m, p_3_5_2_2.m, and p_3_5_2_3.m that are used to develop the book figures. Another special feature we have included is the ***For instance*** and ***In a nutshell*** explanations of selected topics for the non-mathematical reader.

We conclude by acknowledging the many contributors to this text. This naturally includes our instructors, colleagues, collaborators, and students. Among these, we'd like to particularly recognize the contributions of J. Tlusty[1], J. Ziegert, M. Davies, T. Burns, J. Pratt, and G.S. Duncan. We also thank the reviewers of this book for their helpful suggestions: E. Marsh, M. Miller, S. Redkar, G. Stépàn, A. Yi, and J. Ziegert.

University of Florida — Tony L. Schmitz
University of North Carolina at Charlotte — K. Scott Smith
June 2008

[1] The late Prof. Jiri Tlusty established the Machine Tool Research Center at the University of Florida where much of the research in support of this book was completed.

Contents

Chapter 1
Introduction

You have to learn the rules of the game. And then you have to play better than anyone else.

- Albert Einstein

In today's information driven society, it could be argued that the relative significance of a topic is measured by the number of web sites identified by an Internet search. The search term "Super Bowl", for example, yields approximately 63,000,000 sites[1], which tops "World Cup" at 57,300,000. "Nanotechnology" and "machining" score similarly to one another at 12,300,000 and 12,100,000 sites, respectively. Among recently developed manufacturing technologies, selected results include "high speed machining" – 423,000, "laser assisted machining" – 370,000, "laser machining" – 290,000, "friction stir welding" – 101,000, and "selective laser sintering" – 87,400. Refining the high-speed machining search to "high speed machining dynamics" reduces the count by more than four times to 90,900. However, this is almost exactly two times the number of sites (45,400) related to "chinch bugs", a known predator of St. Augustine grass. Apparently, reading this text places you in select company, but not quite as exclusive as those interested in a particular Florida lawn pest!

While this nonscientific survey was performed somewhat tongue in cheek, it does indicate that, although interest in high-speed machining continues to grow (nearly half a million web sites with related content), widespread awareness of the importance of considering the role of the process dynamics in its successful implementation has not yet been achieved. There is still work to be done! Increasing this process dynamics understanding for those interested in "playing the game better" is the intent of this text.

[1] The Internet searches were completed in April, 2008 using the Google™ search engine.

T.L. Schmitz, K.S. Smith, *Machining Dynamics*, DOI 10.1007/978-0-387-09645-2_1,

1.1 The big picture

Discrete part production by machining remains an important manufacturing application. In commercial situations, the focus is naturally on producing accurate parts in the required time frame under conditions of maximized profit. Unfortunately, a number of factors can influence our ability to do so. Important contributors to process efficiency include:

- workpiece loading/unloading from the machine;
- fixturing, including clamping/unclamping the workpiece on the machine;
- machining parameters, such as spindle speed, depth of cut, and feed rate;
- path planning strategies;
- tooling and holder selection;
- tool wear;
- tool changes;
- coolant management;
- chip evacuation;
- tool and workpiece vibrations, including chatter and errors due to the cutting forces (we refer to the latter as surface location errors);
- part measurement (on machine or post process); and
- machine accuracy, including geometric errors in the machine construction, thermally induced errors from heat sources associated with the machining process, and trajectory following errors caused by controller and machine structural dynamics.

We will focus our attention on process parameter selection to enable high material removal rates without introducing significant part errors due to cutting forces and the resulting dynamic tool deflections. The remaining issues, while important, are outside the scope of this text. Figure 1.1.1 displays an overview of our focus areas. There are two critical items upon which our modeling efforts are based: the frequency response function, FRF, and the force model. These are identified in the central portion of the figure (boxes with heavy solid lines). In milling, for example, we require knowledge of the tool-holder-spindle-machine FRF as reflected at the tool point. We can obtain the necessary assembly dynamics through modal testing techniques or by a combination of testing and modeling using the Receptance Coupling Substructure Analysis, RCSA, approach. For the force model, the required coefficients can be determined from cutting tests carried out on a force dynamometer. Given the tool point FRF and force model, we can choose either: 1) a time-domain simulation strategy to predict stability and surface location errors, SLE; or 2) a frequency-domain approach to stability and SLE predictions. For time-domain simulation, an additional step of modal fitting is required to obtain modal parameters that describe the system FRF. We detail each of these topics in Chapters 2 through 7.

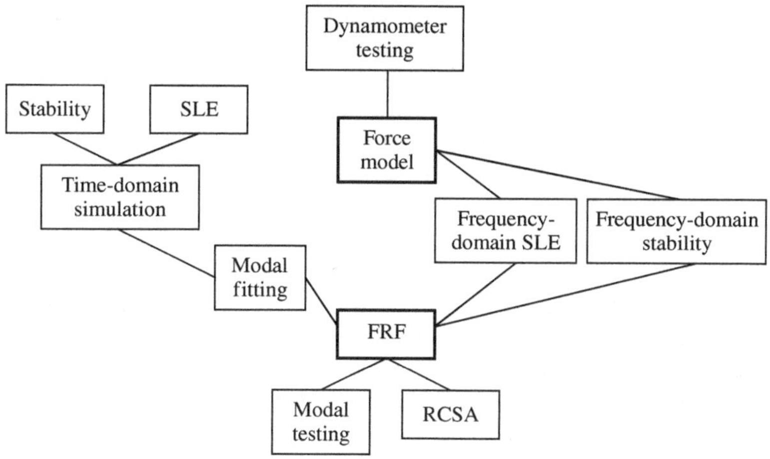

Fig. 1.1.1 Focus areas for study of machining dynamics. The critical frequency response function, FRF, and force model are indicated by the heavy solid lines

1.2 A Brief Review

We cannot hope to capture all the significant contributions to machining science that have been previously reported in the literature and we are wary of accidental omissions. However, we would like to provide a brief review of the foundational work that has led to our modern understanding of machining process dynamics. Clearly, with the process descriptions and models provided in this text, we are "dwarfs standing on the shoulders of giants (*nanos gigantium humeris insidentes*)"[2].

A primary building block for the study of machining is Taylor's "On the Art of Cutting Metals" [2]. This paper established an empirical basis for the relationships between cutting parameters and process performance; contemporary research efforts still rely on variations of Taylor's tool life model, for example. Later, Merchant's work provided a mechanics-based understanding of cutting forces, as well as the corresponding stresses and strains during material removal [3]. Within the broad view of machining encompassed by these and other early efforts, researchers have subsequently studied such basic aspects of machining as chip geometry, shear stresses, friction, and cutting temperatures [4]. In this text, we do not attempt to address these issues, but rather focus on the process dynamics. The contributions of chip formation to turning and milling behavior are indirectly included through the force models, which effectively treat this

[2] This quote is attributed to the 12th century philosopher Bernard of Chartres (*Bernardus Carnotensis*) [1].

complex behavior using "process coefficients" that relate cutting force levels to the uncut chip area [5].

While advances in computer simulation of machining processes continue, the foundation for much of this work can be traced to papers by Tlusty, Tobias, and Merrit [6-9], which, in turn, followed earlier work by Arnold [10] and others. Based on these efforts, an understanding of the regeneration of surface waviness during material removal as the primary mechanism for self-excited vibrations (or chatter) in machining was established. When combined with the effects of forced vibrations during stable cutting, we have the basis for exploring the role of machining dynamics in discrete part production. Comprehensive reviews of subsequent modeling and experimental efforts have been compiled and presented in the literature. We refer the reader to [4, 11, 12, 13, 14, 15, 16, 17], for example.

1.3 Roadmap

Figure 1.3.1 shows a modified version of Fig. 1.1.1, where we have now identified the relevant sections that detail these focus areas. The FRF is defined in Sections 2.2 and 2.4. Its measurement is outlined in Sections 2.5 through 2.7, while its prediction using the RCSA method is detailed in Section 7.6. The cutting force model is described in Section 3.1 for turning and Sections 4.1 and 4.7 for milling. Experimental methods for identifying force model coefficients in milling are covered in Section 4.7 as well. Time-domain simulations are provided in Section 3.5 (turning), Sections 4.4 through 4.6 (milling circular tooth path), and Section 5.3 (milling cycloidal

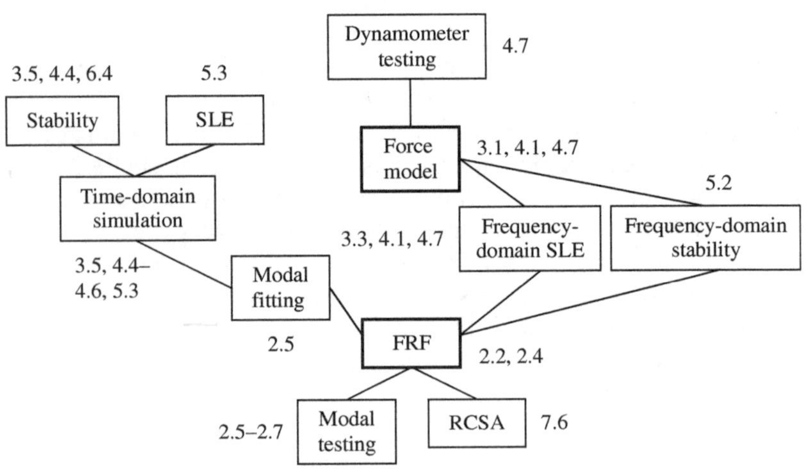

Fig. 1.3.1 A roadmap for the text is shown; section numbers are identified for each focus area

tool path). The necessary modal fitting step for determining the modal mass, damping, and stiffness coefficients used in the time-domain simulation equations of motion is reviewed in Section 2.5. Application of time-domain simulation to turning stability is shown in Section 3.5. Low radial immersion milling stability is described in Section 6.4 and surface location error prediction is included in Section 5.3. Frequency-domain techniques for stability analysis are described in Section 3.3 for turning and Section 4.3 for milling (two methods are included). The frequency-domain analysis for milling surface location error is provided in Section 5.2.

While the content of this book is mathematical by its nature, it is possible to gain an understanding of the basic concepts and some ability to apply them without a detailed understanding of the mathematical formalities. Throughout the text, there are number of explanations/analogies labeled ***For instance*** or ***In a nutshell*** which attempt to convey the essence of the topic to the non-mathematical reader.

References

1. http://www.en.wikipedia.org/wiki/Standing_on_the_shoulders_of_giants, accessed April, 2008.
2. Taylor, F.W., 1906, On the Art of Cutting Metals, Transactions of the ASME, 28: 31-248.
3. Merchant, M.E., 1950, Metal Cutting Research – Theory and Practice, Machining – Theory and Practice, ASM, 5–44.
4. Komanduri, R., 1993, Machining and Grinding: A Historical Review of the Classical Papers, Applied Mechanics Review, 46/3: 80–132.
5. Koenisgberger, F. and Sabberwal, A., 1961, An Investigation into the Cutting Force Pulsations during Milling Operations, International Journal of Machine Tool Design and Research, 1: 15–33.
6. Tlusty, J. and Polacek, M., 1957, Besipiele der behandlung der selbsterregten Schwingung der Werkzeugmaschinen, FoKoMa, Hanser Verlag, Munchen.
7. Tobias, S.A. and Fishwick, W., 1958, Theory of Regenerative Machine Tool Chatter, The Engineer, London, 258.
8. Tlusty, J. and Polaceck, M., 1963, The Stability of Machine Tools Against Self-excited Vibrations in Machining, International Research in Production Engineering, 465–474.
9. Merrit, H., 1965, Theory of Self-Excited Machine Tool Chatter, Journal of Engineering for Industry, 87/4: 447–454.
10. Arnold, R., 1946, The Mechanism of Tool Vibration in the Cutting of Steel, Proceedings of the Institution of Mechanical Engineers, 54: 261–284.
11. Tlusty, J., 1978, Analysis of the State of Research in Cutting Dynamics, Annals of the CIRP, 27/2: 583–589.
12. Smith, S. and Tlusty, J., 1991, An Overview of Modeling and Simulation of the Milling Process, Journal of Engineering for Industry, 113/2: 169–175.
13. Smith, S. and Tlusty, J., 1993, Efficient Simulation Programs for Chatter in Milling, Annals of the CIRP, 42/1: 463–466.
14. Ehmann, K., Kapoor, S., DeVor. R., and Lazoglu, I., 1997, Machining Process Modeling: A Review, Journal of Manufacturing Science and Engineering, 119/4B: 655–663.

15. Smith, S. and Tlusty, J., 1997, Current Trends in High-Speed Machining, Journal of Manufacturing Science and Engineering, 119/4B: 664–666.
16. Merchant, M.E., 1998, An Interpretive Look at 20th Century Research on Modeling of Machining, Machining Science and Technology, 2/2: 157–163.
17. Altintas, Y. and Weck, M., 2004, Chatter Stability of Metal Cutting and Grinding, Annals of the CIRP, 53/2: 619–642.

Chapter 2
Modal Analysis

All these primary impulses, not easily described in words, are the springs of man's actions.

– Albert Einstein

As described in Chapter 1, a critical step in improving machining productivity through a consideration of the process dynamics is identifying the tool point frequency response function. In this chapter, we will briefly review the fundamentals of single and two degree of freedom free and forced vibrations and, in so doing, we will establish notation conventions for a description of modal analysis. Naturally, this review will not replace the information provided in a mechanical vibrations textbook, such as [1], [2], or others, and some of the subtleties of a comprehensive treatment will not be included. However, it will provide us with the basis we need to describe techniques for frequency response function measurement and model development.

After this brief review, the remaining purposes of the chapter are to:

- provide the background necessary to interpret measured frequency response functions;
- detail a modal fitting technique; and
- describe the experimental procedures and equipment used to measure tool point frequency response functions.

In all discussions, we will assume linear vibrations. While nonlinearities can certainly be observed in physical systems, this idealization is generally acceptable in practice.

2.1 Single Degree of Freedom Free Vibration

Our structure of interest is a cutting tool of some geometry (single or multiple cutting edges) connected to either a tool post/turret and lathe or a tool holder, spindle, and milling machine. Both of these cases can be represented as combined bodies that possess both mass and elasticity, or the ability to deform

T.L. Schmitz, K.S. Smith, *Machining Dynamics*, DOI 10.1007/978-0-387-09645-2_2,

without permanently changing shape. The vibration of bodies that exhibit these characteristics can be divided into three main categories: free, forced, and self-excited vibrations.

2.1.1 Free Vibration

Free vibration occurs in the absence of a long term, external excitation force. It is the result of some initial conditions imposed on the system, such as a displacement from the system's equilibrium position, for example. Free vibration produces motion in one or more of the system's natural frequencies and, because all physical structures exhibit some form of damping (or energy dissipation), it is seen as a decaying oscillation with a relatively short duration; see Fig. 2.1.1. Familiar examples include plucking a guitar string or striking a piano string.

2.1.2 Forced Vibration

Forced vibration takes place when a continuous, external periodic excitation produces a response with the same frequency as the forcing function (after the decay of initial transients). While free vibration is often represented in the time-domain, forced vibration is typically analyzed in the frequency-domain. This emphasizes the magnitude and phase dependence on frequency and enables the convenient identification of natural frequencies. A typical source of forced vibration in mechanical systems is rotating imbalance. Large vibrations occur

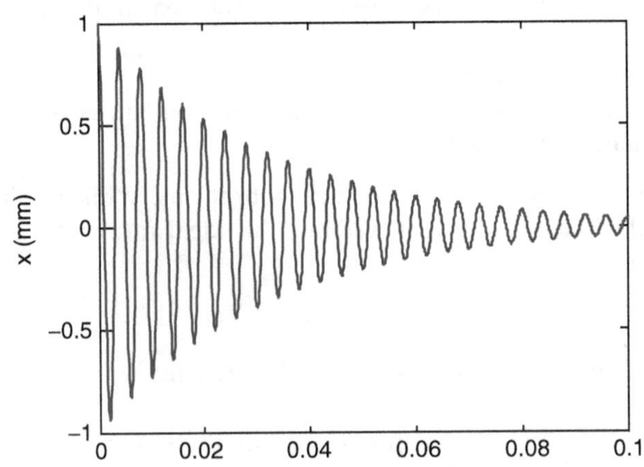

Fig. 2.1.1 Damped free vibration example

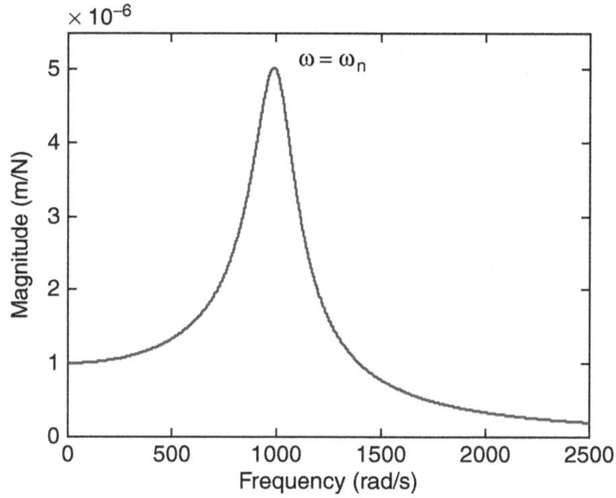

Fig. 2.1.2 Example of forced vibration magnitude

when the forcing frequency, ω, is near a system natural frequency, ω_n, as shown in Fig. 2.1.2. This condition is referred to as resonance and is generally avoided. However, we'll see in Chapters 3 and 4 that operating under resonant conditions can actually improve the stability of cutting processes. While counter intuitive, the mechanism that leads to this behavior (overcutting of the previously machined surface) makes physical sense. We'll also discuss the impact of forced vibrations on machining accuracy in Chapter 5. In this case, we'll see that operating near resonance for lowly damped structures can lead to significant errors in the workpiece geometry although the process remains stable.

2.1.3 Self-Excited Vibration

In self-excited vibration, a steady input force is present, as in the case of forced vibration. However, this input is modulated into vibration at one of the system's natural frequencies, as with free vibration. The physical mechanisms that provide this modulation are varied. Common examples of self-excited vibration include playing a violin, flutter in airplane wings, and chatter in machining. Naturally, our focus is on the latter in Chapters 3 and 4.

Let's begin our discussion of single degree of freedom free vibration with a simple, lumped parameter model. In this model, all the mass is assumed to be concentrated at the coordinate location and the spring that provides the oscillating restoring force is massless. The model is composed of a mass, m, attached to a linear spring, k, that provides a force proportional to its displacement from

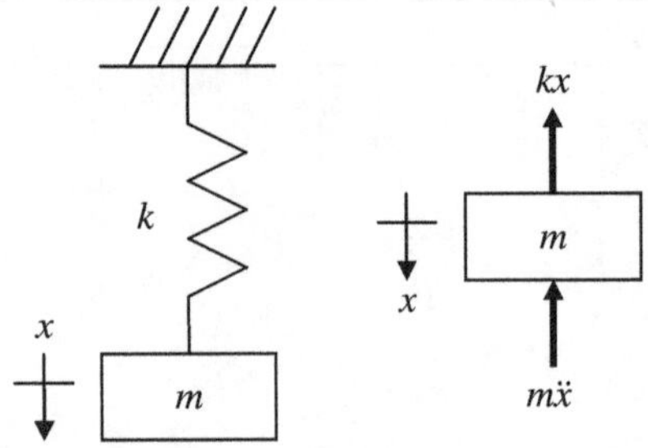

Fig. 2.1.3 Single degree of freedom, undamped lumped parameter model (left); free body diagram (right)

the mass's static equilibrium position. Because the rigid mass is only allowed to move vertically, a single time dependent coordinate, x, is sufficient to describe its motion. In a physical sense, this means that knowledge of the motion of a single coordinate is enough to completely describe the motion of a single degree of freedom system. See Fig. 2.1.3, which includes the free body diagram. Summing the spring and inertial forces in the vertical direction yields the model's equation of motion:

$$m\ddot{x} + kx = 0. \tag{2.1.1}$$

By assuming a harmonic solution of the form $x = Xe^{st}$, where X is a complex coefficient, $s = i\omega$, and ω is the frequency (in rad/s), we can express the velocity as the first time derivative of the displacement, $\dot{x} = sXe^{st} = i\omega Xe^{st}$, and the acceleration as the second time derivative, $\ddot{x} = s^2 Xe^{st} = -\omega^2 Xe^{st}$ (note that $i = \sqrt{-1}$ and $i^2 = -1$). Substitution into Eq. 2.1.1 gives:

$$Xe^{st}\left(ms^2 + k\right) = 0. \tag{2.1.2}$$

In this equation, either Xe^{st} or $\left(ms^2 + k\right)$ is zero. If the first term is zero, this means that no motion has occurred and it is described as the trivial solution. We are interested in the case that the second term is equal to zero. This is referred to as the characteristic equation for the system:

$$ms^2 + k = 0. \tag{2.1.3}$$

Solving for the complex variable s gives the two roots $s = \pm\sqrt{-\frac{k}{m}} = \pm i\sqrt{\frac{k}{m}}$. The vibrating frequency $\sqrt{\frac{k}{m}} = \omega_n$ is the natural frequency for the single degree of freedom system. Typical SI units for k and m are N/m and kg, respectively, which gives units of rad/s for ω_n. Alternately, the natural frequency may be expressed in units of Hz (cycles/s). In this case, we'll use the notation $f_n = \frac{\omega_n}{2\pi}$.

IN A NUTSHELL All mechanical systems have mass and are not infinitely stiff. As a result, they are either at rest (not moving) or they are vibrating. If the vibration energy source is applied only at the beginning of the motion (like plucking a guitar string), then the motion is a free vibration. Free vibrations happen at the natural frequency of the system. The natural frequency is higher if the stiffness is higher or the mass is less and the natural frequency is lower if the stiffness is lower or the mass is greater. In the guitar string example, the high notes (high frequencies) are produced by the strings that are tight (stiff) and thin (low mass). The low notes (low frequencies) are produced by the strings that are loose (not stiff) and thick (high mass). In machine tools, low frequencies are encountered when the vibrating objects are massive (like the column of a machine tool) or flexible (like a long slender workpiece) or both.

The total solution to Eq. 2.1.1 is the sum of the contributions from each of the two roots:

$$x = X_1 e^{i\omega_n t} + X_2 e^{-i\omega_n t}. \tag{2.1.4}$$

The complex coefficients, X_1 and X_2, can be determined from the initial displacement, x_0, and velocity, $\dot{x}_0$, of the single degree of freedom system. Evaluating Eq. 2.1.4 at $t = 0$ gives:

$$x_0 = X_1 + X_2. \tag{2.1.5}$$

The first time derivative of Eq. 2.1.4 is:

$$\dot{x} = i\omega_n X_1 e^{i\omega_n t} - i\omega_n X_2 e^{-i\omega_n t}. \tag{2.1.6}$$

At $t = 0$, Eq. 2.1.6 becomes:

$$\dot{x}_0 = i\omega_n X_1 - i\omega_n X_2. \tag{2.1.7}$$

Using the linear combination approach, Eqs. 2.1.5 and 2.1.7 can be combined to determine the complex conjugate coefficients X_1 and X_2:

$$X_1 = \frac{-i\dot{x}_0 + \omega_n x_0}{2\omega_n} \tag{2.1.8}$$

and

$$X_2 = \frac{i\dot{x}_0 + \omega_n x_0}{2\omega_n}. \tag{2.1.9}$$

These coefficients can then be substituted in Eq. 2.1.4 to determine the time dependent displacement of the mass due to the imposed initial conditions. Alternately, the mass motion can be expressed in exponential form. To use this notation, we first need to identify the real (Re) and imaginary (Im) parts of the complex coefficients:

$$\mathrm{Re}(X_1) = \frac{x_0}{2} \qquad \mathrm{Im}(X_1) = \frac{-\dot{x}_0}{2\omega_n}. \tag{2.1.10}$$

$$\mathrm{Re}(X_2) = \frac{x_0}{2} \qquad \mathrm{Im}(X_2) = \frac{\dot{x}_0}{2\omega_n}. \tag{2.1.11}$$

These real and imaginary parts can then be used to write the coefficients in exponential form:

$$\begin{aligned} X_1 &= Ae^{i\beta} = \sqrt{\mathrm{Re}(X_1)^2 + \mathrm{Im}(X_1)^2}\,\exp\left(i\cdot\tan^{-1}\left(\frac{\mathrm{Im}(X_1)}{\mathrm{Re}(X_1)}\right)\right) \\ X_1 &= \sqrt{\left(\frac{x_0}{2}\right)^2 + \left(\frac{-\dot{x}_0}{2\omega_n}\right)^2}\,\exp\left(i\cdot\tan^{-1}\left(\frac{\frac{-\dot{x}_0}{2\omega_n}}{\frac{x_0}{2}}\right)\right) \\ X_1 &= \sqrt{\frac{x_0^2\omega_n^2 + \dot{x}_0^2}{4\omega_n^2}}\,\exp\left(i\cdot\tan^{-1}\left(-\frac{\dot{x}_0}{x_0\omega_n}\right)\right) \end{aligned} \tag{2.1.12}$$

where the magnitude is $A = \sqrt{\frac{x_0^2\omega_n^2 + \dot{x}_0^2}{4\omega_n^2}}$ and the phase is $\beta = \tan^{-1}\left(-\frac{\dot{x}_0}{x_0\omega_n}\right)$. Similarly, $X_2 = Ae^{-i\beta}$ (same magnitude, but negative phase) because it is the complex conjugate of X_1. We can then rewrite the total solution from Eq. 2.1.4 in the form:

$$x = Ae^{i\beta}e^{i\omega_n t} + Ae^{-i\beta}e^{-i\omega_n t} = A\left(e^{i(\omega_n t+\beta)} + e^{-i(\omega_n t+\beta)}\right). \tag{2.1.13}$$

Finally, by applying the Euler identity $e^{i\theta} + e^{-i\theta} = 2\cos(\theta)$, Eq. 2.1.13 can be rewritten as:

$$x = 2A\cos(\omega_n t + \beta). \tag{2.1.14}$$

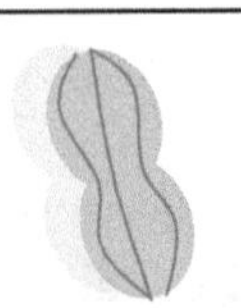

IN A NUTSHELL If we know the mass and stiffness, and we know the initial conditions that started the vibration, then it is possible to describe the motion of the system at any time.

While Eq. 2.1.14 emphasizes the oscillatory nature of the mass motion and the dependence of the magnitude and phase on the initial conditions, we must also include damping in our analysis in order to model physical systems. Damping refers to the "leakage" of the input energy into the vibrating system. In other words, not all of the input energy serves to cause motion. Some of it is dissipated in other ways. A comprehensive model of damping is complicated and not particularly well suited for incorporation into our simple mathematical description of single degree of freedom free vibration. Therefore, one or more of three mathematically simple, but effective, damping models are typically applied.

2.1.4 Viscous Damping

A common assertion is that the retarding damping force is proportional to the mass velocity. You may have experienced this phenomenon if you've attempted to force a body through a fluid, such as pulling your hand through water or sticking your hand out the window of a moving vehicle. You probably observed that increasing the speed of your hand relative to the fluid raised the resistance proportionally. If we write the damping force as:

$$f = c\dot{x} \tag{2.1.15}$$

and substitute the velocity expression $\dot{x} = sXe^{st} = i\omega Xe^{st}$, we see that viscous damping is frequency dependent. When sketching models of lumped parameter systems, the damping element is often illustrated as a fluid dashpot (similar to a car's shock absorber) when the viscous damping model is applied. Typical SI units for c are N-s/m.

FOR INSTANCE In a car, the body provides the mass, the suspension spring provides the stiffness, and the shock absorber provides the damping. Inside the shock absorber is a plate with holes that moves through a fluid and produces a force proportional to the velocity of the motion.

2.1.5 Coulomb Damping

Another effective damping model is Coulomb damping, or dry sliding friction. Here, energy is dissipated due to relative motion between two contacting surfaces. The force magnitude depends on the sliding (kinetic) friction coefficient, μ, and the normal force, N, between the two bodies. See Fig. 2.1.4. Because the friction force always opposes the direction of motion, the resulting equation of motion is nonlinear. A piecewise definition of the Coulomb damping force is [3]:

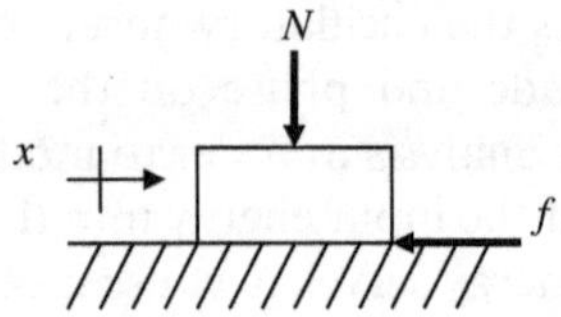

Fig. 2.1.4 Coulomb damping occurs due to dry sliding friction between the two surfaces. The normal and friction forces are shown

$$f=\begin{Bmatrix} -\mu N & \dot{x}>0 \\ 0 & \dot{x}=0 \\ \mu N & \dot{x}<0 \end{Bmatrix}. \tag{2.1.16}$$

FOR INSTANCE Coulomb damping accounts for the energy dissipation by friction between two objects. It converts the kinetic energy from a rotating wheel into heat during braking in a car, for example. The key point regarding Coulomb damping in vibrations is that the Coulomb damping force always opposes the motion like viscous damping, but instead of being proportional to velocity, it is proportional to normal force.

2.1.6 Solid Damping

Even in the absence of an external fluid medium or sliding friction against another surface, the motion of a freely oscillating body decays over time. This is due to energy dissipation internal to the body (perhaps a good mental picture is molecules sliding relative to each other within the body itself during periodic motion and elastic deformation). The energy dissipation during a cycle of motion for this solid or structural damping is taken to be proportional to the square of the vibration magnitude. Using the concept of equivalent viscous damping, solid damping is often incorporated with stiffness to arrive at a complex stiffness term in the differential equation of motion [4]. We implement this approach in Section 7.4 when describing beam models for use in receptance coupling analyses. We use receptance coupling to predict the frequency response of tool-holder-spindle-machine assemblies in Section 7.6.

IN A NUTSHELL While there are certainly several kinds of damping, it is mathematically simpler and sufficiently accurate to consider the damping as viscous.

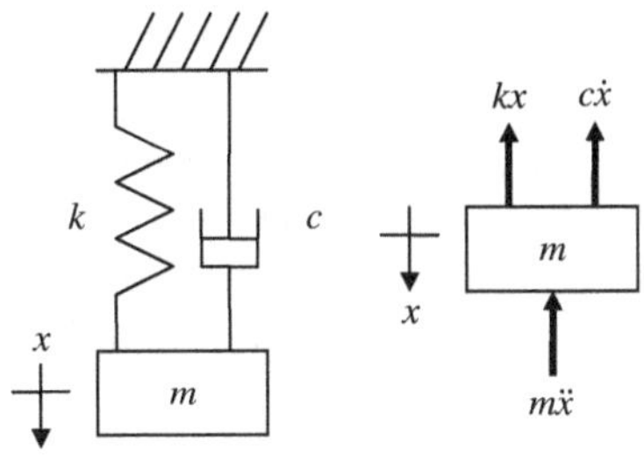

Fig. 2.1.5 Single degree of freedom, damped lumped parameter model (left); free body diagram (right)

For the remainder of this chapter, however, we will use viscous damping to describe energy dissipation in the lumped parameter models. The equation of motion for free vibration of the single degree of freedom spring-mass-damper (Fig. 2.1.5) can then be written as:

$$m\ddot{x} + c\dot{x} + kx = 0. \tag{2.1.17}$$

Again assuming the harmonic solution $x = Xe^{st}$, we obtain the characteristic equation:

$$ms^2 + cs + k = 0, \tag{2.1.18}$$

which can be rewritten as:

$$s^2 + \frac{c}{m}s + \frac{k}{m} = 0. \tag{2.1.19}$$

This equation is quadratic in s^2 and has the two roots:

$$s_{1,2} = -\frac{c}{2m} \pm \sqrt{\left(\frac{c}{2m}\right)^2 - \frac{k}{m}}. \tag{2.1.20}$$

The vibratory behavior of the spring-mass-damper depends on the term under the radical in Eq. 2.1.20. If $\left(\frac{c}{2m}\right)^2 - \frac{k}{m} < 0$, the system is underdamped and vibratory. If $\left(\frac{c}{2m}\right)^2 - \frac{k}{m} = 0$, the system is said to be critically damped and, if $\left(\frac{c}{2m}\right)^2 - \frac{k}{m} > 0$, then the system is overdamped. For these two cases, no vibration occurs. Because the damping is generally low in tool-holder combinations for lathes and milling machines, we will consider only the underdamped option in our analyses. For underdamped systems, Eq. 2.1.20 can be rewritten as:

$$s_{1,2} = -\zeta\omega_n \pm i\omega_d, \tag{2.1.21}$$

where we've introduced the dimensionless damping ratio, $\zeta = \frac{c}{2\sqrt{km}}$, and damped natural frequency, $\omega_d = \omega_n\sqrt{1-\zeta^2}$. Under the viscous damping assumption, we see that the free vibrating frequency is reduced in the presence of damping. However, for typical tool-holder systems, the damping is low enough that the frequency change is negligible. Using the two roots in Eq. 2.1.21, the total solution for the free motion of the single degree of freedom spring-mass-damper is:

$$x = X_1 e^{(-\zeta\omega_n + i\omega_d)t} + X_2 e^{(-\zeta\omega_n - i\omega_d)t} = e^{-\zeta\omega_n t}\left(X_1 e^{i\omega_d t} + X_2 e^{-i\omega_d t}\right). \quad (2.1.22)$$

Like the undamped case, the complex coefficients can be determined from the initial conditions. Taking the time derivative of Eq. 2.1.22, substituting the initial displacement, x_0, and velocity, $\dot{x}_0$, and solving for X_1 and X_2 gives the complex conjugate pair:

$$X_1 = \frac{x_0}{2} - i\frac{\dot{x}_0 + \zeta\omega_n x_0}{2\omega_d} \text{ and } X_2 = \frac{x_0}{2} + i\frac{\dot{x}_0 + \zeta\omega_n x_0}{2\omega_d}. \quad (2.1.23)$$

Using these coefficients, the exponential form can again be developed in a similar way to Eq. 2.1.12 by substituting for the real and imaginary parts. For example, $\mathrm{Re}(X_1) = \frac{x_0}{2}$ and $\mathrm{Im}(X_1) = -\frac{\dot{x}_0 + \zeta\omega_n x_0}{2\omega_d}$ for the coefficient X_1. Note that these terms simplify to Eq. 2.1.10 for the undamped case if ζ is set equal to zero.

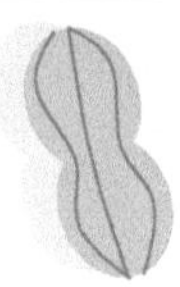

IN A NUTSHELL Free vibration of a single degree of freedom system with damping consists of two parts: a sinusoidal motion at the damped natural frequency and a decaying exponential envelope that bounds it. More damping means that the exponential envelope converges to zero more quickly.

2.2 Single Degree of Freedom Forced Vibration

We will again consider the spring-mass-damper model shown in Fig. 2.1.5. However, a harmonic external force is now applied to the mass. The force is shown as $fe^{i\omega t}$ in Fig. 2.2.1, but may be more properly considered as $\frac{f}{2}(e^{i\omega t} + e^{-i\omega t})$ so that the force is always real valued. This is illustrated in the real-imaginary (or complex) plane as two counter-rotating vectors with magnitude $\frac{f}{2}$ and phase $\pm\omega t$ whose sum is always a scalar value oscillating from f to $-f$ with the forcing frequency ω (Fig. 2.2.2).

$$m\ddot{x} + c\dot{x} + kx = f \quad (2.2.1)$$

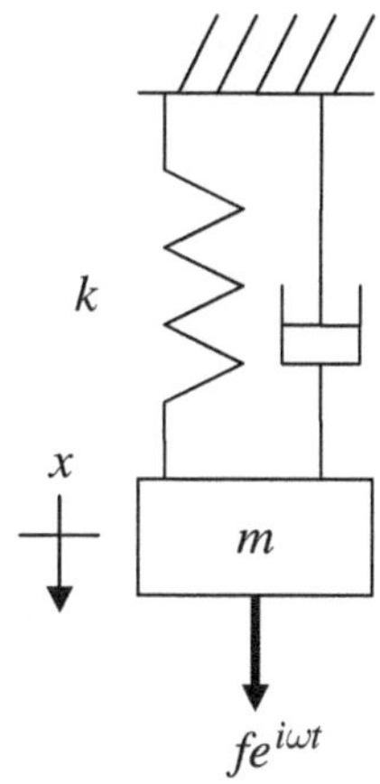

Fig. 2.2.1 Single degree of freedom, lumped parameter model (damped with force)

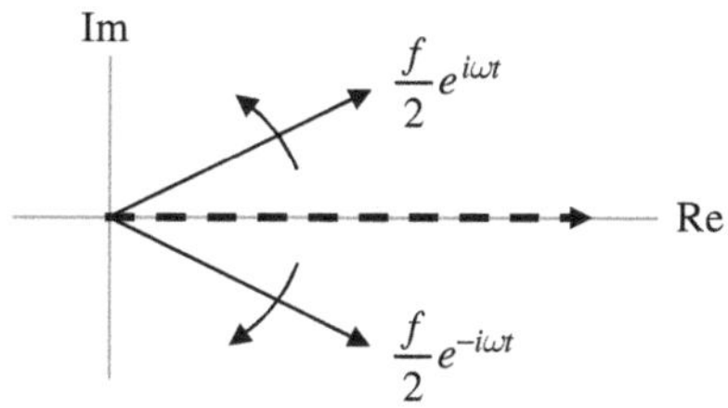

Fig. 2.2.2 Real valued force shown as sum of counter-rotating vectors

IN A NUTSHELL The external force exciting a forced vibration is assumed to be sinusoidal. In fact, any external excitation force can be thought of as a sum of sinusoidal forces (see the Fourier transform described in Section 4.3), so the method is generic to all kinds of external forces.

Although the total solution to Eq. 2.2.1 includes both the homogeneous (transient) and particular (steady state) components, we have already described the damped transient response in the previous section. We will therefore consider only the steady state solution here. Because the motion response has the same frequency as the forcing function, we can assume a solution of the form $x = Xe^{i\omega t}$. The velocity and acceleration can then be written as $\dot{x} = i\omega X e^{i\omega t}$ and $\ddot{x} = -\omega^2 X e^{i\omega t}$. Substituting in Eq. 2.2.1 gives:

$$\left(-\omega^2 m + i\omega c + k\right) X e^{i\omega t} = f e^{i\omega t}. \tag{2.2.2}$$

Rewriting Eq. 2.2.2 gives the complex valued frequency response function. We will use this description of Eq. 2.2.3, rather than transfer function, because we can only consider positive frequencies and a single system configuration (damping and natural frequency) when we perform measurements. The term transfer function refers to the theoretical situation where all frequencies ($-\infty$ to $+\infty$) and $\zeta\omega_n$ combinations are included.

$$\frac{X}{F} = \frac{1}{-m\omega^2 + ic\omega + k} \tag{2.2.3}$$

There are two primary ways to represent the complex function shown in Eq. 2.2.3. The first is to separate the function into its magnitude and phase components and the second is to express the function using its real and imaginary parts. The frequency dependent magnitude and phase are written as:

$$\left|\frac{X}{F}\right| = \frac{1}{k}\sqrt{\frac{1}{\left(1-\left(\frac{\omega}{\omega_n}\right)^2\right)^2 + \left(2\zeta\frac{\omega}{\omega_n}\right)^2}} \tag{2.2.4}$$

and

$$\Phi = \tan^{-1}\left(\frac{\mathrm{Im}\left(\frac{X}{F}\right)}{\mathrm{Re}\left(\frac{X}{F}\right)}\right) = \tan^{-1}\left(\frac{-2\zeta\frac{\omega}{\omega_n}}{1-\left(\frac{\omega}{\omega_n}\right)^2}\right). \tag{2.2.5}$$

IN A NUTSHELL The magnitude and phase parts of the frequency response function emphasize key features of forced vibration: 1) the forced vibration occurs at the frequency of the exciting force; and 2) the size of the motion compared to the force (the magnitude) and the time delay between when the force reaches its maximum and the displacement reaches its maximum (the phase) depend on the frequency of the exciting force and the natural frequency of the system.

When the exciting frequency is low in comparison to the natural frequency, the motion is small and nearly in phase with the force (the force and the displacement reach their peaks at the same time). As the frequency of excitation increases, the size of the motion gets larger and the peak of the displacement begins to happen later than the peak of the force. When the excitation force is very close to the natural frequency, then the displacement is as large as possible and it is zero when the force is maximum (90 deg out of phase). Further increases in the frequency of the excitation force cause the size of the displacement to be reduced. At very high excitation frequencies, the size of the displacement is very small and the displacement reaches its maximum when the force reaches its minimum (180 deg out of phase).

Because Eqs. 2.2.4 and 2.2.5 are somewhat cumbersome, it is common to replace the frequency ratio $\frac{\omega}{\omega_n}$ with another variable, such as r. We will also adopt this convention. The real and imaginary parts of the frequency response function, or FRF, are provided in Eqs. 2.2.6 and 2.2.7.

$$\mathrm{Re}\left(\frac{X}{F}\right) = \frac{1}{k}\left(\frac{1-r^2}{(1-r^2)^2+(2\zeta r)^2}\right) \tag{2.2.6}$$

$$\mathrm{Im}\left(\frac{X}{F}\right) = \frac{1}{k}\left(\frac{-2\zeta r}{(1-r^2)^2+(2\zeta r)^2}\right) \tag{2.2.7}$$

IN A NUTSHELL "Real" and "Imaginary" are not terms that imply "exists" or "does not exist". Rather, they are mathematical terms relating to the solution procedure. It may be more intuitive to think of these terms in the following way.

As we mentioned previously, we can consider the input force to be sinusoidal (sine or cosine). The resulting displacement is therefore also sinusoidal with the same frequency. If the input is a cosine force, then the "real part" of the resulting motion is the part that is also a cosine. The "imaginary part" of the resulting motion is the part that is a sine. The combination of the motion's cosine and sine parts describes both the size (magnitude) of the resulting motion and the phase shift relative to the force.

Example 2.2.1: FRF for single degree of freedom system Let's consider a single degree of freedom spring-mass-damper system with a mass of 1 kg, a spring constant of 1×10^6 N/m, and a viscous damping coefficient of 200 N-s/m. In order to apply Eqs. 2.2.4-2.2.7, we must calculate the (undamped) natural frequency and damping ratio.

$$\omega_n = \sqrt{\frac{k}{m}} = \sqrt{\frac{1 \times 10^6}{1}} = 1000 \text{ rad/s}$$
$$\zeta = \frac{c}{2\sqrt{km}} = \frac{200}{2\sqrt{1 \times 10^6 \cdot 1}} = 0.1$$

Figure 2.2.3 shows the magnitude and phase as a function of the frequency ratio, r. Although a logarithmic magnitude axis (i.e., a semilog plot) is often shown in the literature, we will use a linear convention for plots in this text

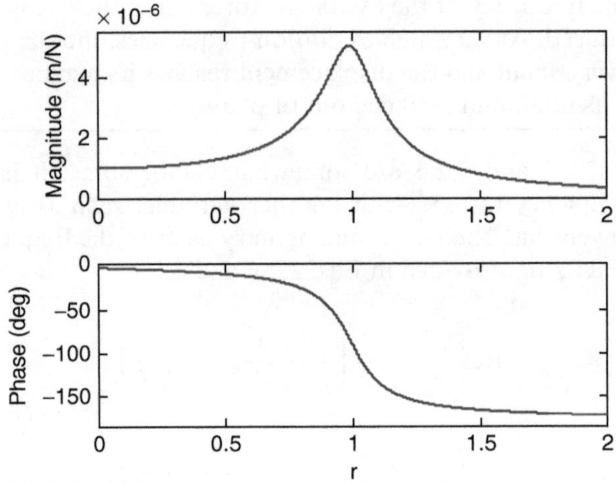

Fig. 2.2.3 Magnitude and phase for single degree of freedom system

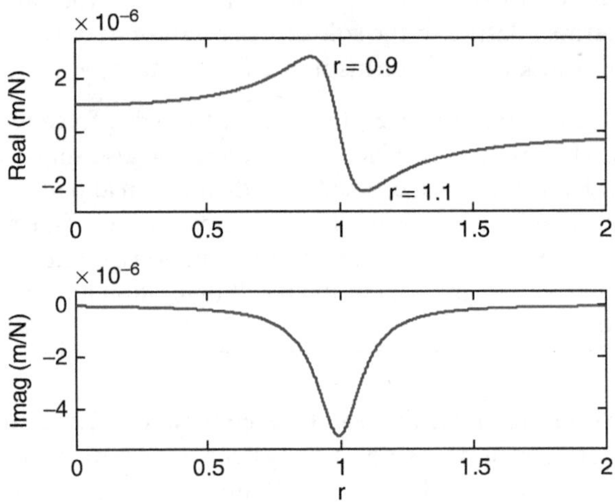

Fig. 2.2.4 Real and imaginary parts for single degree of freedom system

unless specified otherwise. The real and imaginary parts are provided in Fig. 2.2.4. Note that the zero frequency (DC) value for both the real part and magnitude is $\frac{1}{k} = 1 \times 10^{-6}$ m/N. This represents the real valued static deflection of the spring (away from its equilibrium position) under a unit force. We can also see that the magnitude at resonance ($r = 1$ or $\omega = \omega_n$) is significantly larger than the DC deflection. This magnitude is $\frac{1}{2k\zeta} = 5 \times 10^{-6}$ m/N.

IN A NUTSHELL The logarithmic scale is often used to display information covering a broad range in a single graph. The sound pressure levels in acoustics vary over a huge range from a whisper to jet engine noise, for example. The mathematics of acoustics and mechanical vibrations are similar and, for this reason, many textbooks use the logarithmic scales from acoustics to show similar concepts in mechanical vibrations. In our case, focused on mechanical vibrations, the range of the signals is not so large and the linear scale is more intuitive for most people, so we will use it almost exclusively.

The maximum value of the real part occurs at $r = \sqrt{1 - 2\zeta}$, which we will approximate as $r = 1 - \zeta = 0.9$ (this approximation is valid for small ζ values when ζ^2 is negligible; in this case the error is only 0.6%). The minimum real part occurs at $r = \sqrt{1 + 2\zeta}$, approximated as $r = 1 + \zeta = 1.1$. The difference in the real value between these maximum and minimum points is the same as the magnitude peak value $\frac{1}{2k\zeta} = 5 \times 10^{-6}$ m/N. The imaginary minimum is seen at resonance with a value of $\frac{-1}{2k\zeta} = -5 \times 10^{-6}$ m/N.

In addition to the frequency dependent representations of the FRF shown in Figs. 2.2.3 and 2.2.4, the Argand diagram can also be selected. In this case, the real part is graphed versus the imaginary part (i.e., the complex plane) and the same information identified in the previous paragraphs is compactly represented. As we traverse the "circle" clockwise from $r = 0$, where the real part is $\frac{1}{k} = 1 \times 10^{-6}$ m/N and the imagnary part is zero, we sequentially encounter the $r = 1 - \zeta = 0.9$ point where the real part is maximum, the $r = 1$ point where the real part is zero and the imaginary part is most negative, the $r = 1 + \zeta = 1.1$ point where the real part is most negative, and, finally, we approach the $r = \infty$ frequency ratio where both the real and imaginary parts are zero.

Using a vector representation for $\frac{X}{F}$, the magnitude is identified as the length of the vector which extends from the origin to any point (i.e., a desired r value) on our "circle". The phase lag between the displacement and force is the angle between the vector and the positive real axis. The real and imaginary parts are simply the projections of the vector on the real and imaginary axes. The MATLAB® program used to produce Figs. 2.2.4–2.2.6 for this example is provided on the companion CD as p_2_2_1_1.m.

IN A NUTSHELL You can think of vibration as a vector (arrow) rotating in a plane (like a hand on a clock) in the counter-clockwise direction. The length of the vector is the size of the vibration. The speed of rotation is the frequency of the vibration. The projection of the tip of the rotating vector onto the horizontal axis (the real part) gives the vibration. Both the exciting force and the resulting

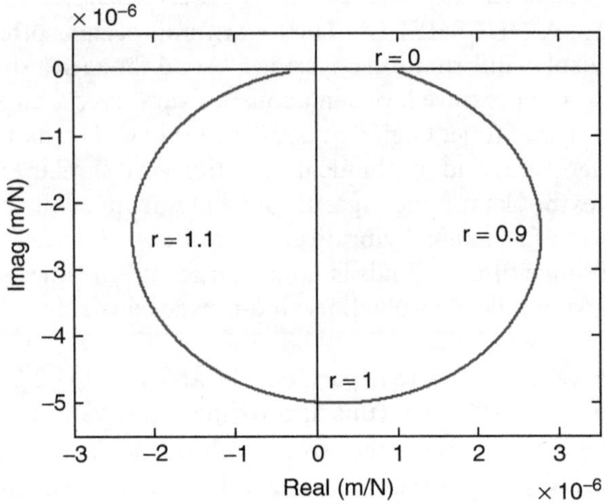

Fig. 2.2.5 Argand diagram for example single degree of freedom system

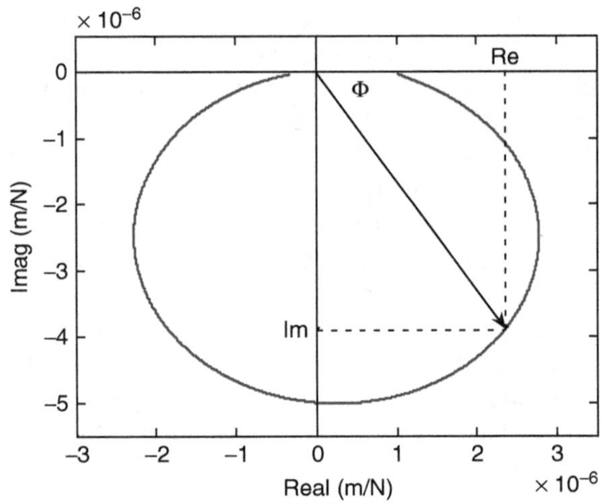

Fig. 2.2.6 Vector representation of FRF in the complex plane

displacement in a forced vibration can be represented this way. They are vectors that rotate together (have the same frequency), although they may have different lengths and they may be shifted, or pointing to different times on the clock.

The Argand, or phase plane, representation summarizes the relationship between the force and the displacement. You can imagine that a picture has been taken when the force vector is horizontal (completely real). The curve on the Argand diagram represents the places where the tip of the resulting displacement vector can be found (as a function of frequency). When the frequency is low, the displacement vector is just below force vector. It is small in size and reaches its maximum at a time only shortly after the force. At resonance (where the exciting frequency is equal to the natural frequency), the displacement vector is large and is located 90 deg behind the force vector. When the excitation frequency is high, the displacement vector is again small and is 180 deg behind the force (it points in the opposite direction).

Of course few vibrating systems can be represented by a single coordinate. We extend the analysis to include a second coordinate in Section 2.3.

2.3 Two Degree of Freedom Free Vibration

We will again use the lumped parameter spring-mass-damper model as the basis for our discussion, but we will now include a second degree of freedom by adding a second spring-mass-damper to the first in a "chain-type" configuration; see Fig. 2.3.1. Using the free body diagrams for the top and bottom masses, where inertial forces are shown in addition to the spring and viscous damper forces, the two equations of motion can be written by equating the sum of the forces in the vertical direction to zero. The equation of motion for the top mass is:

$$m_1\ddot{x}_1 + (c_1 + c_2)\dot{x}_1 + (k_1 + k_2)x_1 - c_2\dot{x}_2 - k_2x_2 = 0 \tag{2.3.1}$$

and the equation of motion for the bottom mass is:

$$m_2\ddot{x}_2 - c_2\dot{x}_1 - k_2x_1 + c_2\dot{x}_2 + k_2x_2 = 0. \tag{2.3.2}$$

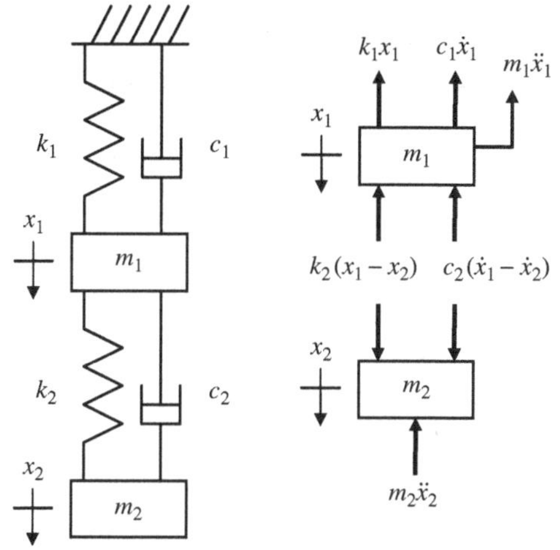

Fig. 2.3.1 Two degree of freedom, damped lumped parameter model (left); free body diagram (right)

IN A NUTSHELL A single degree of freedom system has one equation of motion. A two degree of freedom system has two equations of motion.

The difference we may observe between the single and two degree of freedom situations is that, for the two degree of freedom case, the equations of motion are coupled; they both contain the displacements x_1 and x_2 and velocities $\dot{x}_1$ and $\dot{x}_2$. This complicates the system solution and provides the motivation for modal analysis, which enables us to uncouple these two equations through a coordinate transformation and then use our single degree of freedom solution techniques. Before describing this approach, however, let's continue with our discussion of the chain-type two degree of freedom model.

The equations of motion are compactly expressed using a matrix formulation:

$$\begin{bmatrix} m_1 & 0 \\ 0 & m_2 \end{bmatrix} \begin{Bmatrix} \ddot{x}_1 \\ \ddot{x}_2 \end{Bmatrix} + \begin{bmatrix} c_1 + c_2 & -c_2 \\ -c_2 & c_2 \end{bmatrix} \begin{Bmatrix} \dot{x}_1 \\ \dot{x}_2 \end{Bmatrix} + \begin{bmatrix} k_1 + k_2 & -k_2 \\ -k_2 & k_2 \end{bmatrix} \begin{Bmatrix} x_1 \\ x_2 \end{Bmatrix} = \begin{Bmatrix} 0 \\ 0 \end{Bmatrix}. \tag{2.3.3}$$

The coupling is seen to occur in the symmetric damping and stiffness matrices for this chain-type model due to the nonzero off-diagonal terms in the matrix positions (1,2) and (2,1). If we represent the mass and stiffness matrices as $[M]$ and $[K]$, neglect damping for now, and assume a harmonic solution of the form $x = Xe^{st}$, we can write:

$$\left([M]s^2 + [K]\right)\{X\}e^{st} = \{0\}. \tag{2.3.4}$$

Similar to Eq. 2.1.2, there are two possibilities for the product in Eq. 2.3.4. If $\{X\} = \{0\}$, we obtain the trivial solution. We are therefore interested in the case when $\left([M]s^2 + [K]\right) = \{0\}$. From linear algebra [5], we know that for this matrix of equations to have a non-trivial solution, the determinant must be equal to zero. This represents the characteristic equation for our system.

$$\left|[M]s^2 + [K]\right| = 0 \tag{2.3.5}$$

The determinant of a two row, two column (2×2) matrix can be calculated by finding the difference between the products of the on-diagonal (1,1 and 2,2) terms and the off-diagonal terms. This is expressed generically as:

$$\begin{vmatrix} as^2 + b & cs^2 + d \\ cs^2 + d & es^2 + f \end{vmatrix} = 0 \tag{2.3.6}$$

or

$$\left(as^2 + b\right)\left(es^2 + f\right) - \left(cs^2 + d\right)^2 = 0. \tag{2.3.7}$$

This equation is quadratic in s^2, i.e., $gs^4 + hs^2 + m = 0$, and we can find the roots, s_1^2 and s_2^2, using the quadratic equation. These two roots are the eigenvalues for the two degree of freedom system. The natural frequencies are calculated as:

$$s_1^2 = -\omega_{n1}^2 \text{ and } s_2^2 = -\omega_{n2}^2, \tag{2.3.8}$$

where, by convention, $\omega_{n1} < \omega_{n2}$.

IN A NUTSHELL As you might guess, a single degree of freedom system has a single natural frequency (the frequency at which the system would "like" to vibrate) and a two degree of freedom system has two natural frequencies. Interestingly, vibration in either of the two degree of freedom system's natural frequencies is associated with a characteristic deformation pattern. That pattern is called a mode shape. Mathematically, it is referred to as an eigenvector, a German term that indicates that the natural frequencies (eigenvalues) and corresponding mode shapes (eigenvectors) are fundamental properties of the system.

To find the eigenvectors, or mode shapes, we substitute s_1^2 and s_2^2 into the equation of motion for the top or bottom mass (either will give the same solution because we imposed linear dependence between the two equations when we set the determinant equal to zero in Eq. 2.3.5). The equation of motion for the top mass corresponds to the top row in Eq. 2.3.9; recall that we are ignoring damping for now. See Eq. 2.3.10.

$$\begin{bmatrix} m_1 s^2 + k_1 + k_2 & -k_2 \\ -k_2 & m_2 s^2 + k_2 \end{bmatrix} \begin{Bmatrix} X_1 \\ X_2 \end{Bmatrix} = \begin{Bmatrix} 0 \\ 0 \end{Bmatrix} \tag{2.3.9}$$

$$\left(m_1 s^2 + k_1 + k_2\right) X_1 - k_2 X_2 = 0 \tag{2.3.10}$$

Because the two mode shapes represent the relative magnitude and direction of vibration between the two coordinates in the two degree of freedom system, we want to calculate either the ratio $\frac{X_1}{X_2}$ or $\frac{X_2}{X_1}$. We can choose to normalize the eigenvector to either coordinate x_1 or x_2. In most situations, the coordinate of interest or location of force application is selected. For example, we will normalize to the tool point for our studies of machining gess dynamics. For the chain-type model, if we wish to normalize to coordinate x_1, we require the ratios $\frac{X_1}{X_1} = 1$ and $\frac{X_2}{X_1}$. Using Eq. 2.3.10, we find that $\frac{X_2}{X_1} = \frac{m_1 s^2 + k_1 + k_2}{k_2}$ and the first mode shape is:

$$\psi_1 = \begin{Bmatrix} \frac{X_1}{X_1} \\ \frac{X_2}{X_1} \end{Bmatrix} = \begin{Bmatrix} 1 \\ \frac{m_1 s_1^2 + k_1 + k_2}{k_2} \end{Bmatrix}. \tag{2.3.11}$$

The second mode shape is determined by substitution of s_2^2 in place of s_1^2:

$$\psi_2 = \begin{Bmatrix} \frac{X_1}{X_1} \\ \frac{X_2}{X_1} \end{Bmatrix} = \begin{Bmatrix} 1 \\ \frac{m_1 s_2^2 + k_1 + k_2}{k_2} \end{Bmatrix}. \tag{2.3.12}$$

The first mode shape corresponds to vibration in the first natural frequency ω_{n1}, while the second mode shape is associated with vibration at ω_{n2}. In general, the system will vibrate in a linear combination of both mode shapes/natural frequencies, depending on the initial conditions. If we've followed the convention of $\omega_{n1} < \omega_{n2}$ and normalized to the x_1 coordinate, we'll find that the first mode shape will take the form $\psi_1 = \begin{Bmatrix} 1 \\ a>0 \end{Bmatrix}$, where a is a real number, which indicates that the two masses are vibrating exactly in phase with one another (i.e., they reach their maximum and minimum displacements at the same instants in time). We'll also see that the second mode shape will take the form $\psi_2 = \begin{Bmatrix} 1 \\ b<0 \end{Bmatrix}$, where b is a real number, which means that the mass motions are exactly out of phase with one another (i.e., when one mass reaches its maximum displacement, the other is at its minimum displacement).

IN A NUTSHELL Mode shapes only show relative size. A mode shape could show, for example, that when vibrating at the first natural frequency, the motion of coordinate 1 is twice as large as the motion of coordinate 2. This mode shape could be represented as $\psi_1 = \begin{Bmatrix} 1 \\ 0.5 \end{Bmatrix}$, $\psi_1 = \begin{Bmatrix} 2 \\ 1 \end{Bmatrix}$, or $\psi_1 = \begin{Bmatrix} 40000 \\ 20000 \end{Bmatrix}$. All of these representations are equivalent; they convey the characteristic deformation pattern. The "size" of the motion depends on the initial conditions, but the mode shapes do not.

Example 2.3.1: Free vibration using complex coefficients In this example we will calculate the time response of the system in Fig. 2.3.1 when the mass values are $m_1 = 1$ kg and $m_2 = 0.5$ kg, the stiffness values are $k_1 = 1 \times 10^7$ N/m and $k_2 = 2 \times 10^7$ N/m, the initial displacement of x_1 is $x_{0,1} = 1$ mm and the initial displacement of x_2 is $x_{0,2} = -1$ mm, and the initial velocities are zero. The equations of motion in matrix form are:

$$\begin{bmatrix} 1 & 0 \\ 0 & 0.5 \end{bmatrix} \begin{Bmatrix} \ddot{x}_1 \\ \ddot{x}_2 \end{Bmatrix} + \begin{bmatrix} 1 \times 10^7 + 2 \times 10^7 & -2 \times 10^7 \\ -2 \times 10^7 & 2 \times 10^7 \end{bmatrix} \begin{Bmatrix} x_1 \\ x_2 \end{Bmatrix} = \begin{Bmatrix} 0 \\ 0 \end{Bmatrix}.$$

The characteristic equation is:

$$\begin{vmatrix} 1s^2 + 3 \times 10^7 & -2 \times 10^7 \\ -2 \times 10^7 & 0.5s^2 + 2 \times 10^7 \end{vmatrix} = 0, \text{ or } 0.5s^4 + 3.5 \times 10^7 s^2 + 2 \times 10^{14} = 0.$$

This equation yields the two roots $s_1^2 = -6.277 \times 10^6$ (rad/s)^2 and $s_2^2 = -6.372 \times 10^7$ (rad/s)^2, which give the natural frequencies $\omega_{n1} = \sqrt{-s_1^2} = 2505$ rad/s and $\omega_{n2} = \sqrt{-s_2^2} = 7983$ rad/s. Expressed in units of Hz, these natural frequencies are $f_{n1} = \frac{\omega_{n1}}{2\pi} = 398.8$ Hz and $f_{n2} = \frac{\omega_{n2}}{2\pi} = 1271$ Hz.

Let's normalize the mode shapes to x_2 and arbitrarily select the equation of motion for the top mass to calculate the ratio $\frac{X_1}{X_2} = \frac{2 \times 10^7}{1s^2 + 3 \times 10^7}$. We obtain the first mode shape, which corresponds to vibration in ω_{n1}, by substituting s_1^2 in this ratio:

$$\psi_1 = \begin{Bmatrix} \frac{X_1}{X_2} \\ \frac{X_2}{X_2} \end{Bmatrix} = \begin{Bmatrix} \frac{2 \times 10^7}{-6.277 \times 10^6 + 3 \times 10^7} \\ 1 \end{Bmatrix} = \begin{Bmatrix} 0.8431 \\ 1 \end{Bmatrix}.$$

See Fig. 2.3.2, where the relative deflection amplitudes between coordinates 1 and 2 are identified. The second mode shape, which corresponds to vibration in ω_{n2}, is:

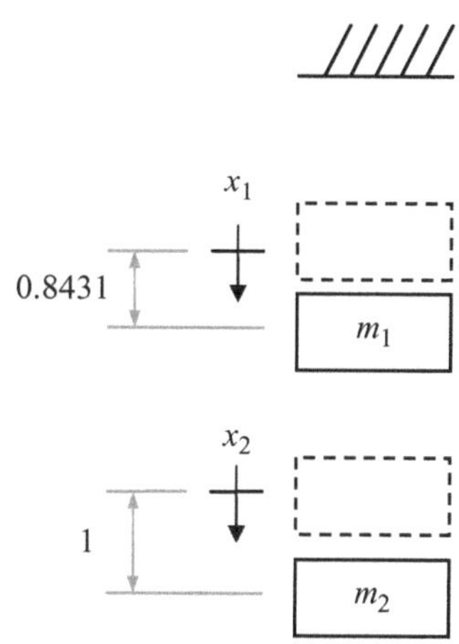

Fig. 2.3.2 Mode shape 1 normalized to coordinate 2

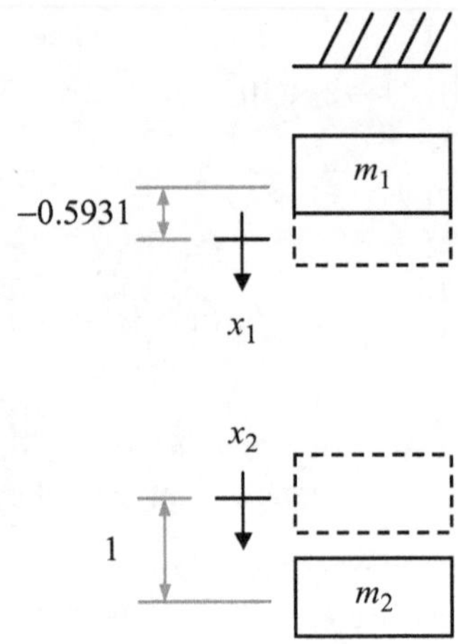

Fig. 2.3.3 Mode shape 2 normalized to coordinate 2

$$\psi_2 = \begin{Bmatrix} \dfrac{X_1}{X_2} \\ \dfrac{X_2}{X_2} \end{Bmatrix} = \begin{Bmatrix} \dfrac{2 \times 10^7}{-6.372 \times 10^7 + 3 \times 10^7} \\ 1 \end{Bmatrix} = \begin{Bmatrix} -0.5931 \\ 1 \end{Bmatrix}.$$

See Fig. 2.3.3, where the deflections are now in opposite directions (out of phase). Similar to Eq. 2.1.4, we can generically write the time-domain solution for the x_1 and x_2 vibrations as:

$$x_1 = X_{11}e^{i2505t} + X_{11}{}^*e^{-i2505t} + X_{12}e^{i7983t} + X_{12}{}^*e^{-i7983t}$$

and

$$x_2 = X_{21}e^{i2505t} + X_{21}{}^*e^{-i2505t} + X_{22}e^{i7983t} + X_{22}{}^*e^{-i7983t}.$$

Here, X_{ij} and $X_{ij}{}^*$ represent a complex conjugate pair, where the subscript i indicates the coordinate number and the subscript j denotes the natural frequency number. This solution suggests the general case that the total vibration is a linear combination of vibration in each of the two modes. The first time derivatives are:

$$\dot{x}_1 = i2505\left(X_{11}e^{i2505t} - X_{11}{}^*e^{-i2505t}\right) + i7983\left(X_{12}e^{i7983t} - X_{12}{}^*e^{-i7983t}\right)$$

and

$$\dot{x}_2 = i2505\left(X_{21}e^{i2505t} - X_{21}{}^*e^{-i2505t}\right) + i7983\left(X_{22}e^{i7983t} - X_{22}{}^*e^{-i7983t}\right).$$

Substitution of the initial conditions leads to a system of four equations with eight unknowns.

$$x_{0,1} = 1 = X_{11} + X_{11}{}^{*} + X_{12} + X_{12}{}^{*}$$
$$x_{0,2} = -1 = X_{21} + X_{21}{}^{*} + X_{22} + X_{22}{}^{*}$$
$$\dot{x}_{0,1} = 0 = i2505(X_{11} - X_{11}{}^{*}) + i7983(X_{12} - X_{12}{}^{*})$$
$$\dot{x}_{0,2} = 0 = i2505(X_{21} - X_{21}{}^{*}) + i7983(X_{22} - X_{22}{}^{*})$$

However, we can apply the mode shape relationships to reduce this to a system of four equations with four unknowns. Using the same definitions for the X_{ij} subscripts, we can write $\frac{X_{11}}{X_{21}} = 0.8431$ and $\frac{X_{11}}{X_{21}} = -0.5931$. After substitution and rewriting in matrix form, we obtain:

$$\begin{bmatrix} 0.8431 & 0.8431 & -0.5931 & -0.5931 \\ 1 & 1 & 1 & 1 \\ i2112 & -i2112 & i4734 & -i4734 \\ i2505 & -i2505 & i7983 & -i7983 \end{bmatrix} \begin{Bmatrix} X_{21} \\ X_{21}{}^{*} \\ X_{22} \\ X_{22}{}^{*} \end{Bmatrix} = \begin{Bmatrix} 1 \\ -1 \\ 0 \\ 0 \end{Bmatrix}, \text{ or} [A]\{X\} = \{b\}.$$

We can determine the coefficients by inverting $[A]$ and premultiplying $\{b\}$ by this result, $\{X\} = [A]^{-1}\{b\}$. This can be accomplished in MATLAB® using the `inv` function. At the command prompt (>>), first define the 4×4 $[A]$ matrix and 4×1 $\{b\}$ vector and then compute $\{X\}$:

```
>> A = [0.8431 0.8431 -0.5931 -0.5931; 1 1 1 1; i*2112
  -i*2112 i*4734 -i*4734; i*2505 -i*2505  i*7983 -i*7983];
>> b = [1 -1 0 0]';
>> X = inv (A)*b
```

The result is $\begin{Bmatrix} X_{21} \\ X_{21}{}^{*} \\ X_{22} \\ X_{22}{}^{*} \end{Bmatrix} = \begin{Bmatrix} 0.1417 \\ 0.1417 \\ -0.6417 \\ -0.6417 \end{Bmatrix}$. Using these values and the mode shape relationships to obtain the remaining four coefficients, we can substitute in the original x_1 and x_2 expressions to determine the time dependent free vibration for our example system.

$$x_1 = 0.1194e^{i2505t} + 0.1194e^{-i2505t} + 0.3805e^{i7983t} + 0.3805e^{-i7983t}$$

$$x_2 = 0.1417e^{i2505t} + 0.1417e^{-i2505t} - 0.6417e^{i7983t} - 0.6417e^{-i7983t}$$

Further, we can use the Euler identity $e^{i\theta} + e^{-i\theta} = 2\cos(\theta)$ to rewrite x_1 and x_2 as a sum of cosines. It is seen that the final motion of each mass is a linear combination of vibration in the two natural frequencies.

$$x_1 = 0.2388\cos(2505t) + 0.7610\cos(7983t)$$

$$x_2 = 0.2834\cos(2505t) + 1.283\cos(7983t)$$

IN A NUTSHELL The prior procedure is long, but the concept is this: the free vibration of a two degree of freedom system happens at the first natural frequency, at the second natural frequency, or a linear combination of the two at the same time. The final behavior depends on the initial conditions. No other frequencies of motion are possible for the two degree of freedom system in free vibration.

A potential problem with this approach is that, for additional degrees of freedom, the size of the matrix varies with the square of the number of coordinates. For example, we inverted a $2^2 \times 2^2$, or 4×4, complex matrix for our two degree of freedom system. For a three degree of freedom model, it would be necessary to invert a $3^2 \times 3^2$, or 9×9, complex matrix, and so on. While computational capabilities continually increase, modal analysis offers an alternative to this approach. The fundamental idea behind modal analysis is that a coordinate transformation is applied to convert from the model, or local, coordinate system into a modal coordinate system. While these modal coordinates do not have physical significance, they lead to uncoupled equations of motion because the off-diagonal terms in the mass and stiffness matrices are zero. The coordinate transformation is a diagonalization process and relies upon the orthogonality of the eigenvectors. Let's rework Example 2.3.1 to demonstrate the modal analysis approach.

IN A NUTSHELL There are many possible choices of coordinates. For example, we used the motion of each mass relative to ground. An equally valid choice would be to use the motion of one mass relative to ground and the motion of the other mass relative to the first mass. Perhaps it is possible that there is a more favorable choice of coordinates – one that would make the math easier. Surprisingly, there is. It is always possible to choose a set of coordinates so that the equations of motion are uncoupled. For a two degree of freedom system, it means that it is possible to choose coordinates so that there are two separate single degree of freedom equations of motion. These special coordinates are called modal coordinates.

Example 2.3.2: Free vibration by modal analysis The first step in the modal analysis approach is typically to find the eigensolution (natural frequencies and mode shapes). However, we have already completed this step in the previous example. Our next task is to define the modal matrix, $[P]$. It is a square matrix whose columns are composed of the mode shapes, $[P] = [\psi_1 \quad \psi_2] = \begin{bmatrix} 0.8431 & -0.5931 \\ 1 & 1 \end{bmatrix}$, where we've continued with the decision to normalize to coordinate x_2 for illustrative purposes. As noted, the orthogonality conditions for eigenvectors (see Appendix A) enable us to diagonalize (i.e., make the off-diagonal terms zero) the mass and stiffness matrices and, therefore, uncouple the two equations of motion. The new mass and stiffness matrices in modal coordinates (identified by the q subscripts) are determined by premultiplying the mass and stiffness matrices in local coordinates by the transpose of the modal matrix and postmultiplying this product by the modal matrix.

$$[M_q] = [P]^T[M][P] = \begin{bmatrix} 0.8431 & 1 \\ -0.5931 & 1 \end{bmatrix} \begin{bmatrix} 1 & 0 \\ 0 & 0.5 \end{bmatrix} \begin{bmatrix} 0.8431 & -0.5931 \\ 1 & 1 \end{bmatrix} = \begin{bmatrix} 1.211 & 0 \\ 0 & 0.8518 \end{bmatrix} \text{kg}$$

$$[K_q] = [P]^T[K][P] = \begin{bmatrix} 0.8431 & 1 \\ -0.5931 & 1 \end{bmatrix} \begin{bmatrix} 3 \times 10^7 & -2 \times 10^7 \\ -2 \times 10^7 & 2 \times 10^7 \end{bmatrix} \begin{bmatrix} 0.8431 & -0.5931 \\ 1 & 1 \end{bmatrix}$$

$$[K_q] = \begin{bmatrix} 7.601 \times 10^6 & 0 \\ 0 & 5.4282 \times 10^7 \end{bmatrix} \text{N/m}$$

These computations can be performed in MATLAB® by first defining $[P]$, $[M]$, and $[K]$, then typing the following commands:

```
>> Mq = P' *M* P;
>> Kq = P' *K* P;
```

Adding the semicolon to the ends of these commands suppresses the echo feature where the answer is displayed as soon as the "Enter" button is pressed. To see the results, either omit the semicolon or follow the first command with >> `Mq` which will display the most recent value of the variable `Mq`. Do not be alarmed if the off-diagonal terms are not identically zero in the modal mass and stiffness matrices. This is due to round-off error in our modal matrix entries. Carrying more significant digits will drive the off-diagonal terms closer to zero.

The two equations of motion can now be written in modal coordinates q_1 and q_2 using the matrix formulation: $[M_q]\{\ddot{q}\} + [K_q]\{q\} = \{0\}$. We see that the two equations are uncoupled and may be treated as separate single degree of freedom systems.

$$1.211\ddot{q}_1 + 7.601 \times 10^6 q_1 = 0$$
$$0.8518\ddot{q}_2 + 5.4282 \times 10^7 q_2 = 0$$

To use the solution techniques we developed in Section 2.1, we also need the initial conditions to be expressed in modal coordinates. Because the relationship between local and modal coordinates is $\{x\} = [P]\{q\}$, we can write $\{q\} = [P]^{-1}\{x\}$. To invert our 2×2 modal matrix, we switch the on-diagonal terms, change the sign of the off-diagonal terms, and divide each term by the scalar determinant, $|P| = P(1,1)P(2,2) - P(1,2)P(2,1)$.

$$[P]^{-1} = \frac{\begin{bmatrix} 1 & 0.5931 \\ -1 & 0.8431 \end{bmatrix}}{0.8431 \cdot 1 - (-0.5931) \cdot 1} = \begin{bmatrix} 0.6963 & 0.4130 \\ -0.6963 & 0.5870 \end{bmatrix}$$

We can then calculate the initial displacements $\{q_0\} = [P]^{-1}\{x_0\} = \begin{Bmatrix} 0.2833 \\ -1.283 \end{Bmatrix}$ mm and velocities $\{\dot{q}_0\} = [P]^{-1}\{\dot{x}_0\} = \begin{Bmatrix} 0 \\ 0 \end{Bmatrix}$ in modal coordinates. These calculations may be carried out in MATLAB® by typing the following commands:

```
>> P = [0.8431 -0.5931; 1 1];
>> x0 = [1 -1]';
>> q0 = inv(P)*x0
>> dx0 = [0 0]';
>> dq0 = inv(P)*dx0
```

We introduce here another general form for the solution of undamped free vibration given the initial displacement and velocity (in addition to the information provided in Section 2.1). The resulting displacement can be written as $x = \frac{\dot{x}_0}{\omega_n}\sin(\omega_n t) + x_0\cos(\omega_n t)$. Using this form, the modal displacements are $q_1 = 0\sin(2505t) + 0.2833\cos(2505t)$, which represents motion in the first natural frequency, and $q_2 = 0\sin(7983t) - 1.283\cos(7983t)$, which describes motion in the second natural frequency. To obtain the motion in local coordinates, we must perform the coordinate transformation $\{x\} = [P]\{q\} = \begin{bmatrix} 0.8431 & -0.5931 \\ 1 & 1 \end{bmatrix}\begin{Bmatrix} q_1 \\ q_2 \end{Bmatrix}$, which provides the relationships:

$$x_1 = 0.8431q_1 - 0.5931q_2$$

and

$$x_2 = q_1 + q_2.$$

It should be emphasized that the x_2 vibration is determined simply by summing the modal displacements, q_1 and q_2. This is a direct outcome of normalizing our mode shapes to x_2 and is an important result for us. We will take advantage of the fact that the local response can be written as a sum of the modal contributions when we perform our modal fitting of measured FRFs. Also, we see that the x_1 motion is a linear combination of q_1 and q_2, where each modal response is scaled by the corresponding mode shape. Substitution of our q_1 and q_2 values into the previous equations for x_1 and x_2 yields the same result we obtained using the technique shown in Example 2.3.1, but the modal analysis approach did not require the inversion of the $2^2 \times 2^2$ complex matrix.

$$
\begin{aligned}
x_1 &= 0.2388\cos(2505t) + 0.7610\cos(7983t) \\
x_2 &= 0.2834\cos(2505t) + 1.283\cos(7983t)
\end{aligned}
$$

IN A NUTSHELL All free vibration of a two degree of freedom systems can be thought of, in the general case, as a linear combination of motion in the two modes. Each mode has a natural frequency and a corresponding mode shape. The resulting combination depends on the initial conditions. In Sections 2.5 and 2.6, we discuss the measurement/determination of the modes that characterize the way a complicated system would "like" to vibrate.

The final consideration in this section is solution of the two degree of freedom free vibration problem in the presence of damping. We've already stated that every physical system dissipates energy, so our analysis should incorporate the viscous damping matrix shown in Eq. 2.3.3. However, this complicates the eigensolution. At this point, we need to introduce the concept of proportional damping. Physically, proportional damping means that all the coordinates pass through their equilibrium (zero) positions at the same instant for each mode shape. For the low damping observed in the tool-holder assemblies we will be considering, this assumption is realistic. For very high damping values, however, it is less reasonable because there may be significant phase differences between the motions of individual coordinates. Mathematically, proportional damping requires that the damping matrix can be written as a linear combination of the mass and stiffness matrices: $[C] = \alpha[M] + \beta[K]$, where α and β are real numbers.

IN A NUTSHELL Many kinds of damping are possible. There is great mathematical advantage in assuming that the damping is proportional and we often assume that it is, even if it is not.

Provided the proportional damping requirement is satisfied, then damping may be neglected in the eigensolution and the modal analysis procedure follows the steps provided in Example 2.3.2. The only modifications are that we must calculate the modal damping matrix $\left[C_q\right] = [P]^T[C][P]$ and the general solution to the uncoupled modal equations of motion $\left[M_q\right]\{\ddot{q}\} + \left[C_q\right]\{\dot{q}\} + \left[K_q\right]\{q\} = \{0\}$ is different. For the underdamped case, we can write $q = e^{-\zeta\omega_n t}\left(\frac{\dot{q}_0+\zeta\omega_n q_0}{\omega_d}\sin(\omega_d t) + q_0\cos(\omega_d t)\right)$. Otherwise, the solution proceeds as before.

2.4 Two Degree of Freedom Forced Vibration

We will use the two degree of freedom lumped parameter spring-mass-damper model shown in Fig. 2.3.1, but will impose external harmonic forces at coordinates x_1 and x_2 for the general case. However, because our analysis is limited to linear systems, we can apply the principle of superposition to consider the forces separately and then sum the individual contributions. For demonstration purposes, we will consider only the $f_2e^{i\omega t}$ force applied to coordinate x_2. The equations of motion in matrix form for this system are:

$$\begin{bmatrix} m_1 & 0 \\ 0 & m_2 \end{bmatrix}\begin{Bmatrix} \ddot{x}_1 \\ \ddot{x}_2 \end{Bmatrix} + \begin{bmatrix} c_1+c_2 & -c_2 \\ -c_2 & c_2 \end{bmatrix}\begin{Bmatrix} \dot{x}_1 \\ \dot{x}_2 \end{Bmatrix} + \begin{bmatrix} k_1+k_2 & -k_2 \\ -k_2 & k_2 \end{bmatrix}\begin{Bmatrix} x_1 \\ x_2 \end{Bmatrix} = \begin{Bmatrix} 0 \\ f_2 \end{Bmatrix}. \tag{2.4.1}$$

By assuming solutions of the form $x_{1,2} = X_{1,2}e^{i\omega t}$ and substituting in Eq. 2.4.1, we obtain:

$$\left(-\omega^2[M] + i\omega[C] + [K]\right)\{X\}e^{i\omega t} = \{F\}e^{i\omega t}. \tag{2.4.2}$$

We have two methods that we can use to determine the steady state forced vibration response for this system. The first is modal analysis, which requires proportional damping, and the second is complex matrix inversion, which places no restrictions on the nature of the system damping. Let's begin with modal analysis.

2.4.1 Modal Analysis

Our first step in the modal analysis approach is to write the system equations of motion in local coordinates as shown in Eq. 2.4.1; we continue to consider the f_2 case in this discussion. Provided proportional damping exists (i.e., $[C] = \alpha[M] + \beta[K]$ is true), then we can ignore damping to find the eigensolution. Note that this solution is also independent of the external force(s). We find the eigenvalues (natural frequencies) and eigenvectors (modes shapes) using Eq. 2.3.4, $\left([M]s^2 + [K]\right)\{X\}e^{st} = \{0\}$. The eigenvalues are determined from the roots of Eq. 2.3.5, $\left|[M]s^2 + [K]\right| = 0$. The natural frequencies are computed from $s_j^2 = -\omega_{nj}^2$, $j = 1$ to 2 (the number of degrees of freedom). We can then use either of the equations of motion to find the 2×1 mode shapes for the two degree of freedom system:

$$\psi_1 = \left\{ \begin{matrix} \frac{X_1}{X_2}\left(s_1^2\right) \\ 1 \end{matrix} \right\} \text{and}\, \psi_2 = \left\{ \begin{matrix} \frac{X_1}{X_2}\left(s_2^2\right) \\ 1 \end{matrix} \right\}, \tag{2.4.3}$$

where we have normalized to the location of the force application (coordinate x_2). Using the mode shapes, we assemble the 2×2 modal matrix $[P] = [\psi_1 \quad \psi_2]$. We can then use the modal matrix to transform into modal coordinates (and uncouple the equations of motion). The diagonal modal mass, damping, and stiffness matrices are: $\left[M_q\right] = [P]^T[M][P] = \begin{bmatrix} m_{q1} & 0 \\ 0 & m_{q2} \end{bmatrix}$, $\left[C_q\right] = [P]^T[C][P] = \begin{bmatrix} c_{q1} & 0 \\ 0 & c_{q2} \end{bmatrix}$, and $\left[K_q\right] = [P]^T[K][P] = \begin{bmatrix} k_{q1} & 0 \\ 0 & k_{q2} \end{bmatrix}$, respectively. We must also transform the local force vector into modal coordinates:

$$\{R\} = \left\{ \begin{matrix} R_1 \\ R_2 \end{matrix} \right\} = [P]^T\{F\} = \begin{bmatrix} \frac{X_1}{X_2}\left(s_1^2\right) & 1 \\ \frac{X_1}{X_2}\left(s_2^2\right) & 1 \end{bmatrix} \left\{ \begin{matrix} 0 \\ f_2 \end{matrix} \right\} = \begin{bmatrix} p_1 & 1 \\ p_2 & 1 \end{bmatrix} \left\{ \begin{matrix} 0 \\ f_2 \end{matrix} \right\} = \left\{ \begin{matrix} f_2 \\ f_2 \end{matrix} \right\}. \tag{2.4.4}$$

The modal equations of motion are:

$$\begin{aligned} m_{q1}\ddot{q}_1 + c_{q1}\dot{q}_1 + k_{q1}q_1 &= R_1 \\ m_{q2}\ddot{q}_2 + c_{q2}\dot{q}_2 + k_{q2}q_2 &= R_2 \end{aligned} \tag{2.4.5}$$

and the corresponding complex FRFs (steady state responses in the frequency-domain) are:

$$\frac{Q_1}{R_1} = \frac{1}{k_{q1}}\left(\frac{\left(1 - r_1^2\right) - i\left(2\zeta_{q1}r_1\right)}{\left(1 - r_1^2\right)^2 + \left(2\zeta_{q1}r_1\right)^2}\right) \text{and}\, \frac{Q_2}{R_2} = \frac{1}{k_{q2}}\left(\frac{\left(1 - r_2^2\right) - i\left(2\zeta_{q2}r_2\right)}{\left(1 - r_2^2\right)^2 + \left(2\zeta_{q2}r_2\right)^2}\right), \tag{2.4.6}$$

where $r_{1,2} = \frac{\omega}{\omega_{n1,2}}$ and $\zeta_{q1,2} = \frac{c_{q1,2}}{2\sqrt{k_{q1,2}m_{q1,2}}}$. We transform into local coordinates using $\{X\} = \left\{ \begin{matrix} X_1 \\ X_2 \end{matrix} \right\} = [P]\{Q\} = \begin{bmatrix} p_1 & p_2 \\ 1 & 1 \end{bmatrix} \left\{ \begin{matrix} Q_1 \\ Q_2 \end{matrix} \right\}$ so that $X_1 = p_1Q_1 + p_2Q_2$ and $X_2 = Q_1 + Q_2$. Dividing each of these equations by F_2 gives the cross and direct FRFs for the f_2 force application, respectively. The cross FRF, which indicates that the force and measurement coordinates are not coincident, is:

$$\frac{X_1}{F_2} = \frac{p_1Q_1 + p_2Q_2}{F_2} = p_1\frac{Q_1}{F_2} + p_2\frac{Q_2}{F_2} = p_1\frac{Q_1}{R_1} + p_2\frac{Q_2}{R_2}, \tag{2.4.7}$$

where we see that the cross FRF is the sum of the modal FRFs scaled by the mode shapes (remember that $R_1 = R_2 = F_2$ from Eq. 2.4.4). The direct FRF, which denotes that the measurement is performed at the force input location, is:

$$\frac{X_2}{F_2} = \frac{Q_1 + Q_2}{F_2} = \frac{Q_1}{F_2} + \frac{Q_2}{F_2} = \frac{Q_1}{R_1} + \frac{Q_2}{R_2}. \tag{2.4.8}$$

We observe the important result that the direct FRF is simply the sum of the modal contributions. This is important for our subsequent analyses. Measurement of the frequency response functions on a physical system enable extraction of the model parameters and visualization of the natural frequencies and mode shapes.

Example 2.4.1: Forced vibration by modal analysis Consider the chain-type, lumped parameter two degree of freedom system shown in Fig. 2.4.1. For the upper spring-mass-damper, the local coordinate constants are: $k_1 = 4 \times 10^5$ N/m, $c_1 = 80$ N-s/m, and $m_1 = 2$ kg. For the lower spring-mass-damper, the local coordinate constants are: $k_2 = 6 \times 10^5$ N/m, $c_2 = 120$ N-s/m, and $m_2 = 1$ kg. A harmonic force $f_2 = 100e^{i\omega t}$ N is applied to the lower mass (at coordinate x_2); we will not consider any force applied to the upper mass, although this force could be considered separately and the result added to the solution of the analysis we will perform here. The local mass, damping, and stiffness matrices are:

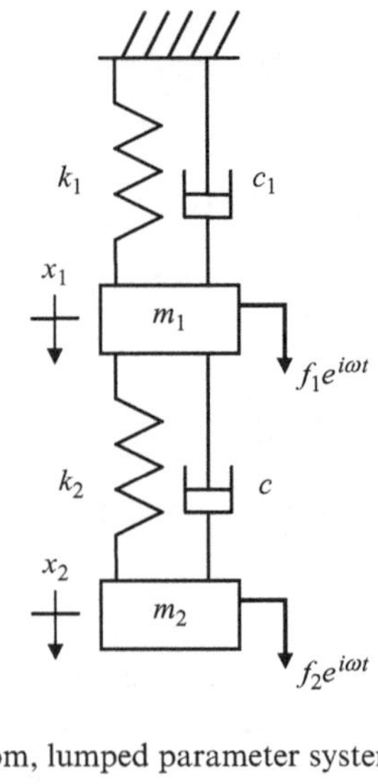

Fig. 2.4.1 Two degree of freedom, lumped parameter system (damped with force)

$M = \begin{bmatrix} 2 & 0 \\ 0 & 1 \end{bmatrix}$ kg, $C = \begin{bmatrix} 200 & -120 \\ -120 & 120 \end{bmatrix}$ N-s/m, and $K = \begin{bmatrix} 1 \times 10^6 & -6 \times 10^5 \\ -6 \times 10^5 & 6 \times 10^5 \end{bmatrix}$ N/m, respectively. To use modal analysis, we must verify that proportional damping exists. For $\alpha = 0$ and $\beta = \frac{1}{5000}$, we see that the relationship $[C] = \alpha[M] + \beta[K]$ is satisfied. We can therefore determine the eigenvalues using:

$$\begin{vmatrix} 2s^2 + 1 \times 10^6 & -6 \times 10^5 \\ -6 \times 10^5 & s^2 + 6 \times 10^5 \end{vmatrix} = 0.$$

The two roots of the determinant are: $s_1^2 = -122799.81$ (rad/s)^2 and $s_2^2 = -977200.19$ (rad/s)^2, which gives the natural frequencies $\omega_{n1} = 350.43$ rad/s and $\omega_{n2} = 988.53$ rad/s ($\omega_{n1} < \omega_{n2}$). To determine the roots using MATLAB®, we first write the characteristic equation: $(2s^2 + 1 \times 10^6)(s^2 + 6 \times 10^5) - (-6 \times 10^5)^2 = 0$, or after simplifying $2s^4 + 2.2 \times 10^6 s^2 + 2.4 \times 10^{11} = 0$. Because this equation is quadratic in s^2, we can find the roots s_1^2 and s_2^2 using the `roots` function.

```
>> roots([2 2.2e6 2.4e11])
```

For the eigenvectors (mode shapes), we normalize to the location of the force application, coordinate x_2. Using the equation of motion for the top mass (arbitrarily selected), we obtain the required ratio $\frac{X_1}{X_2} = \frac{6 \times 10^5}{2s^2 + 1 \times 10^6}$. Substitution of $s_1^2 = -122799.81$ (rad/s)^2 and $s_2^2 = -977200.19$ (rad/s)^2 into Eq. 2.4.3 gives the two mode shapes $\psi_1 = \begin{Bmatrix} 0.7953 \\ 1 \end{Bmatrix}$ and $\psi_2 = \begin{Bmatrix} -0.6287 \\ 1 \end{Bmatrix}$, respectively. We can now construct the modal matrix $[P] = [\psi_1 \quad \psi_2] = \begin{bmatrix} 0.7953 & -0.6287 \\ 1 & 1 \end{bmatrix}$ and transform the local mass, stiffness, and damping matrices into modal coordinates:

$$[M_q] = [P]^T[M][P] = \begin{bmatrix} 2.265 & 0 \\ 0 & 1.790 \end{bmatrix} \text{kg},$$

$$[K_q] = [P]^T[K][P] = \begin{bmatrix} 2.782 \times 10^5 & 0 \\ 0 & 1.750 \times 10^6 \end{bmatrix} \text{N/m},$$

and

$$[C_q] = [P]^T[C][P] = \begin{bmatrix} 55.63 & 0 \\ 0 & 349.9 \end{bmatrix} \text{N} - \text{s/m}.$$

A simple check at this point is to recalculate the natural frequencies using the modal parameters. The results should match the eigenvalue solution. Here, we see that $\omega_{n1} = \sqrt{\frac{k_{q1}}{m_{q1}}} = \sqrt{\frac{2.782\times10^5}{2.265}} = 350.46$ rad/s and $\omega_{n2} = \sqrt{\frac{k_{q2}}{m_{q2}}} = \sqrt{\frac{1.750\times10^6}{1.790}} = 988.76$ rad/s, where the differences are due to round-off error, but the results are essentially the same. We can also determine the modal damping ratios:

$$\zeta_{q1} = \frac{c_{q1}}{2\sqrt{k_{q1}m_{q1}}} = \frac{55.63}{2\sqrt{2.782\times10^5\cdot 2.265}} = 0.035\ (3.5\%\ \text{damping})$$

and

$$\zeta_{q2} = \frac{c_{q2}}{2\sqrt{k_{q2}m_{q2}}} = \frac{349.9}{2\sqrt{1.750\times10^6\cdot 1.790}} = 0.099\ (9.9\%\ \text{damping}).$$

To write our uncoupled equations of motion in modal coordinates, we also need the modal force vector, which we obtain by substitution into Eq. 2.4.4.

$$\{R\} = \begin{bmatrix} 0.7953 & 1 \\ -0.6287 & 1 \end{bmatrix}\begin{Bmatrix} 0 \\ 100 \end{Bmatrix} = \begin{Bmatrix} 100 \\ 100 \end{Bmatrix}\text{N}$$

$$2.265\ddot{q}_1 + 55.63\dot{q}_1 + 2.782\times10^5 q_1 = 100$$

$$1.790\ddot{q}_2 + 349.9\dot{q}_2 + 1.750\times10^6 q_2 = 100$$

The FRFs for the single degree of freedom modal systems are:

$$\frac{Q_1}{R_1} = \frac{1}{2.782\times10^5}\left(\frac{(1-r_1^2) - i(0.070r_1)}{(1-r_1^2)^2 + (0.070r_1)^2}\right)$$

and

$$\frac{Q_2}{R_2} = \frac{1}{1.750\times10^6}\left(\frac{(1-r_2^2) - i(0.198r_2)}{(1-r_2^2)^2 + (0.198r_2)^2}\right),$$

where $r_1 = \frac{\omega}{350.43}$ and $r_2 = \frac{\omega}{988.53}$. The direct and cross FRFs are then $\frac{X_2}{F_2} = \frac{Q_1}{R_1} + \frac{Q_2}{R_2}$ and $\frac{X_1}{F_2} = 0.7953\frac{Q_1}{R_1} - 0.6287\frac{Q_2}{R_2}$, respectively. See Figs. 2.4.2 and 2.4.3. Because motion in the second mode shape, corresponding to $\omega_{n2} = 988.53$ rad/s, exhibits a 180 deg phase shift between the two coordinates (i.e., they are out of phase), the second mode is "inverted" in the cross FRF plot. The MATLAB® program used to produce these figures is provided on the companion CD as p_2_4_1_1.m.

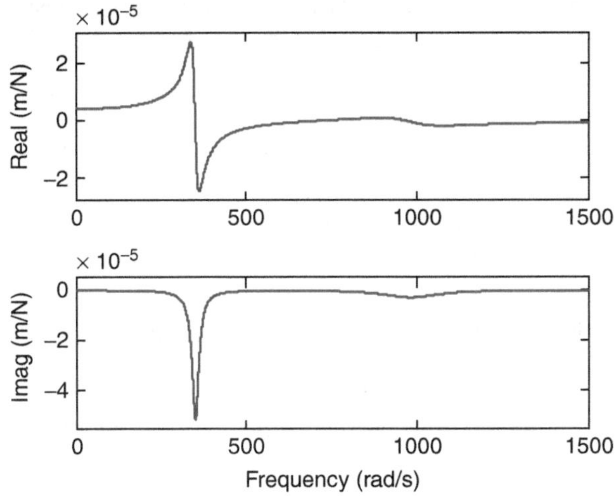

Fig. 2.4.2 The real and imaginary parts of the direct FRF are determined from the sum of the modal contributions

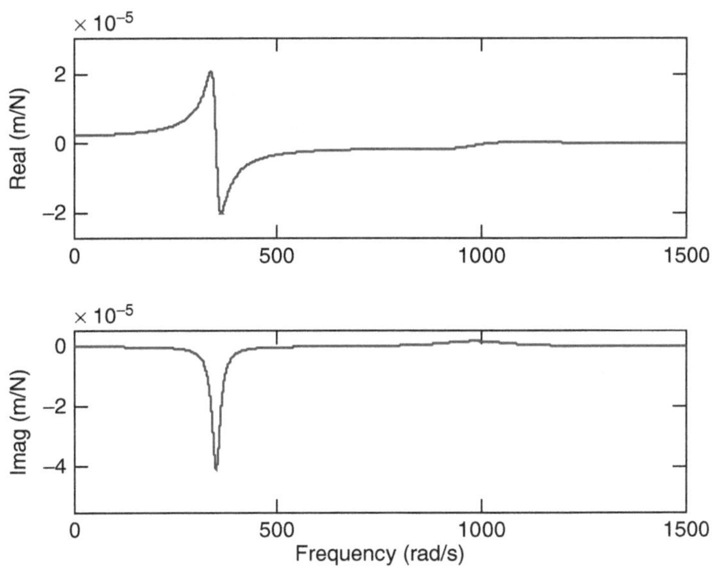

Fig. 2.4.3 The real and imaginary parts of the cross FRF are obtained by scaling the two modes by the corresponding mode shape and summing the results

2.4.2 Complex Matrix Inversion

Our final task of this section is to describe an alternative to modal analysis, referred to as complex matrix inversion. This approach does not require proportional damping, but does include the inversion of a 2×2 frequency dependent, complex matrix for the two degree of freedom system we are considering here. We'll first write Eq. 2.4.2 in the form $[A]\{X\} = \{F\}$, where $[A] = \begin{bmatrix} a_{11} & a_{12} \\ a_{21} & a_{22} \end{bmatrix} = \left(-\omega^2[M] + i\omega[C] + [K]\right)$. The two degree of freedom system has four FRFs that we'd like to determine. First, we have the direct and cross FRFs, $\frac{X_2}{F_2}$ and $\frac{X_1}{F_2}$, due to the force application at coordinate x_2 that we previously determined using modal analysis. Second, we have the direct and cross FRFs, $\frac{X_1}{F_1}$ and $\frac{X_2}{F_1}$, due to the force application at coordinate x_1. We did not explicitly show the modal solution to this case in Section 2.4, but the only differences are that we would normalize the mode shapes to x_1 and the FRFs would then be computed from $\frac{X_1}{F_1} = \frac{Q_1}{R_1} + \frac{Q_2}{R_2}$ and $\frac{X_2}{F_1} = p_1\frac{Q_1}{R_1} + p_2\frac{Q_2}{R_2}$, where $P = \begin{bmatrix} 1 & 1 \\ p_1 & p_2 \end{bmatrix}$ would be used to determine the modal mass, stiffness, and damping matrices.

Rewriting $[A]\{X\} = \{F\}$ as $\{X\}\{F\}^{-1} = [A]^{-1}$provides all four FRFs. They are ordered as: $\begin{Bmatrix} \frac{X_1}{F_1} & \frac{X_1}{F_2} \\ \frac{X_2}{F_1} & \frac{X_2}{F_2} \end{Bmatrix} = \begin{bmatrix} b_{11} & b_{12} \\ b_{21} & b_{22} \end{bmatrix}$, where we've used the b_{ij} notation to indicate the individual terms in the inverted $[A]$ matrix. In our analysis, $[A]$ is symmetric. Therefore, $b_{12} = b_{21}$ and $\frac{X_1}{F_2} = \frac{X_2}{F_1}$. This condition is referred to as reciprocity. Physically, it means that we get the same result if we: 1) excite the system at coordinate x_2 and measure the response at x_1, as if we: 2) excite the system at coordinate x_1 and measure the response at x_2.

For the two degree of freedom system, we can directly write the individual terms in $[A]^{-1}$ as:

$$[A]^{-1} = \frac{\begin{bmatrix} a_{22} & -a_{12} \\ -a_{21} & a_{11} \end{bmatrix}}{a_{11}\cdot a_{22} - a_{12}\cdot a_{21}} = \frac{\begin{bmatrix} -\omega^2 m_2 + i\omega c_2 + k_2 & i\omega c_2 + k_2 \\ i\omega c_2 + k_2 & -\omega^2 m_1 + i\omega(c_1 + c_2) + k_1 + k_2 \end{bmatrix}}{a_{11}\cdot a_{22} - a_{12}\cdot a_{21}}.$$

For example, $\frac{X_1}{F_1} = b_{11} = \frac{-\omega^2 m_2 + i\omega c_2 + k_2}{a_{11}\cdot a_{22} - a_{12}\cdot a_{21}}$. Note that this complex expression is a function of the forcing frequency ω so it must be evaluated over the desired frequency range in order to produce plots equivalent to those obtained for the modal analysis example.

IN A NUTSHELL Complex matrix inversion is a mathematical technique that is used when it is certain that the damping is not proportional.

2.5 System Identification

The previous section describes the modal analysis steps required to obtain the direct and cross FRFs in local coordinates given a system model (we treated the chain-type, lumped parameter case, but other model geometries could be considered as well). This approach required that the mass, damping, and stiffness matrices be known. However, this is not the case for arbitrary tool-holder assemblies. Our actual task is typically to measure the FRFs for the system of interest and then define a model by performing a modal fit to the measured data.

2.5.1 Modal Fitting

Our fitting approach will be a "peak-picking" method where we use the real and imaginary parts of the system FRFs to identify the modal parameters. This approach works well provided the system modes are not closely spaced. However, even if two modeled modes are relatively close in frequency, we can still obtain a reasonable modal fit as we'll see in Example 2.5.1. For additional information on modal testing, see [6].

To demonstrate the fitting steps, consider the direct FRF shown in Fig. 2.5.1. This FRF clearly has two modes within the measurement bandwidth. To determine the modal parameters which populate the 2×2 modal matrices, we must identify three frequencies and one peak value for each mode. [Note that we have automatically assumed proportional damping in using this approach. Additionally, if there were three dominant modes we wished to model, we would obtain 3×3 modal matrices and so on.] The frequencies labeled 1 and 2

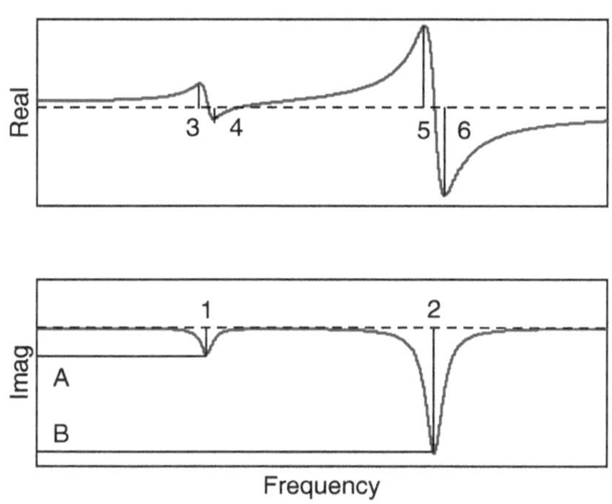

Fig. 2.5.1 Two degree of freedom direct FRF with the frequencies and amplitudes required for peak picking identified

along the horizontal frequency axis in the imaginary part of the direct FRF (Fig. 2.5.1) correspond to the minimum imaginary peaks and provide the two natural frequencies, ω_{n1} and ω_{n2}, respectively. The difference between frequencies 4 and 3, labeled along the frequency axis of the real part of the direct FRF, is used to determine the modal damping ratio for the first mode, ζ_{q1}:

$$\omega_4 - \omega_3 = \omega_{n1}\left(1 + \zeta_{q1}\right) - \omega_{n1}\left(1 - \zeta_{q1}\right) = 2\zeta_{q1}\omega_{n1} \text{ or } \zeta_{q1} = \frac{\omega_4 - \omega_3}{2\omega_{n1}}. \quad (2.5.1)$$

Similarly, the difference between frequencies 6 and 5 is used to determine ζ_{q2}:

$$\zeta_{q2} = \frac{\omega_6 - \omega_5}{2\omega_{n2}}. \quad (2.5.2)$$

The (negative) peak value, A, identified along the vertical axis of the imaginary part of the direct FRF is next used to find the modal stiffness value, k_{q1}:

$$A = \frac{-1}{2k_{q1}\zeta_{q1}} \text{ or } k_{q1} = \frac{-1}{2\zeta_{q1}A}. \quad (2.5.3)$$

Similarly, the peak value B is used to determine k_{q2}:

$$k_{q2} = \frac{-1}{2\zeta_{q2}B}. \quad (2.5.4)$$

At this point, we can directly populate the modal stiffness matrix $\left[K_q\right] = \begin{bmatrix} k_{q1} & 0 \\ 0 & k_{q2} \end{bmatrix}$. However, we must calculate the modal mass and damping values from the additional information we've obtained. We determine the modal masses using the natural frequencies and modal stiffness values:

$$\omega_{n1} = \sqrt{\frac{k_{q1}}{m_{q1}}} \text{ or } m_{q1} = \frac{k_{q1}}{\omega_{n1}^2} \text{ and } m_{q2} = \frac{k_{q2}}{\omega_{n2}^2}. \quad (2.5.5)$$

The modal damping coefficients are computed using the modal damping ratios, stiffness values, and masses:

$$\zeta_{q1} = \frac{c_{q1}}{2\sqrt{k_{q1}m_{q1}}} \text{ or } c_{q1} = 2\zeta_{q1}\sqrt{k_{q1}m_{q1}} \text{ and } c_{q2} = 2\zeta_{q2}\sqrt{k_{q2}m_{q2}}. \quad (2.5.6)$$

We can now write the remaining modal matrices $\left[M_q\right] = \begin{bmatrix} m_{q1} & 0 \\ 0 & m_{q2} \end{bmatrix}$ and $\left[C_q\right] = \begin{bmatrix} c_{q1} & 0 \\ 0 & c_{q2} \end{bmatrix}$. Identification of the modal matrices provides sufficient information for us to complete the time-domain turning and milling simulations we develop in Chapters 3 and 4. However, to complete this discussion, we will detail the steps necessary to define a chain-type, lumped parameter model based on measured FRFs. Before continuing with the model definition, let's complete an example of peak picking to determine the modal matrices.

IN A NUTSHELL Measured frequency response functions contain the information necessary to construct a mathematical model of the system in question.

Example 2.5.1: Peak picking modal fit Figure 2.5.2 shows an example FRF that could be obtained from a tool point measurement. Our task is to perform a modal fit to identify the modal mass, damping, and stiffness matrices. The first step is to decide how many modes we wish to fit. A visual inspection of the FRF shows that a three mode fit is appropriate. The three natural frequencies are identified by locating the three minimum peaks of the imaginary part and recording the associated frequencies. These are identified as 499 Hz, 761 Hz, and 849 Hz in Fig. 2.5.3. We determine the modal damping ratios using the frequencies of the local maximum and minimum values of the real part according to Eq. 2.5.1. These are shown as 460 Hz and 533 Hz for mode 1; 726 Hz and 787 Hz for mode 2; and 827 Hz and 873 Hz for mode 3. The modal damping ratios are then:

$$\zeta_{q1} = \frac{533 - 460}{2 \cdot 499} = 0.073,\ \zeta_{q2} = \frac{787 - 726}{2 \cdot 761} = 0.040,\ \text{and}\ \zeta_{q3} = \frac{873 - 827}{2 \cdot 849} = 0.027.$$

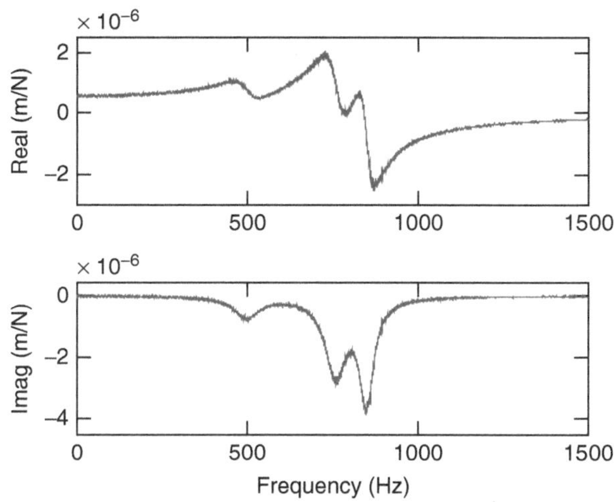

Fig. 2.5.2 Example tool point FRF for peak picking exercise

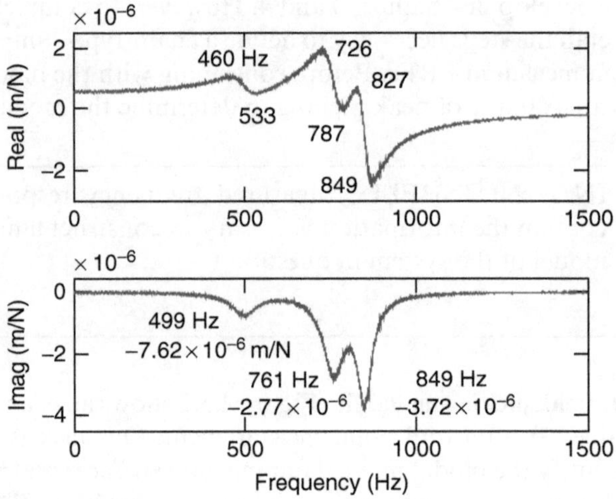

Fig. 2.5.3 Three degree of freedom peak picking example with required frequencies and amplitudes identified

The imaginary part negative peak values for each mode are also listed in Fig. 2.5.3. The modal stiffness values are calculated using Eq. 2.5.3.

$$k_{q1} = \frac{-1}{2 \cdot 0.073 \cdot (-7.62 \times 10^{-7})} = 8.99 \times 10^6 \text{ N/m}$$

$$k_{q2} = \frac{-1}{2 \cdot 0.040 \cdot (-2.77 \times 10^{-6})} = 4.51 \times 10^6 \text{ N/m}$$

$$k_{q3} = \frac{-1}{2 \cdot 0.027 \cdot (-3.72 \times 10^{-6})} = 4.98 \times 10^6 \text{ N/m}$$

We find the modal masses using Eq. 2.5.5. We must be sure to pay special attention to units for these calculations; note that we have switched from frequency units of Hz to rad/s by multiplying by 2π and the stiffness values are expressed in N/m.

$$m_{q1} = \frac{8.99 \times 10^6}{(499 \cdot 2\pi)^2} = 0.914 \text{ kg} \qquad m_{q2} = \frac{4.51 \times 10^6}{(761 \cdot 2\pi)^2} = 0.197 \text{ kg}$$

$$m_{q3} = \frac{4.98 \times 10^6}{(849 \cdot 2\pi)^2} = 0.175 \text{ kg}$$

Finally, the modal damping coefficients are determined using Eq. 2.5.6. Again, units compatibility should be ensured. In the following calculations, stiffness and mass values are expressed in N/m and kg, respectively, to obtain damping coefficient units of N-s/m.

$$c_{q1} = 2 \cdot 0.073\sqrt{8.99 \times 10^6 \cdot 0.914} = 419\,\text{N} - \text{s/m}$$

$$c_{q2} = 2 \cdot 0.040\sqrt{4.51 \times 10^6 \cdot 0.197} = 75.4\,\text{N} - \text{s/m}$$

$$c_{q3} = 2 \cdot 0.027\sqrt{4.98 \times 10^6 \cdot 0.175} = 50.4\,\text{N} - \text{s/m}$$

The 3×3 modal matrices can now be written as:

$$[M_q] = \begin{bmatrix} 0.914 & 0 & 0 \\ 0 & 0.197 & 0 \\ 0 & 0 & 0.175 \end{bmatrix} \text{kg}, \ [C_q] = \begin{bmatrix} 419 & 0 & 0 \\ 0 & 75.4 & 0 \\ 0 & 0 & 50.4 \end{bmatrix} \text{N} - \text{s/m},$$

and

$$[K_q] = \begin{bmatrix} 8.99 \times 10^6 & 0 & 0 \\ 0 & 4.51 \times 10^6 & 0 \\ 0 & 0 & 4.98 \times 10^6 \end{bmatrix} \text{N/m}.$$

The individual modal contributions may be described using Eq. 2.5.6:

$$\frac{Q_j}{R_j} = \frac{1}{k_{qj}} \left(\frac{\left(1 - r_j^2\right) - i(2\zeta_{qj} r_j)}{\left(1 - r_j^2\right)^2 + (2\zeta_{qj} r_j)^2} \right),$$

where $r_j = \frac{\omega}{\omega_{nj}}$, $j = 1$ to 3. The individual modes are plotted, together with the original FRF, in Fig. 2.5.4. As we've discussed, however, the direct FRF in local (physical) coordinates is the sum of the modal contributions so we may simply add the individual modal responses on a frequency by frequency basis to define our final fit. This result is shown in Fig. 2.5.5.

For this contrived example, the original modal parameters used to construct the "measured" FRF are known. Therefore, we can compare our modal approximation to the true values. These results are provided in Table 2.5.1. The MATLAB® program used to produce the example figures is provided on the companion CD as p_2_5_1_1.m.

IN A NUTSHELL The errors between the true and fit parameters arise from limited resolution in the "measurement" and the simplifications used to create the fitting rules.

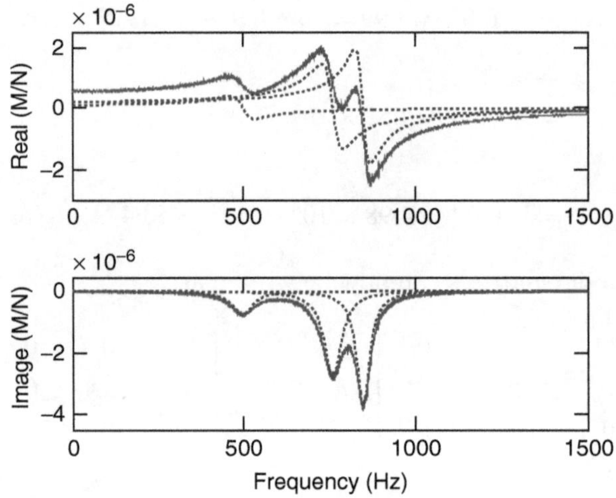

Fig. 2.5.4 Example tool point and three modal coordinate FRFs determined by peak picking approach

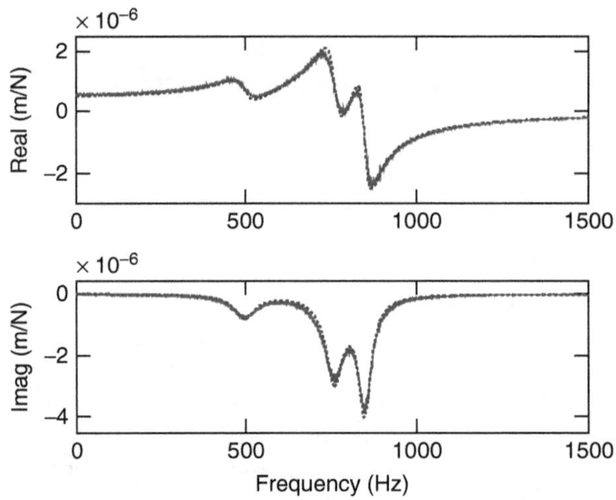

Fig. 2.5.5 Example tool point FRF with three degree of freedom modal fit obtained by peak picking

Table 2.5.1 True modal parameters and values obtained by peak picking modal fit

	Mode 1		Mode 2		Mode 3	
	True	Fit	True	Fit	True	Fit
f_n (Hz)	500	499	760	761	850	849
ζ_q	0.090	0.073	0.050	0.040	0.030	0.027
k_q (N/m)	8.00×10^6	8.99×10^6	4.00×10^6	4.51×10^6	5.00×10^6	4.98×10^6

2.5.2 Model Definition

Once we have determined the modal matrices by peak picking, the next step in defining a model is to use the measured direct and cross FRFs to find the mode shapes and construct the modal matrix. We'll again assume that the measured direct FRF, shown in Fig. 2.5.1, can be approximated with a two mode fit. This means that our model will have two degrees of freedom. As we've seen, for a two degree of freedom model, the mode shapes are 2×1 vectors so that the square modal matrix has dimensions of 2×2. Because the mode shapes have just two entries (one of which is 1), we only require one cross FRF to determine the second entry. As before, we can choose the coordinate to which we normalize our mode shapes for the model shown in Fig. 2.4.1. Let's define the coordinate of interest as x_2 so that the form of the modal matrix is $[P] = \begin{bmatrix} p_1 & p_2 \\ 1 & 1 \end{bmatrix}$. We determine p_1 and p_2 using: 1) the peak imaginary part values denoted C, corresponding to the first mode with the natural frequency ω_{n1}, and D, the second mode with the natural frequency ω_{n2}, in the cross FRF[1] shown in Fig. 2.5.6, together with: 2) the A and B values identified in Fig. 2.5.1.

$$\frac{C}{A} = \frac{\frac{-p_1}{2k_{q1}\zeta_{q1}}}{\frac{-1}{2k_{q1}\zeta_{q1}}} = p_1 \text{ and } \frac{D}{B} = \frac{\frac{-p_2}{2k_{q2}\zeta_{q2}}}{\frac{-1}{2k_{q2}\zeta_{q2}}} = p_2 \tag{2.5.7}$$

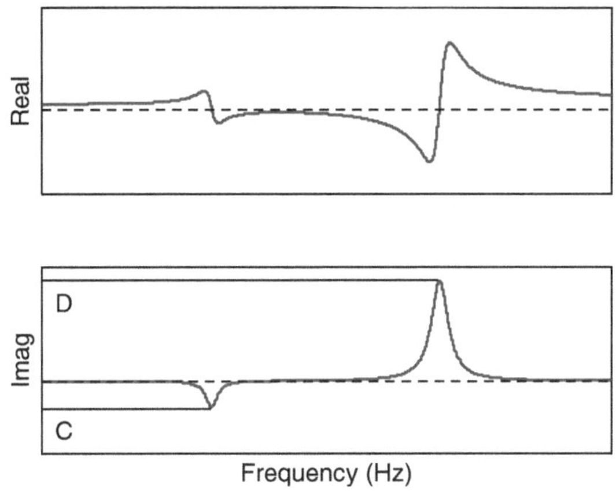

Fig. 2.5.6 Two degree of freedom cross FRF with the amplitudes required for model development identified

[1] We observe that the cross FRF in Fig. 2.5.6 looks very different than the direct FRF in Fig. 2.5.1; the higher frequency mode is "upside down" in Fig. 2.5.6. As we saw in Section 2.4, this is because the two modes are out of phase for the cross FRF, which results in the sign change.

We have used the ratio of the peak of the cross FRF to the direct FRF in each mode to determine the mode shapes because, as we discussed previously, the cross FRF can be expressed as the sum of the modal contributions with each mode scaled by the corresponding system mode shape. See Eq. 2.4.7. Once we have defined the modal matrix, we can determine the model parameters in local coordinates using the transformations (from modal to local coordinates) in Eqs. 2.5.8–2.5.10. The forms of $[M]$, $[C]$, and $[K]$ correspond to the pre-selected two degree of freedom chain-type, lumped parameter model.

$$[P]^{-T}[M_q][P]^{-1} = [M] = \begin{bmatrix} m_1 & 0 \\ 0 & m_2 \end{bmatrix} \tag{2.5.8}$$

$$[P]^{-T}[C_q][P]^{-1} = [C] = \begin{bmatrix} c_1 + c_2 & -c_2 \\ -c_2 & c_2 \end{bmatrix} \tag{2.5.9}$$

$$[P]^{-T}[K_q][P]^{-1} = [K] = \begin{bmatrix} k_1 + k_2 & -k_2 \\ -k_2 & k_2 \end{bmatrix} \tag{2.5.10}$$

IN A NUTSHELL The measured direct and cross frequency response functions provide a way to visualize the mode shape. The relative sizes (and signs) of peaks at the same frequency show the relative motion of the corresponding coordinates at that frequency.

As a final note regarding model definition, it should be emphasized that if the measured direct FRF has three modes that we wish to model, then the square modal matrix will have dimensions of 3×3. To determine the modal matrix, we must measure, at minimum, two cross FRFs to give the two ratios required for the 3×1 mode shapes. Additional cross FRF measurements may be necessary to find measurement locations with good signal to noise ratio (i.e., away from system nodes, or locations of zero vibration amplitude regardless of the force input level).

2.5.3 Modal Truncation

Prior to describing modal testing equipment, there is one remaining issue to highlight regarding modal fitting. Because FRF measurements always have a finite frequency range and elastic bodies possess an infinite number of degrees of freedom, there are necessarily modes that exist outside the measurement range. For tool point tests, we typically measure from zero to a few kHz (perhaps up to 10 kHz for a small mass impact hammer with a steel tip – see Section 2.6). For most machining applications, neglecting (or truncating) modes with natural frequencies of, for example, greater than 5 kHz in our modal fit is reasonable because they are not likely to be excited by the cutting force in any case. However, omitting these higher frequency modes during peak picking does affect the accuracy of the modal fit, particularly the real part of the FRF. Equations 2.2.6 and 2.2.7, which describe the real and imaginary parts of a single degree of freedom FRF, are reproduced here to demonstrate the effect.

$$\mathrm{Re}\left(\frac{X}{F}\right) = \frac{1}{k}\left(\frac{1-r^2}{(1-r^2)^2+(2\zeta r)^2}\right) \tag{2.5.11}$$

$$\mathrm{Im}\left(\frac{X}{F}\right) = \frac{1}{k}\left(\frac{-2\zeta r}{(1-r^2)^2+(2\zeta r)^2}\right) \tag{2.5.12}$$

It is seen that when the frequency ratio $r = \frac{\omega}{\omega_n}$ is large, or the driving frequency ω is very high and outside the measurement range, the denominator within the right parenthetical terms in these two equations becomes very large and the response approaches zero. This is seen at the right hand side of Fig. 2.2.4, for example. However, as r approaches zero, the parenthetical term in the real part approaches one and the parenthetical term in the imaginary part approaches zero. Therefore, the value of the real part approaches $\frac{1}{k}$ as r approaches zero[2]. If there are modes beyond the measurement bandwidth, neglecting these terms and the associated $\frac{1}{k}$ contributions leads to errors in the vertical location of the modal fit's real part. This is demonstrated in Ex. 2.5.2.

Example 2.5.2: High frequency mode truncation during modal fitting A "measured" FRF is provided in Fig. 2.5.7. We will presume that the measurement bandwidth was 2 kHz, although a 5 kHz frequency range is shown for

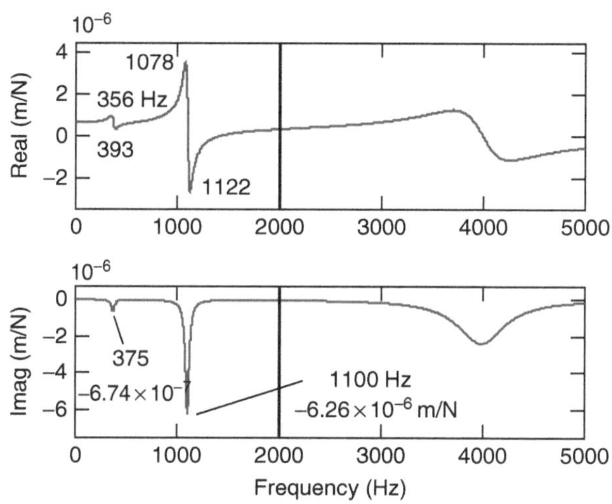

Fig. 2.5.7 "Measured" direct FRF for Ex. 2.5.2. The peak picking values are listed within the 2 kHz measurement bandwidth. A 5 kHz frequency range is provided to show the truncated 4000 Hz mode

demonstration purposes. Within the 2 kHz range, two modes are visible and peak picking can be applied to determine the associated modal parameters. Using the values from the figure, the modal stiffness, mass, and damping matrix terms may be determined as shown in Ex. 2.5.1.

$$\zeta_{q1} = \frac{393 - 356}{2 \cdot 375} = 0.049 \qquad \zeta_{q2} = \frac{1122 - 1078}{2 \cdot 1100} = 0.020$$

$$k_{q1} = \frac{-1}{2 \cdot 0.049 \cdot (-6.74 \times 10^{-7})} = 1.50 \times 10^{7}\,\text{N/m}$$

$$k_{q2} = \frac{-1}{2 \cdot 0.020 \cdot (-6.26 \times 10^{-6})} = 3.99 \times 10^{6}\,\text{N/m}$$

$$m_{q1} = \frac{1.50 \times 10^{7}}{(375 \cdot 2\pi)^2} = 2.70\,\text{kg} \qquad m_{q2} = \frac{3.99 \times 10^{6}}{(1100 \cdot 2\pi)^2} = 0.084\,\text{kg}$$

$$c_{q1} = 2 \cdot 0.049\sqrt{1.50 \times 10^{7} \cdot 2.70} = 624\,\text{N} - \text{s/m}$$

$$c_{q2} = 2 \cdot 0.020\sqrt{3.99 \times 10^{6} \cdot 0.084} = 23.2\,\text{N} - \text{s/m}$$

The fit to the measured direct FRF is determined by summing the two contributions in modal coordinates according to:

$$\frac{X}{F} = \frac{Q_1}{R_1} + \frac{Q_2}{R_2} = \frac{1}{k_{q1}}\left(\frac{(1 - r_1^2) - i(2\zeta_{q1}r_1)}{(1 - r_1^2)^2 + (2\zeta_{q1}r_1)^2}\right) + \frac{1}{k_{q2}}\left(\frac{(1 - r_2^2) - i(2\zeta_{q2}r_2)}{(1 - r_2^2)^2 + (2\zeta_{q2}r_2)^2}\right),$$

where $r_1 = \frac{f}{375}$ and $r_2 = \frac{f}{1100}$ and f is given in Hz. It is seen in Fig. 2.5.8 that, although the shape of the two modes within the 2 kHz bandwidth are correctly identified, there is a noticeable offset in the real part of the fit. It appears too stiff (i.e., it is located below the measured FRF) because the DC compliance due to the 4000 Hz mode has not been considered. Because this mode is outside the measurement frequency range, it is not possible to fit the mode and determine the appropriate modal parameters. However, given the visible offset in Fig. 2.5.8, the combined contributions of truncated modes can be included by adding an effective DC compliance term to the fit. Specifically, for this example, the fit could be rewritten as:

$$\frac{X}{F} = \frac{1}{k} + \frac{Q_1}{R_1} + \frac{Q_2}{R_2},$$

where the $\frac{Q_j}{R_j}$ terms ($j = 1, 2$) are obtained through peak picking as described previously and the $\frac{1}{k}$ value is selected to move the fit to a vertical overlap with the measured FRF. If a value of $k = 3 \times 10^{6}$ N/m is applied

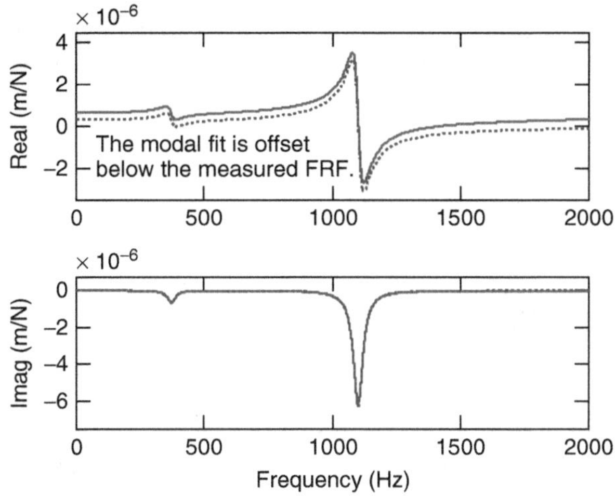

Fig. 2.5.8 Result of modal fitting. An offset in the real part of the fit (dotted line) is observed because the DC compliance of the 4000 Hz mode is not included

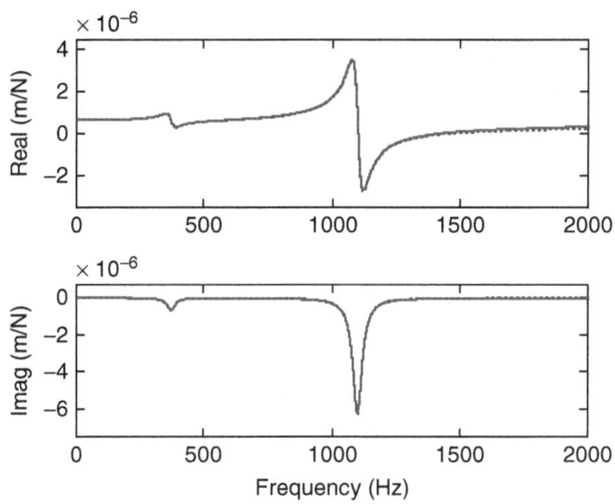

Fig. 2.5.9 Result of modal fitting with the addition of a DC compliance to correct for the truncated mode

here, the fit is improved and the result shown in Fig. 2.5.9 is obtained. Note that this stiffness value is equal to the modal stiffness of the 4000 Hz mode shown in Fig. 2.5.7 (for completeness, the modal damping ratio for this mode is 0.07). The MATLAB® program used to produce the Ex. 2.5.2 figures is provided on the companion CD as p_2_5_2_1.m.

IN A NUTSHELL A measured frequency response function includes the effect of all modes. Some of the modes contributing to the measured FRF may be outside the range of frequencies measured. Nevertheless, they still influence the data in the measured frequency range.

2.6 Modal Testing Equipment

The basic hardware required to measure FRFs is:

- a mechanism for known force input across the desired frequency range;
- a transducer for vibration measurement, again with the required bandwidth; and
- a dynamic signal analyzer to record the time-domain force and vibration inputs and convert these into the desired FRF.

The dynamic signal analyzer includes input channels for the time-domain force and vibration signals and computes the Fourier transform of these signals to convert them to the frequency-domain. It then calculates the ratio of the frequency-domain vibration signal to the frequency-domain force signal; this ratio is the FRF. The form of the FRF depends on the vibration transducer type and can be expressed as:

- receptance/compliance – the ratio of displacement to force (considered in the previous sections);
- mobility – the ratio of velocity to force; and
- inertance/accelerance – the ratio of acceleration to force.

2.6.1 Force Input

Common types of force excitation include:

- fixed frequency sine wave – The complex response is determined one frequency at a time with averaging occurring at each frequency over a short time interval. This is referred to as a sine sweep test.
- random signal – The frequency content of the random signal may be broadband (white noise) or truncated to a desired range (pink noise). Averaging over a fixed period of time is again applied.
- impulse – A short duration impact is used to excite the structure. This approach enables a broad range of frequencies to be excited in a single, short test. Multiple tests are typically averaged in the frequency-domain to improve coherence, or the correlation between the force and vibration signals.

IN A NUTSHELL Hitting the structure with a hammer excites many frequencies at almost the same level at the same time. (Really!)

Common force input hardware includes:

- shaker (similar to a speaker) – These systems include a harmonically driven armature and a base. The armature may be actuated along its axis by a magnetic coil or hydraulic force. The magnetic coil, or electrodynamic, configurations can provide excitation frequencies of tens of kHz with force levels from tens to thousands of N (increased force typically means a lower frequency range). Hydraulic shakers offer high force with the potential for a static preload, but relatively lower frequency ranges. In either case, the force is often applied to the structure of interest through a "stinger", or a slender rod that supports axial tension and compression, but not bending or shear. A load cell is often incorporated in the setup to measure the input force. One consideration is that this load cell adds mass to the system under test, which can alter the FRF for low mass structures. Finally, the shaker must be isolated from the structure to prevent reaction forces due to the shaker motion from being transmitted through the shaker base to the structure.
- impact hammer – An impact hammer incorporates a force transducer in a metal, plastic, or rubber tip to measure the force input during a hammer strike. Because the setup and measurement time is short, it is a popular choice for tool-holder testing (referred to as impact testing). Naturally, the energy input to the structure is a function of the hammer mass; therefore, many sizes are available. Also, the bandwidth of the force input depends on the mass and tip stiffness. Stiffer tips tend to excite a wider frequency range, but also spread the input energy over this wider range. Softer tips concentrate the energy over a lower frequency range. Hard, plastic tips are a common choice for tool testing because they do not damage the cutting edge and generally provide sufficient excitation bandwidth.

2.6.2 Vibration Measurement

Vibration transducers are available in both non-contact and contact types. While non-contact transducers, such as capacitance probes and laser vibrometers, are preferred because they do not influence the system dynamics, contacting types, such as accelerometers, are more convenient to implement. As a compromise, low mass accelerometers are often used for tool point FRF testing. For most tools, the addition of a few grams or less of accelerometer mass does not appreciably alter the response and the accelerometer can be attached using wax and then removed without damaging the cutting tool. Because accelerometers produce a signal which is proportional to acceleration, the inertance FRF is obtained. However, to convert from inertance, or $\frac{A}{F}$, to receptance, $\frac{X}{F}$, we can use the relationship:

$$\frac{X}{F} = -\frac{1}{\omega^2} \cdot \frac{A}{F}, \tag{2.6.1}$$

which follows from the harmonic solution, $x = Xe^{i\omega t}$, and its second time derivative $\ddot{x} = -\omega^2 Xe^{i\omega t} = -\omega^2 x$. Equation 2.6.1 effectively describes double numerical integration in the frequency-domain.

2.7 Measurement Uncertainties

As with any measurement situation, a complete description of the FRF should include both the frequency dependent mean values of the real and imaginary parts, for instance, and the frequency dependent uncertainty in these values. A defensible uncertainty statement for FRF testing results requires a bivariate uncertainty analysis because the FRF is complex valued. In [7], an analysis is provided for tool point impact testing that considers statistical variations, imperfect calibration coefficients that convert the voltage to engineering units for the hammer and vibration transducer, misalignment between the intended and actual force/hammer direction during impact, and mass loading when using an accelerometer. The complex valued FRF is expressed by its real and imaginary parts, which are potentially correlated and this correlation is included in the bivariate analysis [8–9]. An ellipsoid shaped confidence region (at each frequency) is defined in the complex plane; the size and orientation of this region is determined from the individual input uncertainties. The scalar, total uncertainty is then determined using an eigenanalysis of the FRF covariance matrix [10]. While the full analysis is not reproduced here, we highlight some of its findings.

- For the end mill-holder-spindle assembly tested, the calibration coefficient uncertainties (as specified by the manufacturers) comprised >80% of the total uncertainty, which was approximately 2% of the mean tool point direct receptance.
- A small bias is introduced when the hammer force input direction is misaligned with the accelerometer axis. This cosine type bias can be compensated using [11]:

$$FRF_t = \frac{X}{F_t} = \frac{X}{F_t\left(1 - \frac{1}{2}u^2(\beta)\right)} = FRF_m\left(1 - \frac{1}{2}u^2(\beta)\right)^{-1}, \tag{2.7.1}$$

- where the t and m subscripts refer to "true" and "measured", respectively, and $u(\beta)$ is the uncertainty in the misalignment angle, β. This uncertainty is typically a few degrees.
- Accelerometer mass loading also leads to a measurement bias. This bias can be compensated using [12]:

$$FRF_t = \frac{X_t}{F} = \frac{FRF_m}{1 + m_a\omega^2 FRF_m}, \tag{2.7.2}$$

- where m_a is the accelerometer mass, expressed in kg for compatibility with FRF units of m/N and frequency units of rad/s.

Exercises

1. A harmonic motion has an amplitude of 0.2 cm and a period of 15 seconds.
   a) Determine the maximum velocity (m/s) and maximum acceleration (m/s^2) of the periodic motion.
   b) Assume that the motion expresses the free vibration of an undamped single degree of freedom system and that the motion was initiated with an initial displacement and no initial velocity. Express the motion (in units of meters) in each of the following four forms.
      i) $A\cos(\omega_n t + \Phi_c)$
      ii) $A\sin(\omega_n t + \Phi_s)$
      iii) $B\cos(\omega_n t) + C\sin(\omega_n t)$
      iv) $De^{i(\omega_n t)} + Ee^{-i(\omega_n t)}$
2. A single degree of freedom lumped parameter system under free vibration can be modeled with the following mass, stiffness, and damping values: $m =$ 1 kg, $k = 4 \times 10^4$ N/m, and $c = 10$ N-s/m.
   a) Determine the natural frequency, f_n (Hz), and the damping ratio, ζ.
   b) Given an initial displacement of 5 mm and an initial velocity of 0 mm/s, find an expression for the time response of the damped free vibration, $x(t)$, of the form:

   $$x(t) = e^{-\zeta\omega_n t}(A\sin(\omega_d t) + B\cos(\omega_d t)),$$

   where $x(t)$ is expressed in meters. Plot the first 25 cycles of motion.
3. A single degree of freedom lumped parameter system has mass, stiffness, and damping values of 1.2 kg, 1×10^7 N/m, and 364.4 N-s/m, respectively. Generate the following plots of the frequency response function:
   a) magnitude (m/N) vs. frequency (Hz) and phase (deg) vs. frequency (Hz)

b) real part (m/N) vs. frequency (Hz) and imaginary part (m/N) vs. frequency (Hz)
c) Argand diagram, real part (m/N) vs. imaginary part (m/N).

4. For the two degree of freedom, damped lumped parameter system shown in Fig. e.2.4, complete parts a) through f).

$$k_a = 2 \times 10^5\,\mathrm{N/m} \qquad k_b = 5.5 \times 10^4\,\mathrm{N/m}$$

$$c_a = 60\,\mathrm{N-s/m} \qquad c_b = 16.5\,\mathrm{N-s/m}$$

$$m_a = 2.5\,\mathrm{kg} \qquad m_b = 1.2\,\mathrm{kg}$$

$$x_1(0) = 1\,\mathrm{mm} \qquad x_2(0) = 0\,\mathrm{mm}$$
$$\dot{x}_1(0) = 0\,\mathrm{mm/s} \qquad \dot{x}_2(0) = 0\,\mathrm{mm/s}$$

a) Obtain the equations of motion in matrix form and transform them into modal coordinates q_1 and q_2. Normalize your eigenvectors to coordinate x_2. Verify that proportional damping exists.
b) Determine the time responses $q_1(t)$ and $q_2(t)$ in mm. Express your solutions in the form: $q_{1,2}(t) = e^{-\zeta_{q1,2}\omega_{n1,2}t}\left(A\cos\left(\omega_{d1,2}t\right) + B\sin\left(\omega_{d1,2}t\right)\right)$.
c) Transform the modal coordinate solutions, $q_1(t)$ and $q_2(t)$, back into local coordinates, $x_1(t)$ and $x_2(t)$.
d) Plot $x_1(t)$ and $x_2(t)$ (in mm) vs. time (in seconds).
e) Determine the time responses $x_1(t)$ and $x_2(t)$ (in mm) if the initial velocities are zero and the initial displacements are $x_1(0) = 0.312$ mm and $x_2(0) = 1$ mm.

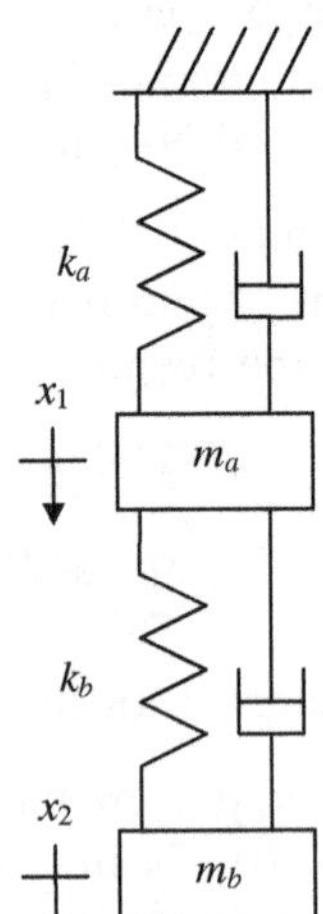

Fig. e.2.4 Two degree of freedom, damped lumped parameter system

5. Assume a harmonic force, $f_2 = f_0 e^{i\omega t}$, is applied to the lower mass (at coordinate x_2) in Fig. e.2.4. Obtain the FRFs $\frac{Q_1}{R_1}$, $\frac{Q_2}{R_2}$, and $\frac{X_2}{F_2}$. Express them in equation form and then plot the real and imaginary parts (in m/N) vs. frequency (in Hz).

References

1. Thomson, W. and Dahleh, M., 1998, Theory of Vibration with Application, 5th Ed., Prentice Hall, Upper Saddle River, NJ.
2. Weaver, Jr., W., Timoshenko, S., and Young, D., 1990, Vibration Problems in Engineering, 5th Ed., John Wiley and Sons, New York, NY.
3. Inman, D., 2001, Engineering Vibration, 2nd Ed., Prentice Hall, Upper Saddle River, NJ, Section 1.10.
4. Thomson, W. and Dahleh, M., 1998, Theory of Vibration with Application, 5th Ed., Prentice Hall, Upper Saddle River, NJ, Section 3.9.
5. Leon, J., 1994, Linear Algebra with Applications, 4th Ed., Prentice Hall, Englewood Cliffs, NJ, Section 5.6.
6. Ewins, D., 2000, Modal Testing: Theory, Practice, and Application, 2nd Ed., Taylor & Francis, London.
7. Kim, H. and Schmitz, T., 2007, Bivariate Uncertainty Analysis for Impact Testing, Measurement Science and Technology, 18: 3565–3571.
8. Jobson J., 1992, Applied Multivariate Data Analysis, Springer, New York, NY.
9. Ridler, N. and Salter, M., 2002, An Approach to the Treatment of Uncertainty in Complex S-parameter Measurements, Metrologia, 39: 295–302.
10. Hall B., 2004, On the Propagation of Uncertainty in Complex-valued Quantities, Metrologia, 41: 173–177.
11. International Standards Organization (ISO), 1993, Guide to the Expression of Uncertainty in Measurement (Corrected and Reprinted 1995).
12. Ashory, M., 1999, High Quality Modal Testing Methods, Ph.D. Dissertation, Imperial College of Science, Technology, and Medicine, London.

Chapter 3
Turning Dynamics

Make everything as simple as possible, but not simpler.

– Albert Einstein

In Chapter 2, we discussed how to use modal analysis to describe the tool point dynamics for tool-holder combinations. In this chapter, we'll discuss regenerative chatter in turning and see how we can apply the dynamics information to develop stability lobe diagrams that describe the limiting chip width (to avoid chatter) as a function of spindle speed. We'll also detail a time-domain simulation for predicting cutting force and tool displacement which also enables us to determine stable and unstable cutting conditions.

3.1 Turning Description

Turning operations are generally carried out on a lathe where a workpiece is rotated in a spindle past a tool mounted on a two axis slide in order to give the desired shape to the axisymmetric part; see Fig. 3.1.1. The final shape can include both internal and external features. The lathe may be manual, where a machinist controls the slide positions during material removal, or computer numerically controlled (CNC). In this case, automatic control is used to command the slide positions to follow the path described by the part program. The part program is based on the desired workpiece dimensions and is typically developed using computer aided design/computer aided manufacturing (CAD/CAM) software.

During turning, a sharp cutting edge is used to remove material in the form of a chip. Many studies have been performed to better understand chip formation and the associated mechanics, but our focus is a broader view of the resulting cutting force and corresponding vibrations of the tool. Therefore, we will not focus so much on the chip itself as on the corresponding system behavior. Texts that provide more information regarding metal cutting

T.L. Schmitz, K.S. Smith, *Machining Dynamics*, DOI 10.1007/978-0-387-09645-2_3,

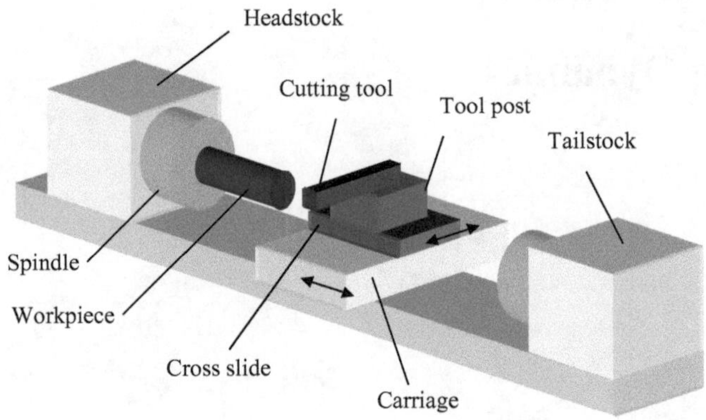

Fig. 3.1.1 Schematic of manual lathe. The workpiece is clamped in the rotating spindle, but may also be supported at its free end using the tailstock. The cutting tool, which is clamped to the tool post, is moved relative to the workpiece by adjusting the positions of the carriage (axial direction) and cross slide (radial direction)

fundamentals, heat generation and heat transfer during cutting, and related technology include, for example, [1–4].

To begin, let's consider the tool and workpiece to be rigid and develop expressions for the cutting force, F. Figure 3.1.2 shows an "orthogonal cutting" operation, where only the normal, F_n, and tangential, F_t, components of the force are considered. In general, the cutting force vector includes the third component along the part axis, but the orthogonal treatment is sufficient for us to describe the process dynamics. The figure also identifies: 1) the mean chip thickness, h_m, or commanded feed per revolution for the facing operation pictured; and 2) the force angle, β, between F and F_n. The side view of this operation (inset in Fig. 3.1.2) shows the chip width, b. Together, the chip

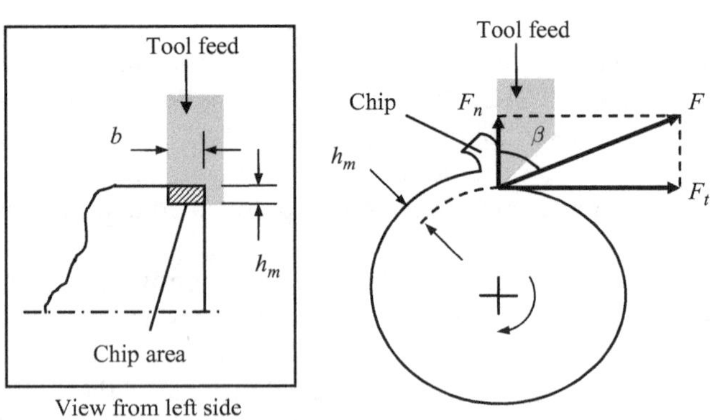

Fig. 3.1.2 Orthogonal cutting operation showing the cutting force with its normal and tangential components

thickness and chip width define the area of material to be removed, $A = bh_m$. We approximate the cutting force as the product of this chip area and an empirical coefficient. This process dependent coefficient is referred to as the specific (or per unit chip area) force, K_s, in [2] and depends on the workpiece material, tool geometry, and, to a lesser extent, the cutting speed (peripheral velocity of the rotating workpiece) and chip thickness.

$$F = K_s A = K_s b h_m \tag{3.1.1}$$

The normal and tangential components, F_n and F_t, can be expressed using F and the force angle:

$$F_n = \cos(\beta)F = \cos(\beta)K_s b h_m = k_n b h_m \text{ and} \tag{3.1.2}$$

$$F_t = \sin(\beta)F = \sin(\beta)K_s b h_m = k_t b h_m, \tag{3.1.3}$$

where we've defined the cutting force coefficients, k_n and k_t, which incorporate both K_s and β. Although efforts continue to calculate these coefficients based on elastic and plastic material properties, a common approach used to characterize these process dependent values is to prescribe known cutting conditions and measure the force components directly. If the tool is mounted on a cutting force dynamometer as shown in Fig. 3.1.3 and the b and h_m values are known, then the measured force component values can be used to determine the coefficients by rewriting Eqs. 3.1.2 and 3.1.3, $k_n = \frac{F_n}{bh_m}$ and $k_t = \frac{F_t}{bh_m}$. Typical units for k_n, k_t, and K_s are N/mm^2.

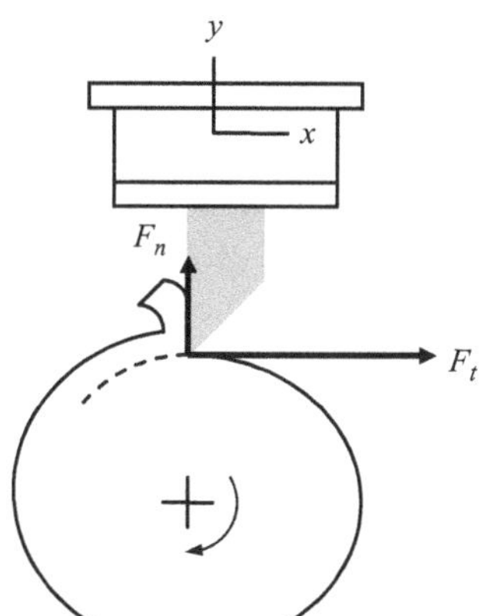

Fig. 3.1.3 Force measurement during turning using a cutting force dynamometer. The tangential and normal cutting forces are measured as the x and y dynamometer force components

Table 3.1.1 Representative K_s values for selected workpiece materials [2]

Material	K_s (N/mm^2)	Material	K_s (N/mm^2)
Gray cast iron	1500	Ni-based Inconel X	3400
1020 carbon steel	2100	Ni-based Udimet 500	3500
1035 carbon steel	2300	Co-based L605	3500
1045 carbon steel	2600	Ti (6Al, 4 V)	2000
302 stainless steel	2700	Al 7075-T6	850
4140/5140 alloy steel	2800	Al 6061-T6	750

Selected K_s values are provided in [2, Tables 7.1 and 8.1]. These have been reproduced in Table 3.1.1, but should be considered to be representative values and not necessarily specific to a particular application.

IN A NUTSHELL The force produced by the cutting operation is proportional to the frontal area of the chip through the coefficient K_s. Deriving K_s from first principles is quite difficult and, in this way, it is similar to the elastic modulus, E. Tabulated values are approximate and often good enough. High precision applications, on the other hand, may require careful measurement of the cut geometry and resulting forces using the intended tooling and workpiece materials, F.

3.2 Regenerative Chatter in Turning

If we remove the assumption of a rigid tool, then it is clear that the cutting force will cause deflections of the cutting tool. Because the tool has stiffness and mass, it can vibrate. If the tool is vibrating as it removes material, these vibrations are imprinted on the workpiece surface as a wavy profile. Figure 3.2.1 shows an exaggerated view, where the initial impact with the workpiece surface causes the tool to begin vibrating and the oscillations in the normal direction to be copied onto the workpiece. When the workpiece begins its second revolution, the vibrating tool encounters the wavy surface produced during the first revolution. Therefore, the chip thickness at any instant depends both on the tool deflection at that time and the workpiece surface from the previous revolution. Vibration of the tool therefore leads to a variable chip thickness which, according to Eq. 3.1.1, will give a variable cutting force since the force is proportional to the chip thickness. The cutting force governs the current tool deflection and, subsequently, the system exhibits feedback. In other words, the current behavior depends on previous behavior – the system has a "memory".

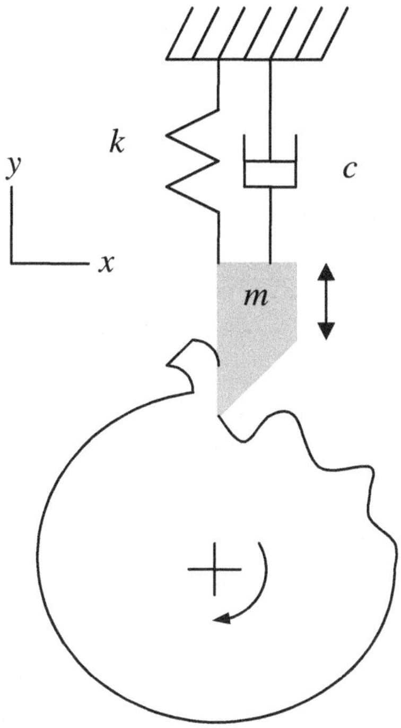

Fig. 3.2.1 Description of regenerative chatter in turning. Initial tool deflections are copied onto the workpiece surface and are encountered in subsequent revolutions. This varies the chip thickness and cutting force which, in turn, affects the resulting tool deflections

---

IN A NUTSHELL Cutting produces a force that is proportional to the chip thickness and chip width. While the tool is stiff in comparison to objects in our everyday lives, it is not infinitely stiff. For this reason, varying cutting forces (from transients such as the initial contact of the tool with the workpiece) produce vibrations of the tool. The vibrating tool changes the chip thickness and leaves a wavy surface. Variable forces cause vibrations... vibrations cause wavy surfaces...wavy surfaces produce variable forces. The reality is that the cutting operation is only partially governed by the selected geometry. Dynamics, the response of a flexible system to varying forces, plays a strikingly powerful role.

From a modeling standpoint, this "regeneration of waviness" appears as a time-delayed term in the chip thickness equation. Figure 3.2.2 shows an unwrapped view of the turning operation, where the surface on the left was produced in the previous revolution and the surface to the right of the tool (offset by the mean feed per revolution) was just cut away by the oscillating tool.

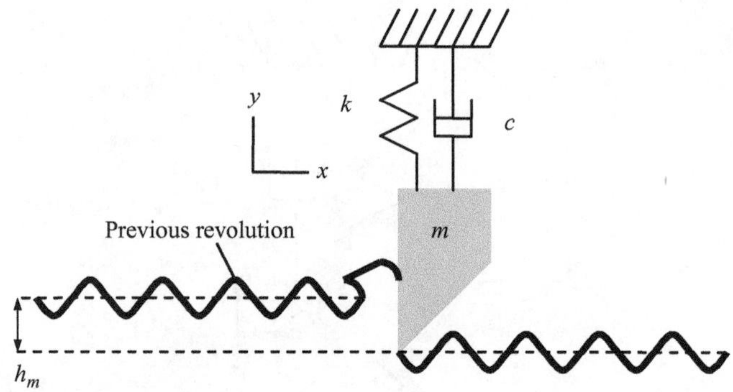

Fig. 3.2.2 Depiction of turning where the surface from the previous revolution, shown to the left of the tool, is removed by the vibrating cutter to produce a new wavy surface to the right of the tool

We will consider only vibrations in the normal direction, y (positive direction out of the cut), which has the most direct influence on the chip thickness.

The time dependent, instantaneous chip thickness, $h(t)$, is determined using Eq. 3.2.1. It is seen that larger positive vibration during the previous revolution, $y(t-\tau)$, where τ is the time for one rotation, gives an increased chip thickness (i.e., less material was removed so the current chip is thicker). Larger positive current vibration, $y(t)$, on the other hand, yields a thinner chip. See Fig. 3.2.3.

$$h(t) = h_m + y(t-\tau) - y(t) \tag{3.2.1}$$

The relative phasing between the surface waviness from one pass to the next determines the level of force variation and whether the operation is stable or

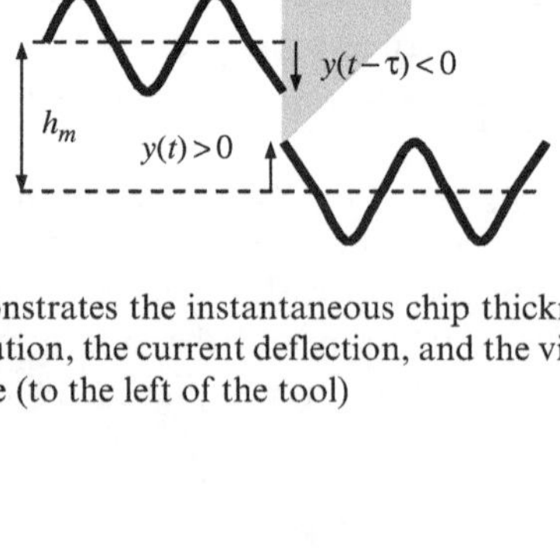

Fig. 3.2.3 The figure demonstrates the instantaneous chip thickness calculation. It depends on the mean feed per revolution, the current deflection, and the vibration during the previous revolution of the workpiece (to the left of the tool)

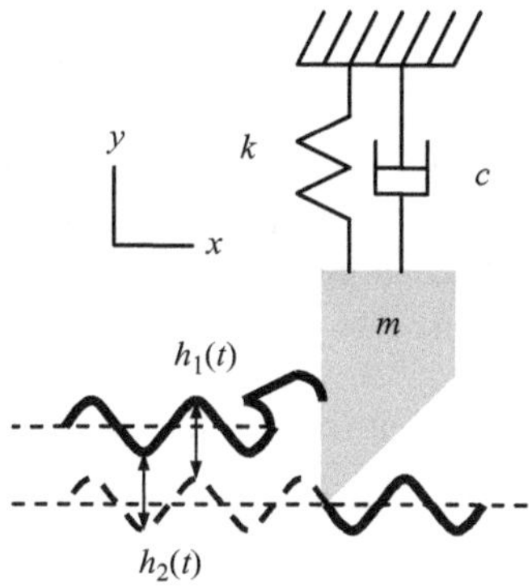

Fig. 3.2.4 The surface waviness between revolutions is in phase. Negligible chip thickness variation is obtained

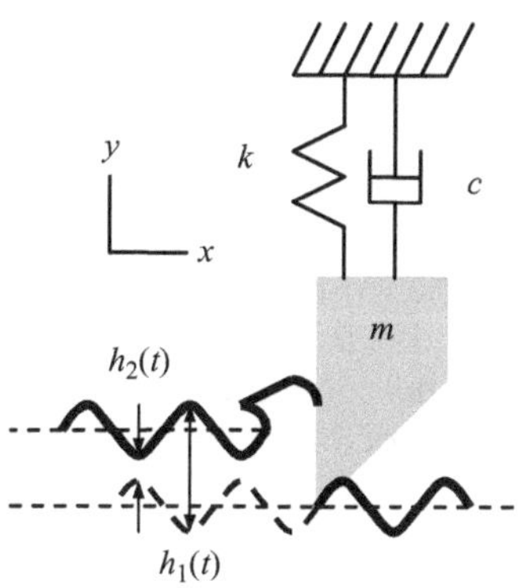

Fig. 3.2.5 Less favorable phase relationship between revolutions yields significant chip thickness variation

unstable (chatter occurs). Figures 3.2.4 and 3.2.5 show two possibilities. In Fig. 3.2.4, the wavy surfaces between two revolutions are in phase. Therefore, even though vibration is present during material removal, the chip thickness variation (vertical distance between the two curves) is negligible and there is no appreciable force variation. This enables stable cutting at larger chip widths. Considering that the tool tends to vibrate at its natural frequency, it is intuitive that matching the workpiece rotating frequency (spindle speed) to the tool's natural frequency will lead to this preferred "in phase" situation. However, this is counter intuitive based on our traditional understanding of resonance where we avoid driving the system at its natural frequency. Figure 3.2.5 shows a less favorable phase relationship where there is significant variation in the chip thickness. This leads to unstable cutting at smaller chip widths than the previous case due to the force variations and subsequent tool deflections.

IN A NUTSHELL The cutting force is very sensitive to the revolution-to-revolution alignment of the current tool motion to the previous motion of the tool imprinted on the surface. A "good" alignment (in phase) results in negligible chip thickness variation. A "bad" alignment (180 deg out of phase) produces a strong chip thickness variation. The worst case occurs when significant variation in the chip thickness coincides with a large magnitude in the frequency response function. It turns out that the worst case is about 270 deg out of phase.

3.3 Stability Lobe Diagrams

Depending on the feedback system "gain", or chip width b, and spindle speed, Ω, the turning operation will either be stable or exhibit chatter (unstable cutting), which causes large vibrations and forces and leads to poor surface finish and, potentially, tool/workpiece damage. In stable machining, the vibrations diminish from revolution to revolution. In unstable machining, the vibrations grow from revolution to revolution until limited in some way. Surprisingly, the vibrations may become large enough that the tool jumps out of the cut, losing contact with the workpiece. The vibrations in unstable cutting may be at least as large as the chip thickness and it is not surprising these large vibrations may result in damage to the machine, tool, and workpiece. The governing relationships for this behavior are provided in Eqs. 3.3.1 through 3.3.3 [2].

$$b_{\lim} = \frac{-1}{2K_s\cos(\beta)\mathrm{Re}[FRF]} \tag{3.3.1}$$

$$\frac{f_c}{\Omega} = N + \frac{\varepsilon}{2\pi} \tag{3.3.2}$$

$$\varepsilon = 2\pi - 2\tan^{-1}\left(\frac{\mathrm{Re}[FRF]}{\mathrm{Im}[FRF]}\right) \tag{3.3.3}$$

In these equations, $b_{\lim}$ is the limiting chip width to avoid chatter, f_c is the chatter frequency (should it occur), N is the integer number of waves of vibration imprinted on the workpiece surface in one revolution, and $\frac{\varepsilon}{2\pi}$ is any additional fraction of a wave, where ε is the phase (in rad) between current and previous tool vibrations. Note that for units consistency in Eq. 3.3.2, if f_c is expressed in Hz, then Ω must be specified in rev/s. Figure 3.3.1 shows an example stability lobe diagram where the Ω versus $b_{\lim}$ family of curves (N = 0, 1, 2, ...) separate the space into two regions. Any (Ω, $b_{\lim}$) pair that appears above the collective boundary indicates unstable behavior, while any pair below the boundary is presumed to be stable. We'll next discuss the foundation of the

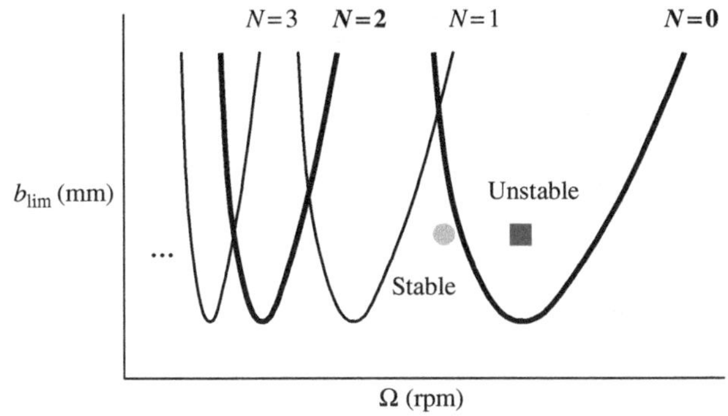

Fig. 3.3.1 Example stability lobe diagram. The stability boundary separates stable chip width-spindle speed combinations (below the boundary, marked as a circle) from unstable pairs (above, marked as a square)

relationships, provided in Eqs. 3.3.1 through 3.3.3, that are applied to construct this diagram. We'll base this discussion on the normal force equation and the vector representation of tool deflections in the complex plane.

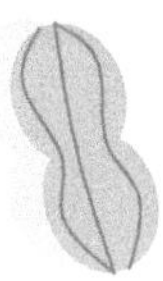

IN A NUTSHELL Whether we know it or not, whether we like it or not, every cutting operation has a picture like the one shown in Fig. 3.3.1. If we choose the cutting conditions at random, or at least without considering the applicable stability lobe diagram, then we sometimes choose stable cutting and sometimes not. Sometimes speeding up helps and other times it makes things worse. It appears to be random and many machine shops struggle with this issue every day. If we have the diagram and choose the cutting conditions accordingly, then it is possible to avoid the unstable conditions (chatter) and increase productivity.

Equation 3.2.1 shows that the instantaneous chip thickness depends on the commanded chip thickness, the normal direction vibration one revolution earlier, and the current vibration in the normal direction. If we substitute this $h(t)$ for h_m in the normal force expression provided in Eq. 3.1.2, we obtain:

$$F_n = k_n bh(t) = k_n b(h_m + y(t - \tau) - y(t)). \tag{3.3.4}$$

This force equation has both a constant part, $k_n bh_m$,, and a variable part, $k_n b(y(t - \tau) - y(t))$. The constant part does not influence the stability of the linear system. We are therefore interested in the behavior of the variable part. To explore the value of b as a function of k_n and the tool vibrations, let's consider a unit value of the variable force, or $1 = k_n b(y(t - \tau) - y(t))$ from Eq. 3.3.4. This equation can be solved for b:

$$b = \frac{1}{k_n(y(t-\tau) - y(t))}. \tag{3.3.5}$$

Regarding the vibration levels from one revolution to the next in Eq. 3.3.5, we can state the following:

- if $y(t) > y(t-\tau)$, then the vibrations are growing from one revolution to the next and unstable behavior results;
- $y(t) = y(t-\tau)$ indicates the limit of stability – the vibration level is neither increasing or decreasing; and
- $y(t) < y(t-\tau)$, then the vibrations are decaying from one revolution to the next and stable behavior is achieved.

Let's draw the $\vec{y}(t)$ and $\vec{y}(t-\tau)$ vectors in the complex plane for the limiting case that $\vec{y}(t) = \vec{y}(t-\tau)$[1]. We will require that the following vector sum is satisfied:

$$\vec{y} + (\vec{y}(t-\tau) - \vec{y}) = \vec{y}(t-\tau). \tag{3.3.6}$$

It is also necessary that the difference $\vec{y}(t-\tau) - \vec{y}(t)$, which represents the variable part of the chip thickness, is real valued. In other words, the instantaneous chip thickness is a scalar quantity; it has no imaginary part. In the complex plane it must therefore be horizontal. Figure 3.3.2 shows the vectors $\vec{y}(t)$ and $\vec{y}(t-\tau)$ as well as the real valued difference $\vec{y}(t-\tau) - \vec{y}(t)$ for a single degree of freedom system. Note the similarity to Fig. 2.2.6. The real valued unit

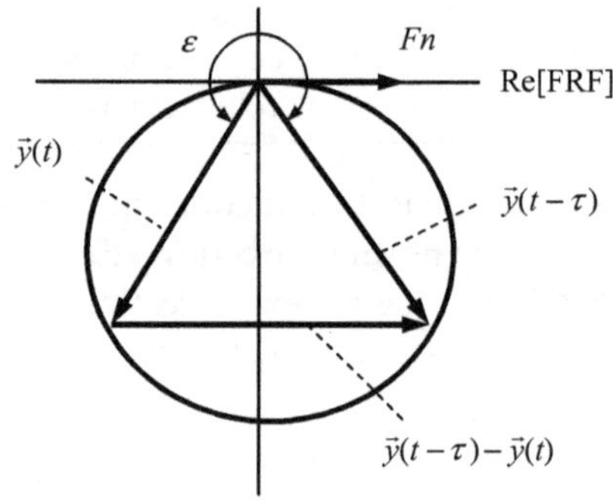

Fig. 3.3.2 Vector representation of unit normal force and tool deflections (current and previous revolutions) for limit of stability

[1] We represent $\vec{y}(t)$ and $\vec{y}(t-\tau)$ as vectors because they have both a magnitude and phase relative to the force, F_n. The force and both displacement vectors are displayed in Fig. 3.3.2.

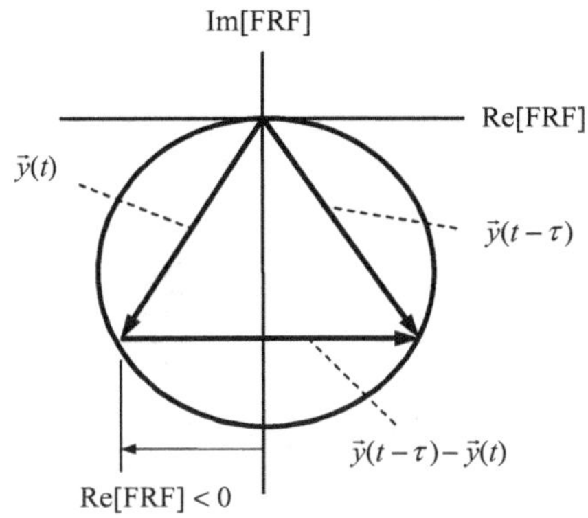

Fig. 3.3.3 Representing the length of the vector difference using the negative real part of the tool point FRF

normal force and the phase between tool vibrations in subsequent revolutions, ε, are also shown. The geometry seen in this figure satisfies the following requirements: 1) the magnitude and phase of the tool deflections depend on the forcing frequency (spindle speed) and the tool's direct FRF as measured at the cutting edge; 2) the amplitudes for $\vec{y}(t)$ and $\vec{y}(t-\tau)$ are equal (limit of stability); and 3) the difference $\vec{y}(t-\tau)-\vec{y}(t)$ is horizontal.

We can now rewrite Eq. 3.3.5 by substituting for $\vec{y}(t-\tau)-\vec{y}(t)$. Figure 3.3.3 shows that, due to the approximate symmetry, the length of the vector difference can be written as twice the negative real (Re) part of the tool's direct FRF. Equation 3.3.7 shows the new relationship, where the negative sign is included in order to obtain positive (limiting) chip width values. This equation matches Eq. 3.3.1, where k_n is substituted for the product $K_s\cos(\beta)$:

$$b_{\text{lim}} = \frac{-1}{k_n(2\text{Re}[FRF])}. \tag{3.3.7}$$

IN A NUTSHELL It is the chip width (and not the chip thickness) that controls whether or not the cutting operation is stable. The real part of the frequency response function defines the limiting chip width. The higher the system's dynamic stiffness (more stiffness and more damping), the larger the chip width that can be obtained without chatter.

Given Eq. 3.3.7, we can determine the smallest and largest potential values of b_{lim} based on the range of magnitudes possible for the negative real part of the tool point FRF. The smallest value is obtained for the minimum, or most negative, value of Re[FRF], when the absolute value of the denominator is

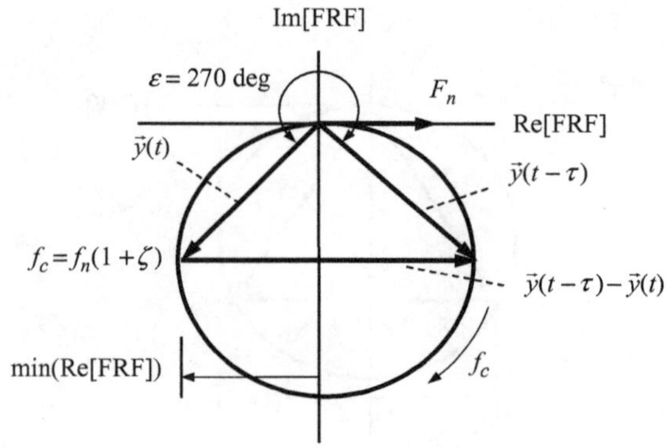

Fig. 3.3.4 Vector representation of $b_{\mathrm{lim,crit}}$ case ($f_c = f_n(1+\zeta)$ and $\varepsilon = 270$ deg)

largest. We will refer to this minimum b_{lim} value as the critical value, $b_{\mathrm{lim,crit}}$. See Eq. 3.3.8. The chatter frequency for this case (i.e., the min(Re[FRF])) is $f_c = f_n(1+\zeta)$ as shown in Fig. 2.2.5. The situation is pictured in Fig. 3.3.4. From the figure, we also see that ε is $\frac{3\pi}{2}$ rad or 270 deg. A representation of the time dependent tool deflections with this phase relationship is shown in Fig. 3.3.5. Again, the chip thickness variation is determined from the vertical distance between the two curves. It is interesting to note that the worst case (smallest chip width) is not obtained when $\varepsilon = 180$ deg, or the vibrations from one revolution to the next are exactly out of phase.

$$b_{\mathrm{lim,crit}} = \frac{-1}{k_n(2 \cdot \min(\mathrm{Re}[FRF]))} \tag{3.3.8}$$

The largest b_{lim} value is obtained when $\varepsilon = 360$ deg. This is the "in phase" situation of negligible chip thickness variation. As seen in the complex plane representation given in Fig. 3.3.6, the real part of the tool point FRF is zero and the chatter frequency is f_n. Substitution in Eq. 3.3.7 would suggest an infinite chip thickness, $b_{\mathrm{lim}} = \frac{-1}{k_n(2 \cdot 0)} = \infty$. However, we obtain finite b_{lim} values for the

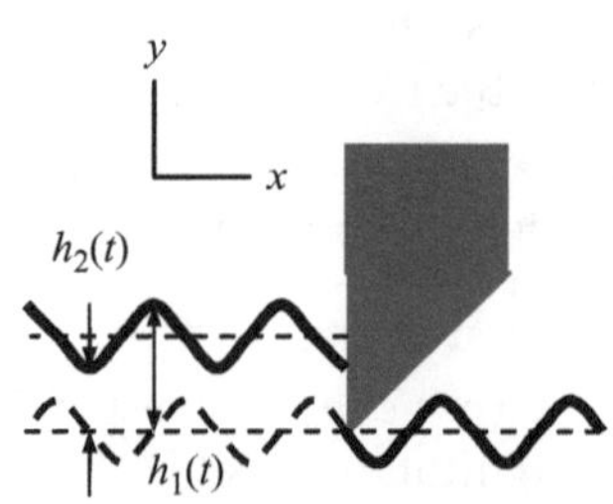

Fig. 3.3.5 Chip thickness variation for the $b_{\mathrm{lim,crit}}$ case when $\varepsilon = 270$ deg

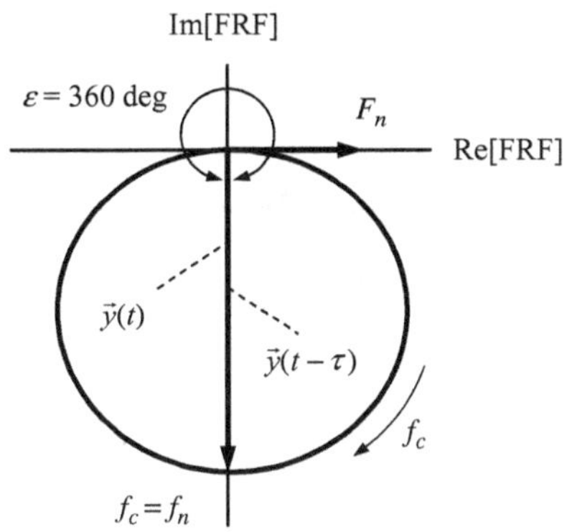

Fig. 3.3.6 Vector representation of largest $b_{\lim}$ case with favorable phase relationship (ε = 360 deg) between subsequent workpiece revolutions

left end of the $N = 0$ curve, even at a chatter frequency equal to f_n, because the adjacent stability curve with one more wave per revolution ($N = 1$) intersects the original curve and truncates it. See Fig. 3.3.7.

We see a similar situation as the chatter frequency approaches infinity, $f_c \to \infty$. In this case (Fig. 3.3.8), even though the revolution to revolution phase relationship is unfavorable ($\varepsilon \to 180°$, deg, or exactly out of phase), the response amplitude approaches zero, Re(FRF) $\to$ 0.. Again, substitution in Eq. 3.3.7 would suggest an infinite chip thickness, $b_{\lim} = \frac{-1}{k_n(2\cdot 0)} = \infty$. However, as seen in Fig. 3.3.7, the left side of the $N = 0$ curve serves to limit the right hand side of the $N = 1$ curve where $f_c \to \infty$.

We'll now see how each individual stability curve is actually a mapping of Re[FRF] onto the (Ω, $b_{\lim}$) diagram. We'll first consider the $N = 0$ curve, which

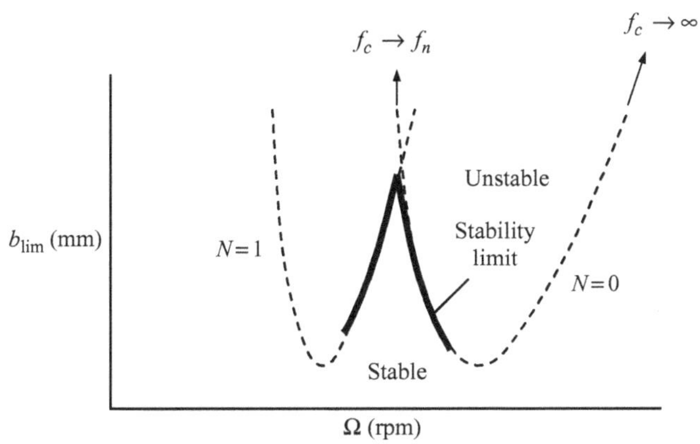

Fig. 3.3.7 Stability lobe diagram exhibiting truncation of $N = 0$ stability curve by $N = 1$ curve to obtain finite $b_{\lim}$ values

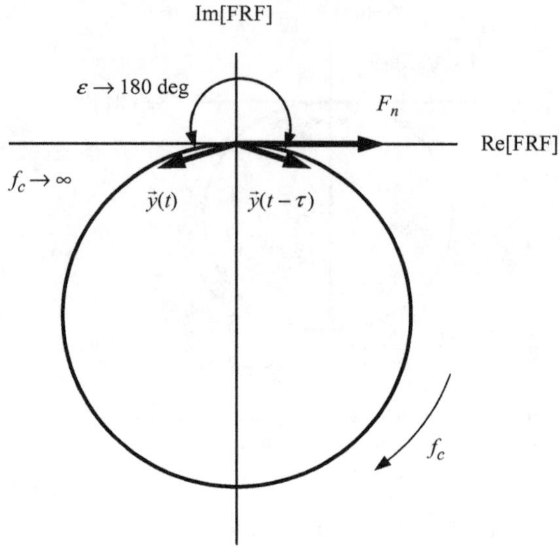

Fig. 3.3.8 Vector representation of $\varepsilon = 180$ deg phase relationship between subsequent workpiece revolutions ($f_c \rightarrow \infty$)

means that less than one wave is imprinted on the surface per revolution. We've already discussed the b_{lim} values, so we'll now focus on the spindle speeds for the left end (labeled as 1 in Fig. 3.3.9), the minimum (2), and the right end (3). As we saw in Fig. 3.3.6, when the chatter frequency is equal to f_n, the $\varepsilon = 360 \text{ deg} = 2\pi$ rad phase relationship is obtained and $b_{lim} = \infty$ because the real part is zero. Point 1 therefore has a spindle speed of:

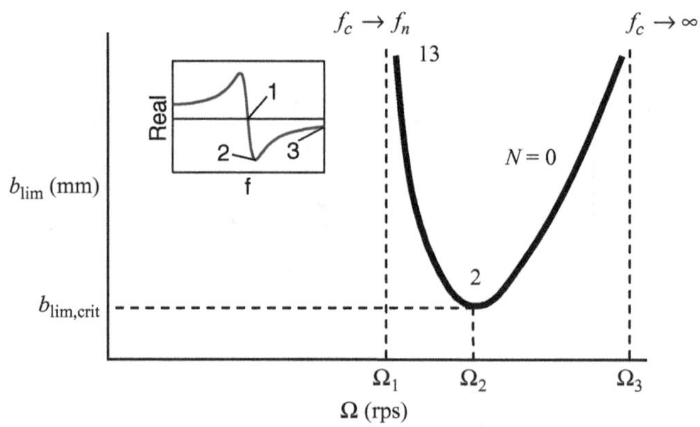

Fig. 3.3.9 Mapping of the real part of the tool point FRF onto the (Ω, b_{lim}) diagram for the $N = 0$ lobe. Spindle speeds provided in Eqs. 3.3.9 through 3.3.11

$$\Omega_1 = \frac{f_c}{N + \frac{\varepsilon}{2\pi}} = \frac{f_n}{0 + \frac{2\pi}{2\pi}} = f_n \tag{3.3.9}$$

from Eq. 3.3.2. (Note that units of Hz for f_n gives equivalent units of rev/s, or rps, for spindle speed.) This corresponds to point 1 in the inset showing the real part of the tool point FRF.

IN A NUTSHELL The chip width can be very large without the occurrence of chatter if the spindle speed is set close to the natural frequency of the most flexible vibration mode. This surprising result comes from favorable alignment of the waviness from one revolution to the next at this speed.

Recall that we are only considering the negative portion of the real part so the applicable frequency range (for this single degree of freedom response) is f_n to ∞. At point 2, which has the minimum chip width, $b_{\text{lim,crit}}$, we saw in Fig. 3.3.4 that the chatter frequency is $f_n(1+\zeta)$ and $\varepsilon = 270$ deg $= \frac{3\pi}{2}$ rad. The corresponding spindle speed is:

$$\Omega_2 = \frac{f_n(1+\zeta)}{0 + \frac{\frac{3\pi}{2}}{2\pi}} = \frac{4}{3} f_n(1+\zeta) \tag{3.3.10}$$

At point 3 the chatter frequency approaches ∞ and ε approaches 180 deg $= \pi$ rad. rad. The spindle speed is:

$$\Omega_3 = \frac{\infty}{0 + \frac{\pi}{2\pi}} = \infty. \tag{3.3.11}$$

Let's next consider the $N = 1$ stability curve (or lobe). In this case, there is at least one wave of vibration per revolution. See Fig. 3.3.10. Although the b_{lim} values are a function of the chatter frequency (via the tool point FRF), they do not depend on the lobe number, N. Therefore, these values do not change relative to the $N = 0$ calculations. Similarly, the ε values are independent of the lobe number and do not change. The spindle speed equation (Eq. 3.3.2), however, is a function of the lobe number. Using the same labeling convention for the left end, minimum, and right end points of Re[FRF], we obtain the three spindle speeds:

$$\Omega_1 = \frac{f_n}{1 + \frac{2\pi}{2\pi}} = \frac{f_n}{2}, \tag{3.3.12}$$

$$\Omega_2 = \frac{f_n(1+\zeta)}{1 + \frac{\frac{3\pi}{2}}{2\pi}} = \frac{4}{7} f_n(1+\zeta), \text{ and} \tag{3.3.13}$$

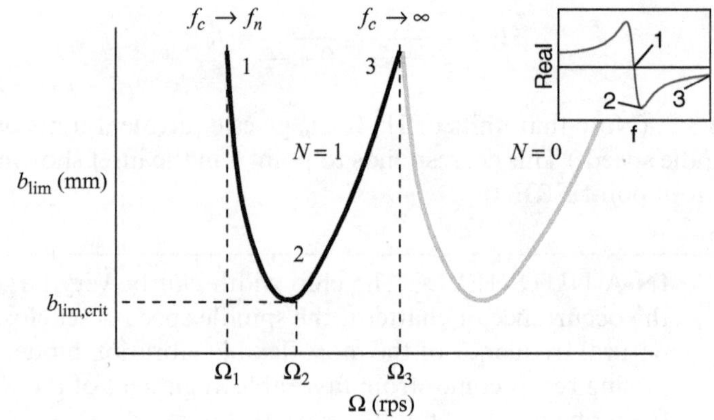

Fig. 3.3.10 Mapping of the real part of the tool point FRF onto the (Ω, b_{lim}) diagram for the $N = 1$ lobe. Spindle speeds provided in Eqs. 3.3.12 through 3.3.14

$$\Omega_3 = \frac{\infty}{1 + \frac{\pi}{2\pi}} = \infty. \tag{3.3.14}$$

If we plot multiple lobes ($N = 0, 1, 2, \ldots$), we obtain a result similar to Fig. 3.3.11. We see that all lobes exhibit the same minimum value, $b_{lim,crit}$, and the peak values are located approximately at integer fractions of the tool point natural frequency. We can therefore write a "best speeds" equation which identifies these spindle speeds (in rev/min, or rpm). See Eq. 3.3.15, where f_n is expressed in Hz. As noted, any (Ω, b_{lim}) pair located below the stability boundary leads to stable operation, while combinations above the boundary result in chatter.

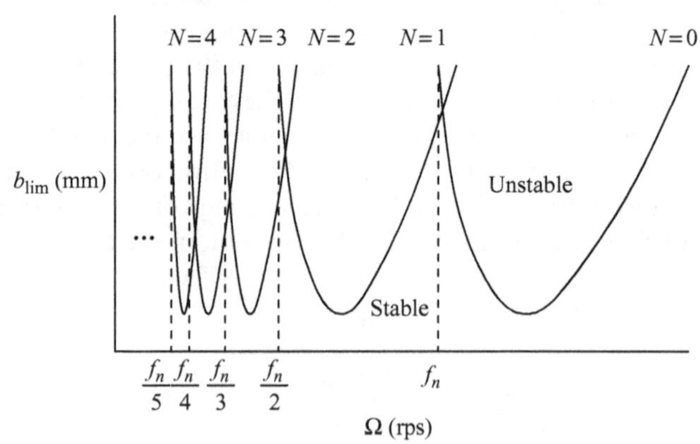

Fig. 3.3.11 Depiction of multiple stability lobes which, taken together, form the stability limit. It is seen that the peaks of the stable zone become less pronounced as N increases

$$\Omega_{best} = \frac{f_n \cdot 60}{N+1}\,(\text{rpm}) \tag{3.3.15}$$

IN A NUTSHELL We get the same favorable alignment of waviness at many different spindle speeds; these multiple speeds correspond to different numbers of integer waves of vibration imprinted on the workpiece surface per revolution. The increase in allowable chip width without chatter gets larger as the spindle speed gets higher (less waves of vibration per revolution). At lower speeds the lobes overlap to such an extent that the limiting chip width approaches a constant value.

Similarly, we can write a "worst speeds" equation that provides the spindle speeds where the $b_{\text{lim,crit}}$ values are encountered; see Eq. 3.3.16. Equations 3.3.15 and 3.3.16 provide good approximations for dynamic systems that can be modeled as single degree of freedom. Multiple degree of freedom systems, on the other hand, are generally best described using the stability lobe diagram itself.

$$\Omega_{worst} = \frac{f_n(1+\zeta) \cdot 60}{N+\frac{3}{4}}\,(\text{rpm}) \tag{3.3.16}$$

Before discussing the concept of the "oriented FRF" introduced by Tlusty [2], let's take a moment to explore which factors affect b_{lim} and the applicability of stability lobe diagrams to typical turning operations. From Eq. 3.3.1, we see that b_{lim} depends on K_s, β, and the negative real part of the tool point FRF, which we'll describe using the stiffness, k, and damping ratio, ζ. (Again, we are assuming single degree of freedom dynamics.) As K_s and β decrease ($0 \leq \beta \leq 90\,\text{deg}$), b_{lim} increases. Considering Table 3.1.1, we can see that, all other conditions being equal, we would obtain approximately a three times increase in the allowable chip width if we compared an aluminum alloy with a low alloy steel due to the corresponding K_s values. Considering only the Re[FRF], it becomes less negative as k and ζ are increased (Section 2.2). See the depiction in Fig. 3.3.12, where the increase in k yields not only a less negative real part, but also an increase in the natural frequency. If the workpiece material/tool geometry are unchanged (so that K_s and β can be considered constant), this Re[FRF] change leads to an increase in b_{lim}, but also a shift in the stability lobes. Naturally this would need to be considered in implementation by a spindle speed adjustment.

Example 3.3.1: Best spindle speeds for a single degree of freedom system Let's consider a setup where the tool behaves like a single degree of freedom system with $f_n = 700\,\text{Hz}$ and corresponding k and ζ values determined from a curve

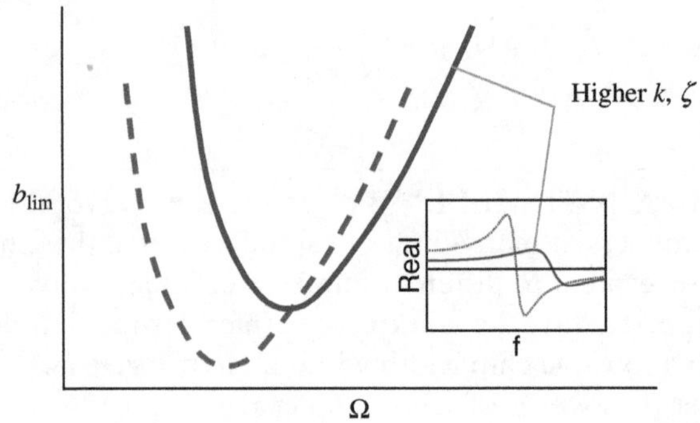

Fig. 3.3.12 Influence of changes in k and ζ on $b_{\lim}$

fit to the direct FRF as measured at the tool point in the y direction. We can apply Eq. 3.3.15 to determine our "best" spindle speed for an increased limiting chip width. If we let $N = 0$ (this gives the right-most peak in the stability lobe diagram), we obtain:

$$\Omega_{best} = \frac{700 \cdot 60}{0 + 1} = 42000 \text{ rpm}.$$

The problem is that conventional lathe spindles don't turn this fast. Let's say the top spindle speed for a selected lathe is 3600 rpm. Which peak, and corresponding best spindle speed, must we then pick? We would require that:

$$\Omega_{best} = \frac{700 \cdot 60}{N + 1} \leq 3600 \text{ rpm}.$$

If we select $N = 11$ (i.e., the 12th peak counting from right to left in the diagram), the best spindle speed would be 3500 rpm and the maximum spindle speed constraint would be satisfied. However, as seen in Fig. 3.3.11, as N increases the relative improvement in $b_{\lim}$ afforded by the stability lobe peaks decreases dramatically. For $N = 11$, the improvement is generally negligible and $b_{\lim} \approx b_{\lim,crit}$. For this reason, stability lobe diagrams are typically more successfully implemented in high speed milling, as opposed to traditional turning operations.

3.4 The Oriented FRF

Using the Tlusty model [2], Eq. 3.3.1 can be rewritten as:

$$b_{\lim} = \frac{-1}{2K_s\mu\text{Re}[FRF]}, \tag{3.4.1}$$

where $\mu = \cos(\beta)$ is referred to as the "directional orientation factor" and the product of μ and Re[FRF] is the real part of the "oriented FRF". The concept is to first project the cutting force, F, into the direction of the system dynamics and, second, project this result into the surface normal, y. The projection into the surface normal is necessary because we are only considering the effects of tool vibrations in this sensitive direction. The two projection steps are required because it may be that the system dynamics are not known in the surface normal direction as we have assumed in our previous descriptions.

Consider, for example, an external turning operation where a square tool bar, whose sides are not aligned with x and y, is used to reduce the diameter of a workpiece; see Fig. 3.4.1. The impact tests described in Section 2.6 would naturally be performed along the tool bar faces (directions u_1 and u_2 in the figure), which would not provide FRFs in the y direction. Here μ would account for these misalignments as well as the force angle and, when combined with the measured FRFs, would provide the system dynamic stiffness in the direction of the surface normal. This is the oriented FRF.

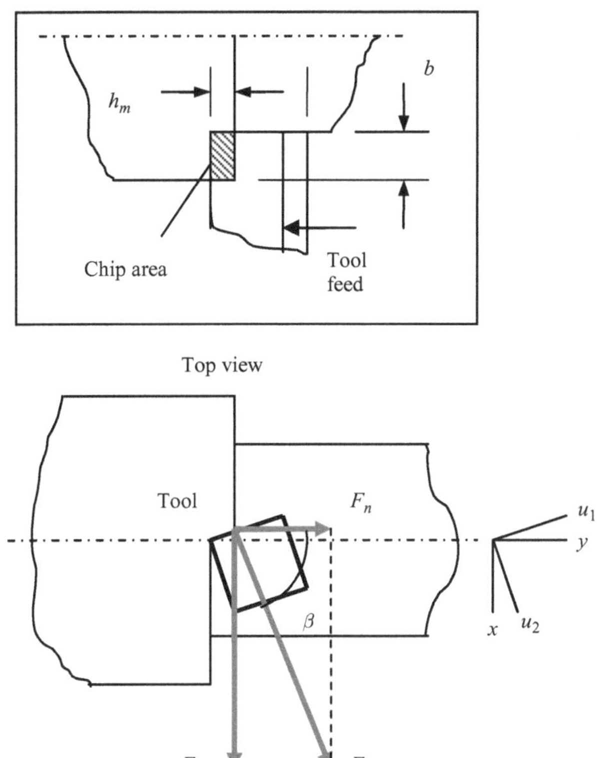

Fig. 3.4.1 Oriented FRF description using external turning operation

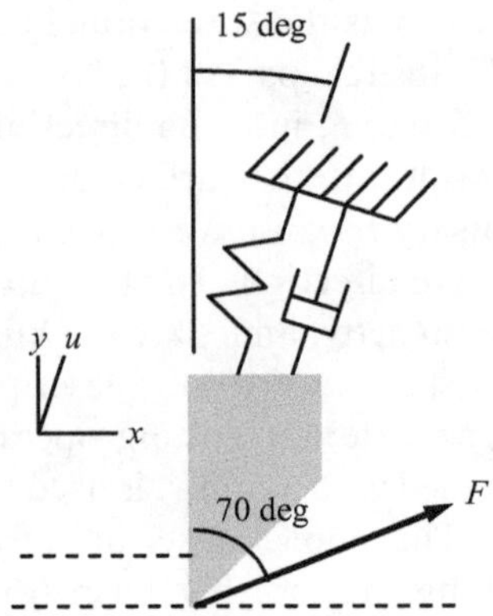

Fig. 3.4.2 Single degree of freedom system with vibration mode that is not aligned with the surface normal

Example 3.4.1: Single and two degree of freedom oriented FRF Let's consider an example single degree of freedom system where the vibration mode direction is not aligned with the surface normal as shown in Fig. 3.4.2. In order to calculate the directional orientation factor, the force projection onto the mode direction u is first determined:

$$F_u = F\cos(70 - 15) = F\cos(55).$$

This result is then projected onto the surface normal (y direction).

$$F_n = F_u\cos(15) = F\cos(55)\cos(15)$$

The directional orientation factor is then $\mu = \cos(55)\cos(15) = 0.55$ and the oriented FRF is the product of μ and the FRF measured in the u direction. Because μ is less than one, the oriented FRF appears stiffer than the u direction FRF; see Fig. 3.4.3. Physically, this indicates that only a portion (55%) of the force/flexibility leads to vibration in the sensitive direction.

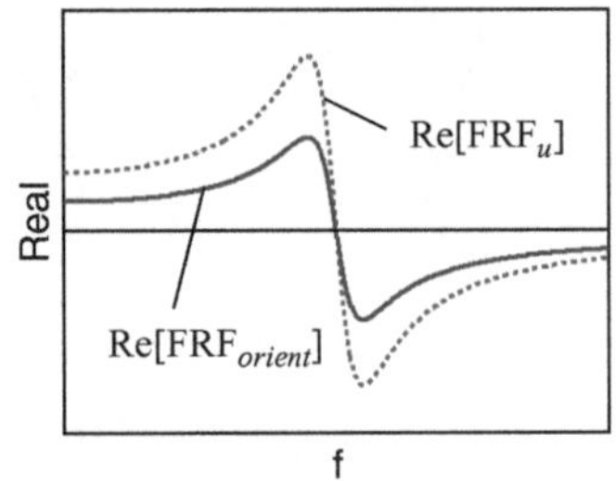

Fig. 3.4.3 Re[FRF] in u direction compared to real part of oriented FRF

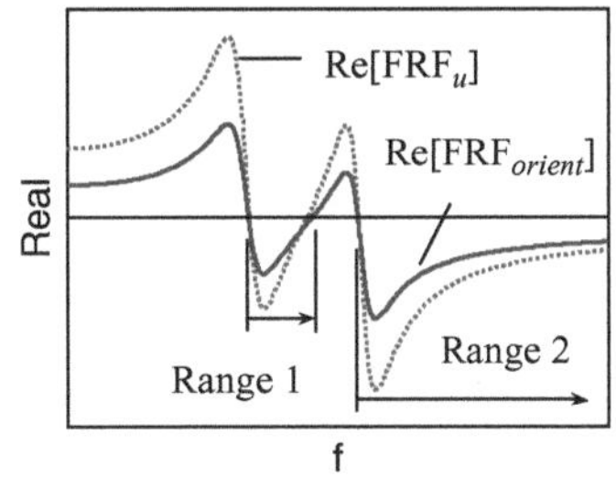

Fig. 3.4.4 Re[FRF] in u direction compared to real part of oriented FRF for two degree of freedom system. Two valid chatter frequency ranges are seen

Next, consider the same u direction as shown in Fig. 3.4.2, but now with a second degree of freedom in this direction. The real part of an example two degree of freedom direct FRF at the tool point, as well as the oriented FRF real part, is shown in Fig. 3.4.4. As discussed in Section 3.3, the negative real part is used in the $b_{\lim}$ calculation and defines the valid chatter frequency range. Figure 3.4.4 shows two valid chatter frequency ranges, unlike the previous single degree of freedom examples we've considered. This yields two stability boundary sections for each N value as shown in Ex. 3.4.2.

Example 3.4.2: Competing lobes for two degree of freedom oriented FRF Consider the model shown in Fig. 3.4.5. A two degree of freedom system is aligned with the surface normal and the force angle is 70 deg. The directional orientation factor therefore only requires the projection of the force into the mode

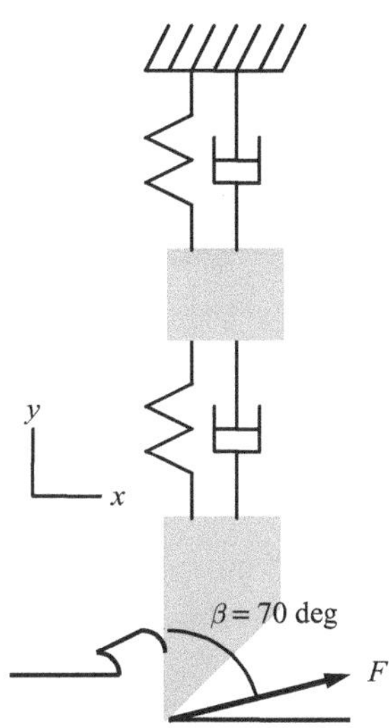

Fig. 3.4.5 Two degree of freedom model for turning stability evaluation

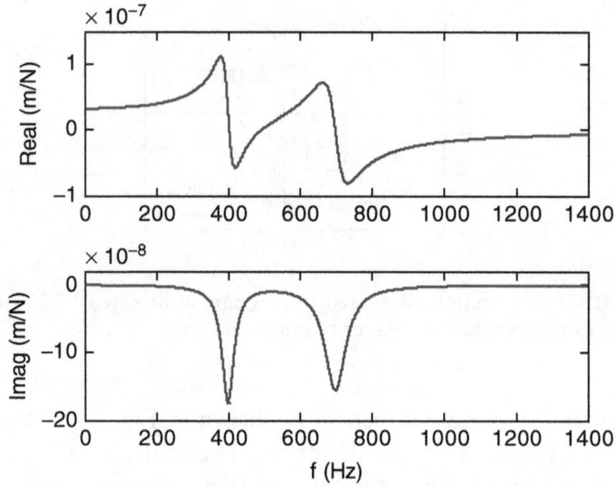

Fig. 3.4.6 Real and imaginary parts of the oriented FRF for the system in Ex. 3.4.2

direction, $\mu = \cos(70)$. The modal parameters are: $f_{n1} = 400\,\text{Hz}$, $k_{q1} = 2 \times 10^7\,\text{N/m}$, $\zeta_{q1} = 0.05$, $f_{n2} = 700\,\text{Hz}$, $k_{q2} = 2.2 \times 10^7\,\text{N/m}$,, and $\zeta_{q2} = 0.05$. We will determine the stability behavior for this turning operation.

The real and imaginary parts of the oriented FRF for this system are shown in Fig. 3.4.6. Two distinct modes with 400 Hz and 700 Hz natural frequencies are observed. As seen previously in Fig. 3.4.4, there are two valid chatter frequency ranges associated with this two mode system. They occur where the real part is less than zero and are pictured in the top panel of Fig. 3.4.7. In the

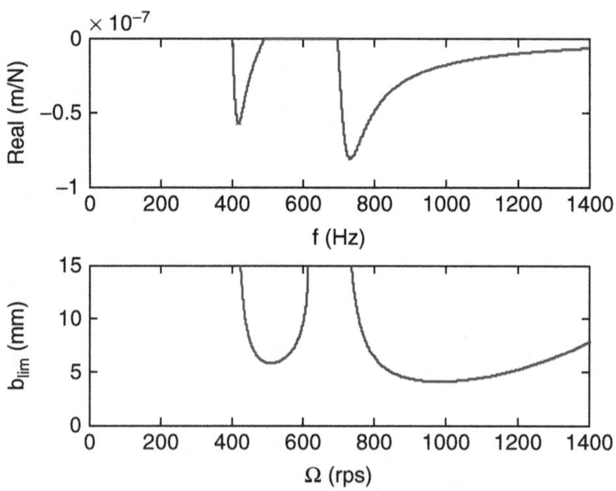

Fig. 3.4.7 (Top panel) Negative real part of oriented FRF. (Bottom panel) Corresponding $N = 0$ stability limit. The two lobes sections are present due to the two portions of the negative real part in the top panel

bottom panel, the Ω and $b_{\lim}$ values, which are both a function of the FRF and therefore the chatter frequency (as seen in Eqs. 3.1–3.3), are plotted against one another to define the stability limit ($K_s = 1500$ N/mm^2). It is seen that a distinct lobe section is associated with each of the two chatter frequency ranges. Only the $N = 0$ pair is shown in this figure. However, it is the whole family of curves, $N = 0, 1, 2, \ldots$, that defines the overall stability boundary. The $N = 0$, 1, and 2 lobes are shown in Figure 3.4.8. Because these lobes can interfere with each other and limit the stable chip width, they may be considered as "competing lobes". For example, it is seen that the right portion of the $N = 2$ lobe (corresponding to the 700 Hz mode) truncates the stable zone between the left (400 Hz mode) and right portions of the $N = 1$ lobe near 20000 rpm. The MATLAB® program used to produce these figures is provided on the companion CD as p_3_4_2_1.m.

As seen in Figure 3.4.1, it is sometimes not sufficient to consider the system flexibility in one direction only. In this case, modes in both the u_1 and u_2 directions should be included in the stability evaluation described in the previous example. The model shown in Fig. 3.4.9 depicts a single degree of freedom aligned with both u_1 and u_2, neither of which are coincident with y, the surface normal. The oriented FRF now depends on the contributions of both modes. We therefore calculate two directional orientation factors, each of which projects the force into the appropriate mode and then this result into the surface normal. Let's consider the u_1 mode. The first step is to project F into the u_1 direction:

$$F_{u_1} = F\cos(\beta - \alpha_1). \tag{3.4.2}$$

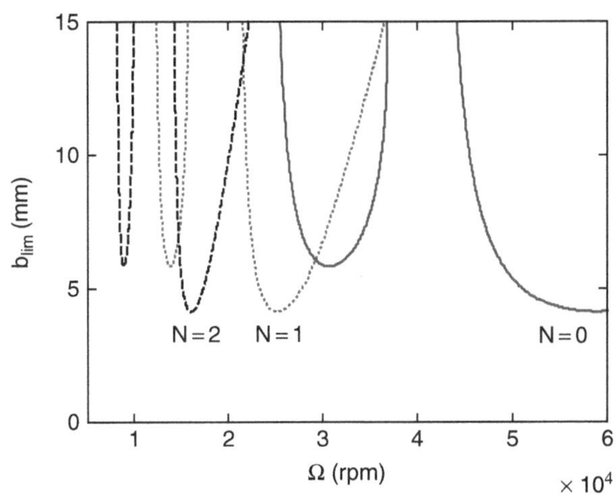

Fig. 3.4.8 Competing $N = 0$, 1, and 2 stability lobes for two degree of freedom system

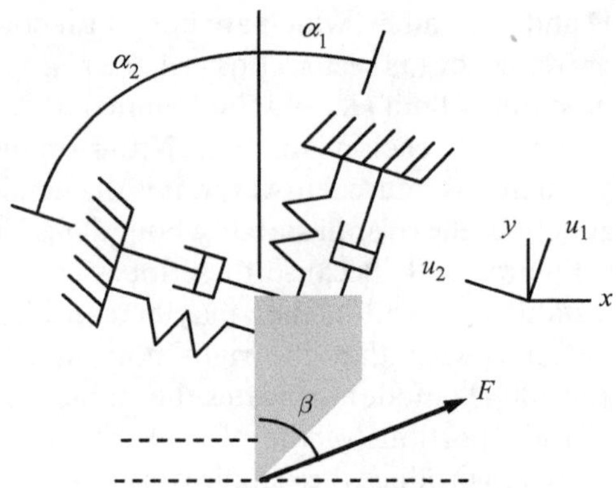

Fig. 3.4.9 Turning model with a single degree of freedom in both the u_1 and u_2 directions

This result is then projected into the surface normal:

$$F_n = F_{u_1}\cos(\alpha_1) = F\cos(\beta - \alpha_1)\cos(\alpha_1). \tag{3.4.3}$$

The u_1 directional orientation factor is, therefore, $\mu_1 = \cos(\beta - \alpha_1)\cos(\alpha_1)$. Similarly, the steps in determining μ_2 are:

$$F_{u_2} = F\cos(\beta + \alpha_2)\text{ and} \tag{3.4.4}$$

$$F_n = F_{u_{12}}\cos(\alpha_2) = F\cos(\beta + \alpha_2)\cos(\alpha_2) \tag{3.4.5}$$

so that $\mu_2 = \cos(\beta + \alpha_2)\cos(\alpha_2)$. The oriented FRF is then calculated as a linear combination of the contributions of both modes/cutting force using these directional orientation factors. Note that this treatment is not limited to a single degree of freedom in either direction; it is generic to any number of degrees of freedom in the two perpendicular directions u_1 and u_2.

$$FRF_{orient} = \mu_1 FRF_{u_1} + \mu_2 FRF_{u_2} \tag{3.4.6}$$

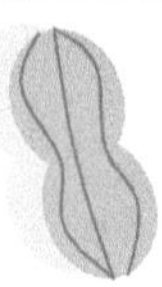

IN A NUTSHELL It is possible for a multiple degree of freedom system to chatter in any of its modes. The effect of each mode is determined by its dynamic characteristics, the alignment of the force with the direction of flexibility of the mode, and the ability of the deflection of the mode to imprint on the surface of the workpiece. These effects are combined in the "directional orientation factor".

Example 3.4.3: Turning model with modes in two perpendicular directions We'll next determine the stability for the model in Fig. 3.4.9 with the following parameters: $\alpha_1 = 30$ deg, $\alpha_2 = 60$ deg, $\beta = 70$ deg, and $K_s = 2000$ N/mm^2. The dynamics are defined by $f_{n1} = 421$ Hz,, $k_1 = 2.8 \times 10^7$ N/m, and, and $\zeta_1 = 0.05$ for the u_1 direction and $f_{n2} = 491$ Hz,, $k_2 = 3.81 \times 10^7$ N/m, and $\zeta_2 = 0.05$ for the u_2 direction. The directional orientation factors are calculated using Eqs. 3.4.2–3.4.5.

$$\mu_1 = \cos(\beta - \alpha_1)\cos(\alpha_1) = \cos(70 - 30)\cos(30) = 0.663$$

$$\mu_2 = \cos(\beta + \alpha_2)\cos(\alpha_2) = \cos(70 + 60)\cos(60) = -0.321$$

The oriented FRF, as well as its components $\mu_1 FRF_{u_1}$ and $\mu_2 FRF_{u_2}$, are shown in Fig. 3.4.10. It is seen that: 1) the minimum real part of FRF$_{orient}$ occurs at 443 Hz with a value of 1.493×10^{-4} mm/N; and 2) the real part crosses through zero amplitude at a frequency of 418 Hz (which corresponds to the natural frequency for a single degree of freedom system). Although this is not a single degree of freedom system, the real part of the oriented FRF bears some similarity to a single degree of freedom FRF real part. As an approximation, we can therefore calculate $b_{\lim,crit}$ and the best and worst spindle speeds using the single degree of freedom equations presented in Section 3.3. The best and worst spindle speeds are summarized in Table 3.4.1.

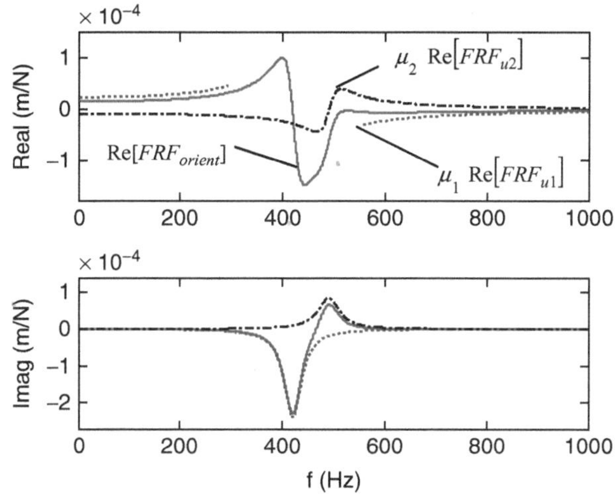

Fig. 3.4.10 Oriented FRF for Ex. 3.4.2. The two components of the oriented FRF are also shown

Table 3.4.1 Approximate best and worst spindle speeds for Ex. 3.4.2

N	Ω_{best} (rps)	Ω_{best} (rpm)	Ω_{worst} (rps)	Ω_{worst} (rpm)
0	418	25080	591	35460
1	209	12540	253	15180
2	139	8340	161	9665
3	105	6300	118	7088

$$b_{\text{lim,crit}} = \frac{-1}{2K_s \cdot \min(\text{Re}[FRF_{orient}])} = \frac{-1}{2 \cdot 2000 \cdot (-1.493 \times 10^{-4})} = 1.7\,\text{mm}$$

$$\Omega_{best} = \frac{418}{N+1}\ (\text{rps})$$

$$\Omega_{worst} = \frac{443}{N+\frac{3}{4}}\ (\text{rps})$$

Figure 3.4.11 shows one valid chatter frequency range associated with the oriented FRF. As before, it occurs where the real part is less than zero (top panel). In the bottom panel, the $N = 0$ Ω and b_{lim} values are plotted against one another to define the stability limit. It is seen that the best speed of 418 rps is a reasonable approximation of the actual behavior. Figure 3.4.12 shows the combined stability boundary for $N = 0$ to 3. The MATLAB® program used to produce these figures is provided on the companion CD as p_3_4_3_1.m. Note that ε in Eq. 3.3.3 must now be calculated using FRF_{orient}.

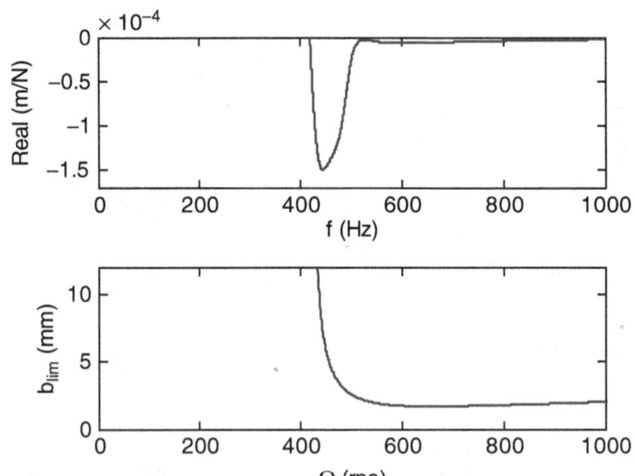

Fig. 3.4.11 (Top panel) Negative real part of oriented FRF for Ex. 3.4.3. (Bottom panel) Corresponding $N = 0$ stability limit

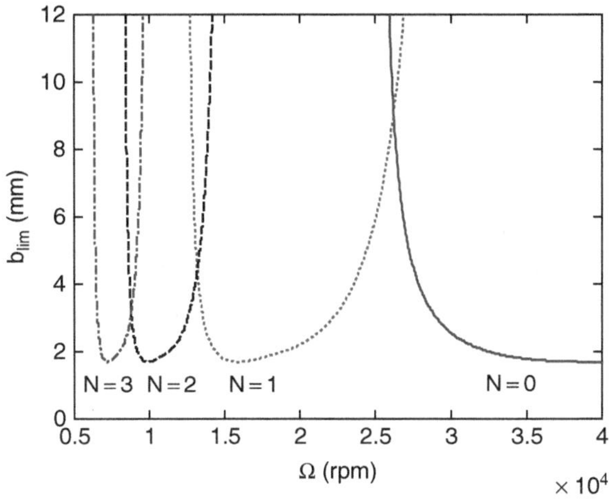

Fig. 3.4.12 Stability lobes for Ex. 3.4.3 ($N = 0$ to 3)

3.5 Turning Time-Domain Simulation

In this section, a time-domain simulation is described that solves the turning equation of motion by numerical integration. This is in contrast to Tlusty's analytical, frequency-domain solution detailed in the previous sections. The stability lobe diagram determined from the analytical solution gives a "global" picture of the stability behavior for a particular turning setup, but does not provide information regarding the cutting force or tool vibrations. On the other hand, time-domain simulation gives "local" force and vibration levels for the selected cutting conditions, but not the same global view. The simulation proceeds as follows: 1) the instantaneous chip thickness is determined using the current and previous tool vibrations; 2) the cutting force is calculated; 3) the force is used to find the new displacement; and 4) the process is repeated in small time steps. The simulation model is the same as was presented in Fig. 3.4.9.

3.5.1 Chip Thickness Calculation

As shown in Eq. 3.2.1, the instantaneous chip thickness depends, at minimum, on the mean chip thickness (or feed per revolution), the current normal (y) direction vibration, and the vibration one revolution earlier, $h(t) = h_m + y(t - \tau) - y(t)$. We used Fig. 3.2.3 as an aid in visualizing this equation. In reality, the situation is a little more complicated because, depending on the vibration levels, the instantaneous chip thickness may depend on the current vibration and the vibration two revolutions earlier, for example. Consider the vibrations in two subsequent

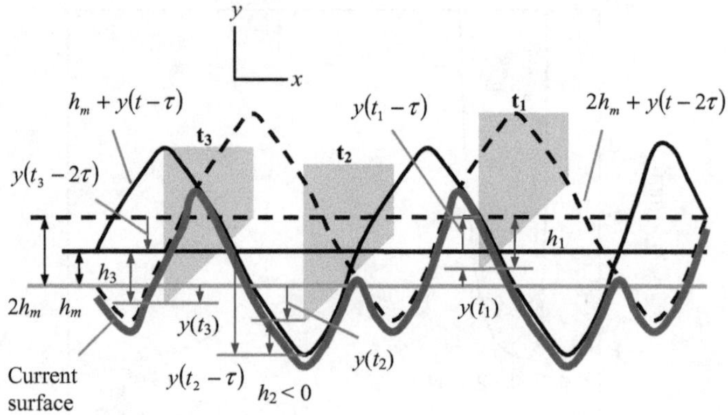

Fig. 3.5.1 Chip thickness determination at three different instants in time

revolutions shown in Fig. 3.5.1. The current vibration state is defined by the tool position at three different times, t_1, t_2, and t_3. The vibration one revolution earlier is represented by the solid, sinusoidal line and is offset from the current commanded mean chip thickness line by h_m. The vibration two revolutions earlier is shown as a dashed line and is offset by $2h_m$. The current surface is given by the heavy solid line. It is seen that the current surface depends on the previous revolution at certain times (including t_1 and t_2) and the vibration two revolutions earlier at other times (such as t_3). At t_1, the chip thickness is determined in the traditional manner, $h_1 = h_m + y(t_1 - \tau) - y(t_1)$. At t_2, the tool point is actually above the current surface. In other words, no cutting is taking place. Here, $h_3 = 2h_m + y(t_3 - 2\tau) - y(t_3)$ so that $h_2 < 0$. Clearly, a negative chip thickness does not make physical sense, but this nonlinearity is conveniently incorporated in the simulation. At t_3, the chip thickness is determined using $h_3 = 2h_m + y(t_3 - 2\tau) - y(t_3)$ At this instant, we must consider the vibrations two revolutions prior.

The simulation is carried out in small time steps, *dt*. Because we are numerically integrating the system equation of motion to determine the tool vibrations, care must be exercised in selecting *dt*. If the value is too large, inaccurate results are obtained. As a rule of thumb, it is generally acceptable to set *dt* at least 10 times smaller than the period corresponding to the highest natural frequency in the system's dynamic model.

Example 3.5.1: Numerical integration time step selection Consider a turning model with two natural frequencies, $f_{n1} = 800$ Hz and $f_{n2} = 1000$ Hz. The period of the higher natural frequency is:

$$\tau_2 = \frac{1}{f_{n2}} = 1 \times 10^{-3}\ \text{s}$$

and the maximum *dt* value should be:

$$dt = \frac{\tau_2}{10} = 1 \times 10^{-4}\ \text{s}.$$

Smaller values are naturally acceptable (dividing the time constant by 50 or 100, for example), but there is a tradeoff between improved numerical accuracy and execution time for the simulation.

Once the time step is selected, the simulation time can be determined using:

$$t_n = n \cdot dt, \tag{3.5.1}$$

where n is the simulation counter ($n = 1, 2, \ldots$). The number of time steps per spindle revolution, or *steps_rev*, is related to dt (in sec) and the selected spindle speed, Ω (in rpm), as:

$$steps\_rev = \frac{60}{dt \cdot \Omega}. \tag{3.5.2}$$

This value is critical because it enables us to keep track of the vibration state from one revolution to the next. At some discrete time t_n in the simulation, the behavior one revolution earlier ($t - \tau$ in a continuous sense) is $t_{n\text{-}steps\_rev}$ (in a discrete sense). In practice, because *steps_rev* is used as part of the index for the simulation variables, only integer values are allowed. This is accomplished in MATLAB® using the `round` function. For example:

```
steps_rev= round(60/(dt*omega));
```

where `omega` represents the spindle speed in rpm. In simulation, the dependence of the actual surface on more than the most previous revolution can be incorporated by calculating the chip thickness according to:

$$h = y_{\min} - y_{n-1}, \tag{3.5.3}$$

where $y_{\min}$ is the smallest value of $\{\{h_m + y_{n-steps\_rev}, 2h_m + y_{n-2\cdot steps\_rev}, \ldots\}$ which, according to Fig. 3.5.1, defines the current workpiece surface. The index (n-1) on y in Eq. 3.5.3 is used because the vibration in the current time step (with index n) is only known after the force is determined and the equation of motion is solved.

3.5.2 Force Calculation

Once the chip thickness is computed using Eq. 3.5.3, the force in the current time step is determined using the selected chip width, b, and specific force:

$$F = K_s bh. \tag{3.5.4}$$

As described previously, it is possible that the calculated chip thickness will be negative if the current tool vibration is larger than the surface location (refer to Fig. 3.5.1 at t_2). In this case, no cutting is occurring. Therefore, the cutting force should be zero. We incorporate this nonlinearity by setting the force equal to zero if the result from Eq. 3.5.4 is less than zero (due to a negative chip thickness). Additionally, the surface is updated to reflect the actual location, y_{min}, rather than the tool vibration: $y_{n-1} = y_{min}$ Once the cutting force is known, it is resolved into the mode directions, u_1 and u_2, as depicted in Fig. 3.4.9.

$$F_{u_1} = F\cos(\beta - \alpha_1) \text{ and } F_{u_2} = F\cos(\beta + \alpha_2) \tag{3.5.5}$$

3.5.3 Displacement Calculation

Considering the single degree of freedom models shown in the u_1 and u_2 directions in Fig. 3.4.9, the corresponding equations of motion are:

$$m_1\ddot{u}_1 + c_1\dot{u}_1 + k_1u_1 = F_{u_1} \text{ and } m_2\ddot{u}_2 + c_2\dot{u}_2 + k_2u_2 = F_{u_2}. \tag{3.5.6}$$

The mode direction accelerations in the current time step due to the two force components are determined by rewriting Eq. 3.5.6:

$$\ddot{u}_1 = \frac{F_{u_1} - c_1\dot{u}_1 + k_1u_1}{m_1} \text{ and } \ddot{u}_2 = \frac{F_{u_2} - c_2\dot{u}_2 + k_2u_2}{m_2}, \tag{3.5.7}$$

where the velocities, $\dot{u}_1$ and $\dot{u}_2$, and positions, u_1 and u_2, from the previous time step are used (they are set equal to zero initially). The velocities and positions for the current time step are then determined by numerical (Euler) integration:

$$\dot{u}_1 = \dot{u}_1 + \ddot{u}_1dt \text{ and } \dot{u}_2 = \dot{u}_2 + \ddot{u}_2dt \tag{3.5.8}$$

$$u_1 = u_1 + \dot{u}_1dt \text{ and } u_2 = u_2 + \dot{u}_2dt, \tag{3.5.9}$$

where the velocities on the right hand side of the equal signs in Eq. 3.5.8 are retained from the previous time step and used to update the current values. These current values are then applied to determine the current displacements in Eq. 3.5.9. Again, the displacements on the right hand side of Eq. 3.5.9 are those from the previous time step. Once u_1 and u_2 are known, they are projected into the normal (y) direction:

$$y_n = u_1\cos(\alpha_1) + u_2\cos(\alpha_2), \tag{3.5.10}$$

where the n subscript on y indicates the time step. Note that this value represents y_{n-1} in the next time step and is applied in Eq. 3.5.3 to calculate the updated chip thickness. The process of computing the force and resulting displacements is then repeated.

Example 3.5.2: Stability evaluation by time-domain simulation Consider the turning model shown in Fig. 3.5.2. The dynamic constants are: $m_1 = 1\text{kg}$, $c_1 = 450\text{N} - \text{s/m}$, and $k_1 = 2 \times 10^7\,\text{N/m}$ for the u_1 direction and $m_2 = 1\,\text{kg}$, $c_2 = 650\text{N} - \text{s/m}$, and $k_2 = 3 \times 10^7\,\text{N/m}$ for the u_2 direction. The other parameters are: $\alpha_1 = 35\,\text{deg}$, $\alpha_2 = 55\,\text{deg}$, $\beta = 65\,\text{deg}$, $h_m = 0.1\,\text{mm}$, and $K_s = 2000\,\text{N/mm}^2$. Prior to determining the time-domain force and displacements using the simulation described in the previous paragraphs, let's calculate $b_{\text{lim,crit}}$, the approximate best and worst speeds, and the stability lobe diagram for this model.

As seen in Ex. 3.4.3, the oriented FRF must be determined using the directional orientation factors prior to calculating $b_{\text{lim,crit}}$. The directional orientation factors are:

$$\mu_1 = \cos(\beta - \alpha_1)\cos(\alpha_1) = \cos(65 - 35)\cos(35) = 0.709 \text{ and}$$

$$\mu_2 = \cos(\beta + \alpha_2)\cos(\alpha_2) = \cos(65 + 55)\cos(55) = -0.287.$$

The oriented FRF is displayed in Fig. 3.5.3. The minimum real part of occurs at 748 Hz with a value of $1.993 \times 10^{-4}\,\text{mm/N}$ and the real part crosses through zero amplitude at a frequency of 709 Hz. Similar to Ex. 3.4.3, although this is not a single degree of freedom system, the real part of the oriented FRF resembles single degree of freedom system behavior. We can therefore approximate $b_{\text{lim,crit}}$ and the best and worst spindle speeds using the single degree of freedom equations presented in Section 3.3. The best and worst spindle speeds are summarized in Table 3.5.1.

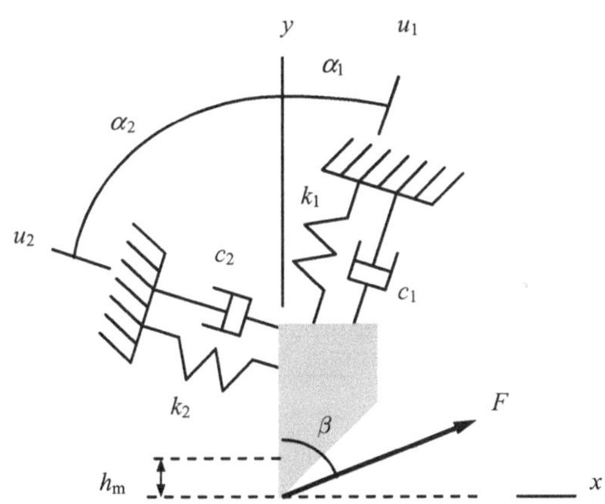

Fig. 3.5.2 Model for turning time-domain simulation in Ex. 3.5.2

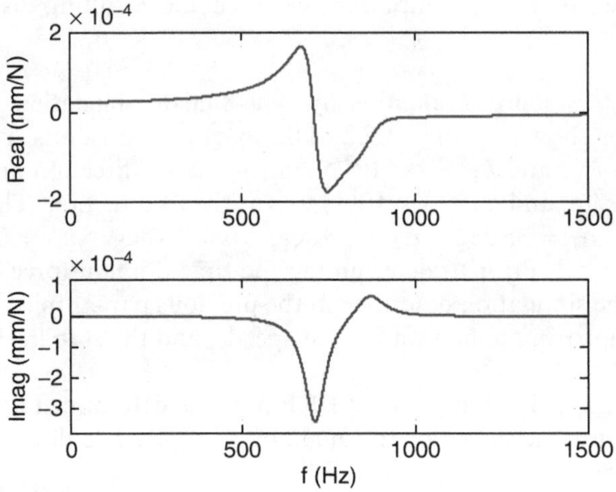

Fig. 3.5.3 Oriented FRF for Ex. 3.5.2

$$b_{\text{lim,crit}} = \frac{-1}{2K_s \cdot \min(\text{Re}[FRF_{orient}])} = \frac{-1}{2 \cdot 2000 \cdot (-1.993 \times 10^{-4})} = 1.3\,\text{mm}$$

$$\Omega_{best} = \frac{709}{N+1}\ (\text{rps})$$

$$\Omega_{worst} = \frac{748}{N+\frac{3}{4}}\ (\text{rps})$$

The top panel of Fig. 3.5.4 shows a single valid chatter frequency range (real part < 0) for the oriented FRF. In the bottom panel, it is observed that the best speed of 709 rps is an acceptable approximation of the actual behavior, although this is not a reasonable spindle speed (42540 rpm) for typical turning applications. Figure 3.5.5 shows the combined stability boundary for $N = 0$ to 4. The MATLAB® programs used to produce these figures are provided on the companion CD as p_3_5_2_1.m and p_3_5_2_2.m.

As discussed before, Fig. 3.5.5 provides a global view of the process stability, but does not provide local information, such as force and tool displacement,

Table 3.5.1 Approximate best and worst spindle speeds for Ex. 3.5.2

N	Ω_{best} (rps)	Ω_{best} (rpm)	Ω_{worst} (rps)	Ω_{worst} (rpm)
0	709	42540	997	59840
1	355	21270	427	25646
2	236	14180	272	16320
3	177	10635	199	11968
4	142	8508	157	9448

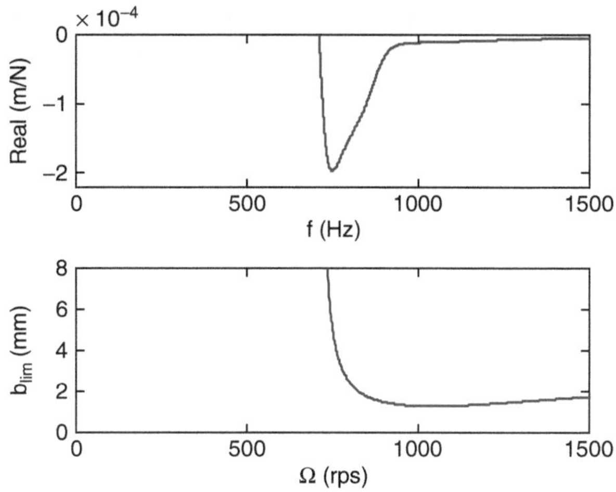

Fig. 3.5.4 (Top panel) Negative real part of oriented FRF for Ex. 3.5.2. (Bottom panel) Corresponding $N = 0$ stability limit

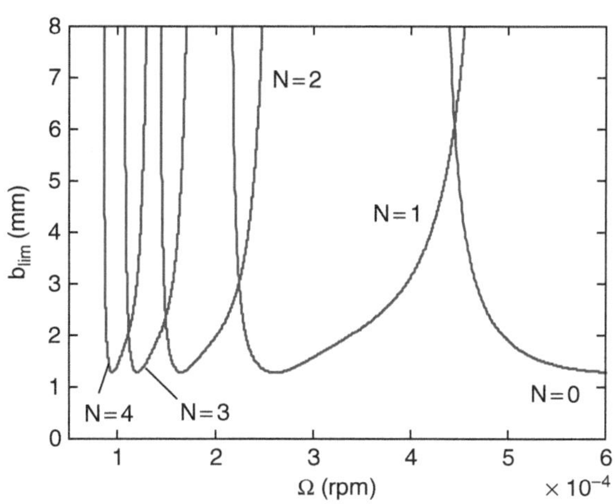

Fig. 3.5.5 Stability lobes for Ex. 3.5.2 ($N = 0$ to 4)

about particular operating parameter combinations. For example, the stability boundary would suggest that 44490 rpm is a good operating speed for b values up to 6 mm [this spindle speed is somewhat higher than the simple approximation of 42540 rpm provided in Table 3.5.1]. It also shows that the maximum allowable chip width at 26140 rpm is slightly less than 1.3 mm. We will now use the time-domain simulation to determine the force and tool displacement values near these stability thresholds. Figure 3.5.6 provides the results for $\Omega = 44490$ rpm and $b = 5$ mm, where the total simulation time

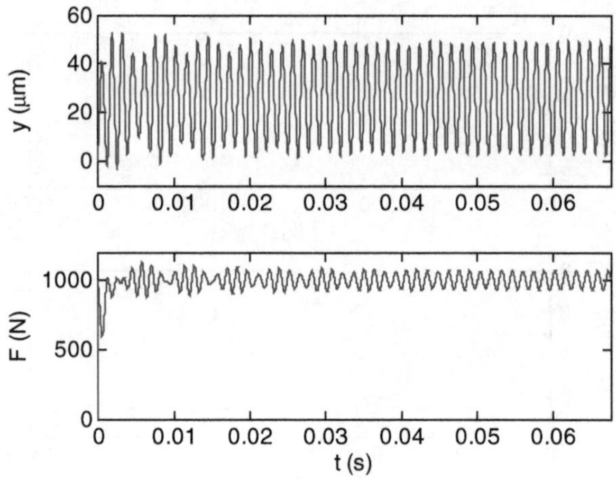

Fig. 3.5.6 Example 3.5.2 time-domain force and displacement for $\Omega = 44490$ rpm and $b = 5$ mm

corresponds to 50 revolutions, $\frac{50}{\frac{44490}{60}} = 0.067$s. It is seen that these conditions are actually near the stability limit since the force and displacement levels are neither growing nor diminishing. In this way, time-domain simulation can be used to refine the analytical stability limit results. We can also check the simulation mean values against Eqs. 3.5.11 and 3.5.12. Equation 3.5.11 gives the mean cutting force, F_m, based on the commanded chip area and specific force. Equation 3.5.12 provides the mean normal direction displacement, y_m, where the sum of the $\frac{\mu}{k}$ ratios gives the static compliance of the oriented FRF. As expected, the simulation results in Fig. 3.5.6 agree with the analytical expressions in both cases. At $\Omega = 44490$ rpm and $b = 6$ mm, the cut is clearly unstable; the force and displacement values are growing significantly with time. See Fig. 3.5.7.

$$F_m = K_s b h_m = 2000 \cdot 5 \cdot 0.1 = 1000\,\text{N} \tag{3.5.11}$$

$$y_m = F_m\left(\frac{\mu_1}{k_1} + \frac{\mu_2}{k_2}\right) = 1000\left(\frac{0.709}{2 \times 10^7} + \frac{-0.287}{3 \times 10^7}\right) = 2.6 \times 10^{-5}\,\text{m} = 26\,\mu\text{m} \tag{3.5.12}$$

At the less favorable spindle speed of $\Omega = 26140$ rpm, chatter is observed at $b = 2$ mm in Fig. 3.5.8. Not only are the force and displacement values growing with time, but the force nonlinearity is also observed (beginning at approximately $t = 0.025$ s) when the tool vibration grows large enough that there is no cutting intermittently and $F = 0$. The total simulated time for 50 revolutions is now: $\frac{50}{\frac{26140}{60}} = 0.067$s. If the chip width is reduced to 1 mm, stable operation is obtained (Fig. 3.5.9). Here it is seen that the initial transient response quickly

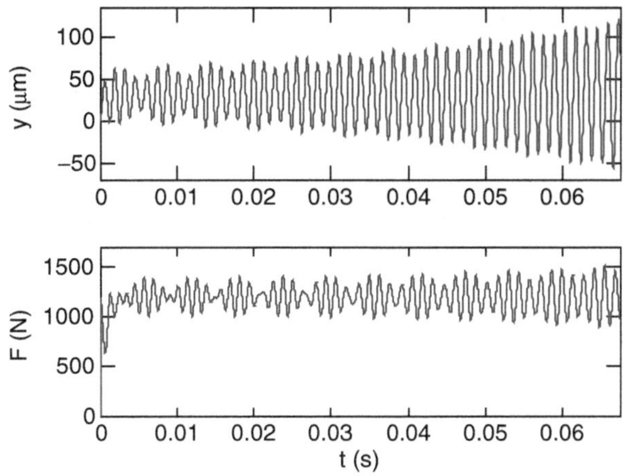

Fig. 3.5.7 Example 3.5.2 time-domain force and displacement for Ω = 44490 rpm and $b = 6$ mm

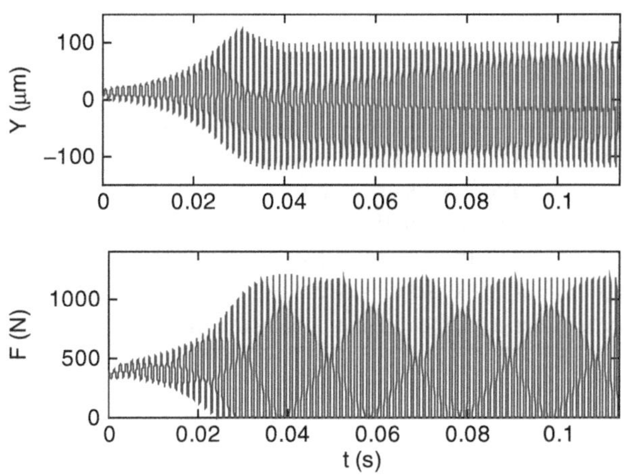

Fig. 3.5.8 Example 3.5.2 time-domain force and displacement for Ω = 26140 rpm and $b = 2$ mm

attenuates and the steady state force and displacement are obtained. The MATLAB® program used to produce these figures is provided on the companion CD as p_3_5_2_3.m.

$$F_m = K_s b h_m = 2000 \cdot 1 \cdot 0.1 = 200\,\text{N}$$

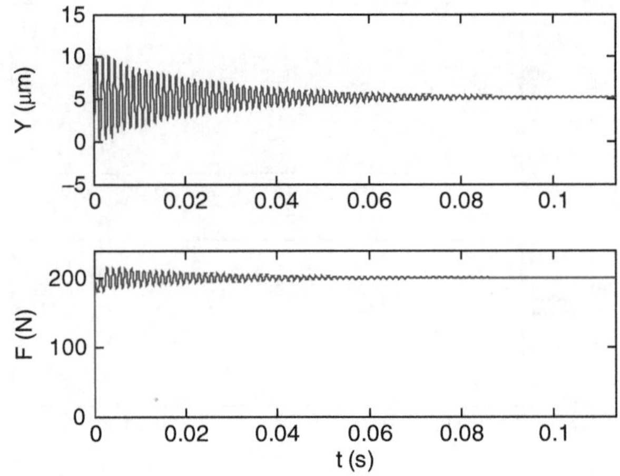

Fig. 3.5.9 Example 3.5.2 time-domain force and displacement for $\Omega = 26140$ rpm and $b = 1$ mm

$$y_m = F_m\left(\frac{\mu_1}{k_1} + \frac{\mu_2}{k_2}\right) = 200\left(\frac{0.709}{2 \times 10^7} + \frac{-0.287}{3 \times 10^7}\right) = 5 \times 10^{-6}\,\text{m} = 5\,\mu\text{m}$$

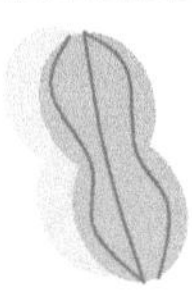

IN A NUTSHELL Time-domain simulation eliminates many of the simplifying assumptions required to obtain the global analytical solution. For that reason, time-domain simulation is more accurate than the analytical solution. However, it provides information on a case-by-case basis only. Using time-domain simulation to produce a stability lobe diagram requires many repeated executions.

3.5.4 Multiple Degree of Freedom Modeling

As a final point of consideration in this section, let's discuss how to extend the time-domain simulation to include multiple degrees of freedom in the two mode directions u_1 and u_2. The important point to remember from Chapter 2 is that, provided we have measured the direct FRF at the location of interest (the tool point), the response can be expressed as a sum of the modal contributions. Therefore, once we have completed the curve fitting exercise to obtain the modal parameters (Section 2.5), we can treat each degree of freedom for both mode directions separately and determine the individual solutions using Euler integration as shown in Eqs. 3.5.7 through 3.5.9. After the displacements in each mode have been calculated, they are simply summed in each direction. For example, assume that two modes were identified and curve fitted using peak picking for the u_1 direction and three modes were selected for the u_2 direction. The modal parameters m_{q1,u_1}, c_{q1,u_1}, k_{q1,u_1}, m_{q2,u_1}, c_{q2,u_1}, and k_{q2,u_1} would be obtained for u_1, while m_{q1,u_2}, c_{q1,u_2}, k_{q1,u_2}, m_{q2,u_2}, c_{q2,u_2}, k_{q2,u_2}, m_{q3,u_2}, c_{q3,u_2}, and k_{q3,u_2} would be obtained for u_2. The modal accelerations in the u_1 direction would now include:

$$\ddot{u}_{1,q1} = \frac{F_{u_1} - c_{q1,u_1}\dot{u}_{1,q1} + k_{q1,u_1}u_{1,q1}}{m_{q1,u_1}} \text{ and } \ddot{u}_{1,q2} = \frac{F_{u_1} - c_{q2,u_1}\dot{u}_{1,q2} + k_{q2,u_1}u_{1,q2}}{m_{q2,u_1}},$$

where the local u_1 direction force is used because the modal force vector is composed of identical local force values for each modeled mode (see Eq. 2.4.4.). The modal coordinate velocities and positions would then be determined according to:

$$\dot{u}_{1,q1} = \dot{u}_{1,q1} + \ddot{u}_{1,q1}dt \quad \text{and} \quad \dot{u}_{1,q2} = \dot{u}_{1,q2} + \ddot{u}_{1,q2}dt \quad \text{and}$$
$$u_{1,q1} = u_{1,q1} + \dot{u}_{1,q1}dt \quad \text{and} \quad u_{1,q2} = u_{1,q2} + \dot{u}_{1,q2}dt.$$

Finally, the u_1 local displacement would be calculated using: $u_1 = u_{1,q1} + u_{1,q2}$. Similarly, the u_2 direction modal accelerations would be:

$$\ddot{u}_{2,q1} = \frac{F_{u_2} - c_{q1,u_2}\dot{u}_{2,q1} + k_{q1,u_2}u_{2,q1}}{m_{q1,u_2}}, \ddot{u}_{2,q2} = \frac{F_{u_2} - c_{q2,u_2}\dot{u}_{2,q2} + k_{q2,u_2}u_{2,q2}}{m_{q2,u_2}}, \text{ and}$$

$$\ddot{u}_{2,q3} = \frac{F_{u_2} - c_{q3,u_2}\dot{u}_{2,q3} + k_{q3,u_2}u_{2,q3}}{m_{q3,u_2}}.$$

The modal coordinate velocities and positions in the u_2 direction would then be determined using:

$$\dot{u}_{2,q1} = \dot{u}_{2,q1} + \ddot{u}_{2,q1}dt, \dot{u}_{2,q2} = \dot{u}_{2,q2} + \ddot{u}_{2,q2}dt, \quad \text{and} \quad \dot{u}_{2,q3} = \dot{u}_{2,q3} + \ddot{u}_{2,q3}dt$$
$$u_{2,q1} = u_{2,q1} + \dot{u}_{2,q1}dtu_{2,q2} = u_{2,q2} + \dot{u}_{2,q2}dt \quad \text{and} \quad u_{2,q3} = u_{2,q3} + \dot{u}_{2,q3}dt$$

and the u_2 local displacement would be calculated using: $u_2 = u_{2,q1} + u_{2,q2} + u_{2,q3}$. As with the single degree of freedom case, all initial values of the modal velocities and displacements should be set equal to zero prior to simulation execution.

IN A NUTSHELL Modal analysis provides a convenient way to summarize the dynamic characteristics of complicated vibration systems. When applying time-domain simulation, the compact modal data representation facilitates determination of the appropriate parameters for stable machining.

Exercises

1. For the turning schematic shown in Fig. e.3.1., complete parts a) through f). For the single degree of freedom dynamics, the mass is 2 kg, the damping ratio is 0.05, and the stiffness is 2×10^7 N/m. The u direction is oriented at an angle, α, of 35 deg relative to the surface normal, y. The force model parameters are $K_s = 1500$ N/mm^2 and $\beta = 70$ deg.

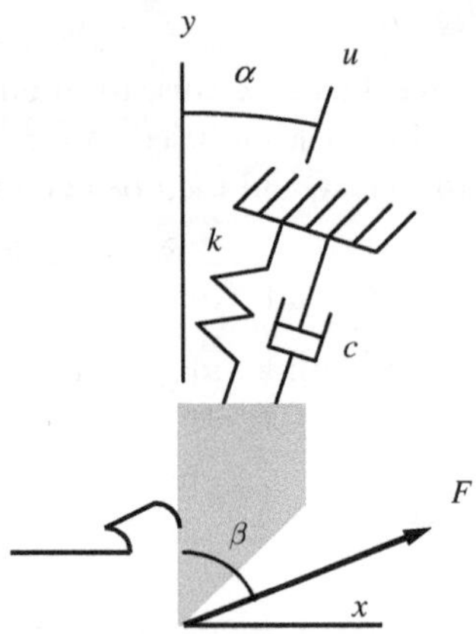

Fig. e.3.1 Turning model with flexible tool

a) Calculate the directional orientation factor. Using this value, compute and plot the real and imaginary parts (in m/N) of the oriented frequency response function vs. frequency (in Hz).
b) Determine the minimum value of the real part of the oriented frequency response function and the corresponding chatter frequency. Calculate $b_{lim,crit}$.
c) Determine the spindle speed (in rpm) corresponding to the stability peak defined by the intersection of the $N = 0$ and $N = 1$ stability lobes.
d) Find the spindle speed (in rpm) corresponding to the minimum stability limit for the $N = 0$ lobe.
e) Determine the spindle speed (in rpm) corresponding to the stability peak defined by the intersection of the $N = 3$ and $N = 4$ stability lobes.
f) Plot the first four stability lobes ($N = 0$ to 3) for this system. Use b_{lim} units of mm and spindle speed units of rpm.

2. Using the turning schematic shown in Fig. e.3.2., complete parts a) through d). For the u_1 direction, the mass is 10 kg, the damping is 170 N-s/m, and the stiffness is 7×10^6 N/m. The u_1 direction is oriented at an angle, α_1, of 60 deg relative to the surface normal, y. For the u_2 direction, the mass is 12 kg, the damping is 1700 N-s/m, and the stiffness is 5×10^7 N/m. The u_2 direction is oriented at an angle, α_2, of 30 deg relative to the y direction. The force model parameters are $K_s = 2000$ N/mm^2 and $\beta = 60$ deg.

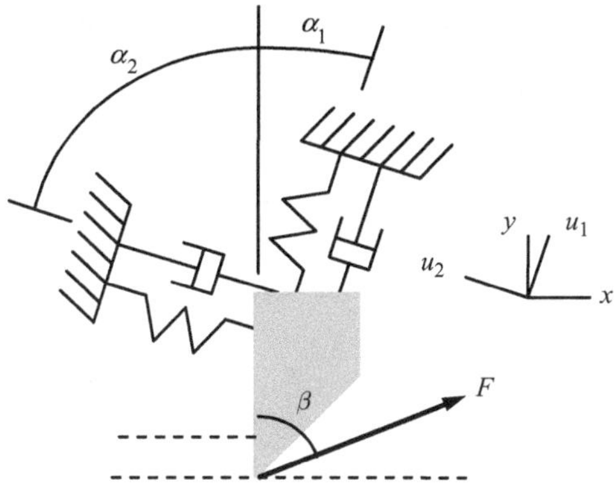

Fig. e.3.2 Turning model with a single degree of freedom in both the u_1 and u_2 directions

a) Compute the directional orientation factors, μ_1 and μ_2. Plot the real and imaginary parts (in m/N) of the oriented frequency response function vs. frequency (in Hz).
b) Determine the minimum value of the real part of the oriented frequency response function and the corresponding chatter frequency. Calculate $b_{\text{lim,crit}}$.
c) Find the spindle speed (in rpm) corresponding to the minimum stability limit for the $N = 2$ lobe.
d) Plot the first five stability lobes ($N = 0$ to 4) for this system. Use b_{lim} units of mm and spindle speed units of rpm.

3. Considering the turning model shown in Fig. e.3.3, determine the critical stability limit if $K_s = 750$ N/mm^2. For both lumped parameter degrees of freedom, the mass is 1 kg, the stiffness is 7×10^6 N/m, and the damping is 200 N-s/m.

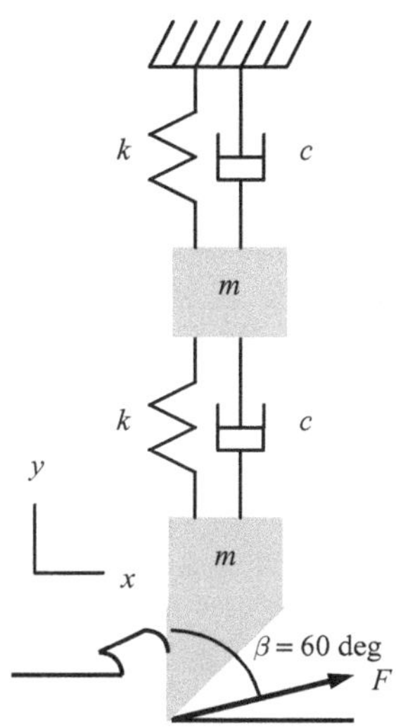

Fig. e.3.3 Two degree of freedom turning model

4. Complete time-domain simulations for the turning model described in Exercise 2. Evaluate the following points for stable or unstable behavior. Use a mean chip thickness (feed per revolution) of 0.15 mm and carry out your simulations for 25 revolutions.

Ω (rpm)	b (mm)
2150	0.1
2150	0.5
2500	0.1
2500	0.5
2930	0.1
2930	0.5
3750	0.1
3750	0.5
4600	0.1
4600	0.5

Superimpose your results on the stability lobe diagram from Exercise 2, part d). Use a circle for stable operating points and an 'x' for unstable points.

References

1. Trent, E. and Wright, P., 2000, Metal Cutting, 4th Ed., Butterworth-Heinemann, Boston, MA.
2. Tlusty, G., 2000, Manufacturing Equipment and Processes, Prentice-Hall, Upper Saddle River, NJ.
3. Altintas, Y., 2000, Manufacturing Automation: Metal Cutting Mechanics, Machine Tool Vibrations, and CNC Design, Cambridge University Press, Cambridge.
4. Stephenson, D. and Agapiou, J., 1997, Metal Cutting Theory and Practice, Marcel Dekker, Inc., New York, NY.

Chapter 4
Milling Dynamics

If we knew what it was we were doing, it would not be called research, would it?

–Albert Einstein

In Chapter 3, we applied knowledge of the tool point dynamics from modal analysis (Chapter 2) to predict regenerative chatter in turning using stability lobe diagrams and time-domain simulation. This chapter follows a similar format, but our focus is milling. We again develop force expressions, discuss surface regeneration, and present the relevant equations and examples for stability lobe diagrams. This is followed by a description of time-domain simulation that predict forces and displacements during milling for selected operating parameters. The chapter concludes with a description of the experimental determination of cutting force coefficients.

4.1 Milling Description

In milling, a rotating tool with defined cutting edges is moved relative to a workpiece in order to remove material and obtain the desired workpiece geometry and dimensions. The tool is typically mounted in a holder which is attached to the spindle. The spindle provides the tool's rotational speed, torque, and power. Multiple axes are then used to manipulate the tool-holder-spindle relative to the workpiece. At minimum, three linear axes are generally arranged in a mutually perpendicular configuration; these linear motions are traditionally labeled x, y, and z with the latter indicating the tool axis. However, milling machines are also available with additional rotational axes to provide contouring capabilities for non-prismatic parts. In this case, the rotational degrees of freedom are typically labeled A, B, and C, which indicate rotations about the x, y, and z axes, respectively. As with lathes, milling machines may be manual or computer numerically controlled. Figure 4.1.1 shows an example representation of a three axis milling machine. A vertical spindle configuration is shown, although horizontal spindle geometries are also available. The latter is often

T.L. Schmitz, K.S. Smith, *Machining Dynamics*, DOI 10.1007/978-0-387-09645-2_4,

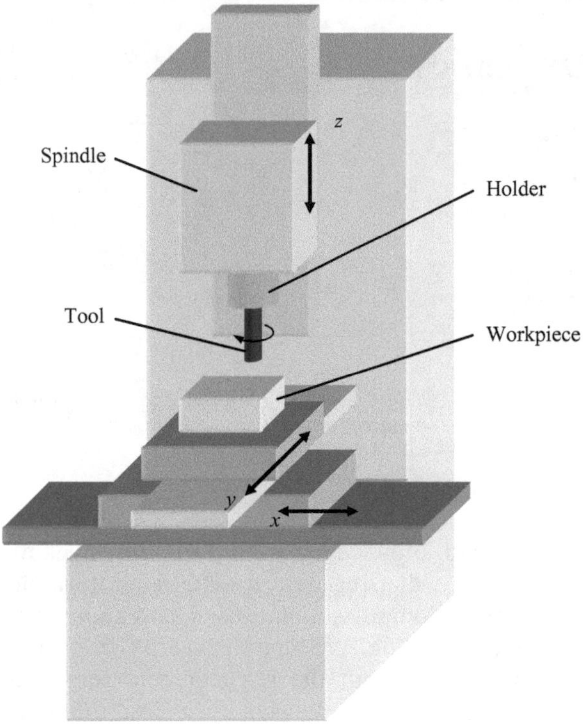

Fig. 4.1.1 Schematic of vertical spindle milling machine. The tool is clamped in a holder which is attached to the rotating spindle. The tool-holder-spindle is moved relative to the workpiece using the three orthogonal axes to remove material in the subtractive milling process

preferred in high speed milling applications because the chips fall from the workpiece for later collection.

Cutting tools and holders are available in many varieties that are tailored to specific applications, such as peripheral, end, contour, and face milling. For analysis purposes, we will focus on peripheral and end milling operations, although the concepts can be extended to other operations as well. Endmills may be loosely categorized according to their free end geometry, including square, ball nose, and bull nose. Square endmills have a cylindrical shape with a small end radius so that the profile is rectangular. Ball nose endmills are hemispherical at their free end to enable contouring of internal cavities and external, three dimensional non-prismatic features. Endmills with a bull nose geometry have larger end radii than square endmills, but are not fully hemispherical. Similar to ball nose endmills, they also allow contouring. See Fig. 4.1.2.

Endmills may have the cutting teeth ground directly into the body or may have replaceable inserts clamped to a cylinder. Typical materials for ground cutters include high speed steel and sintered carbide, while inserted cutters

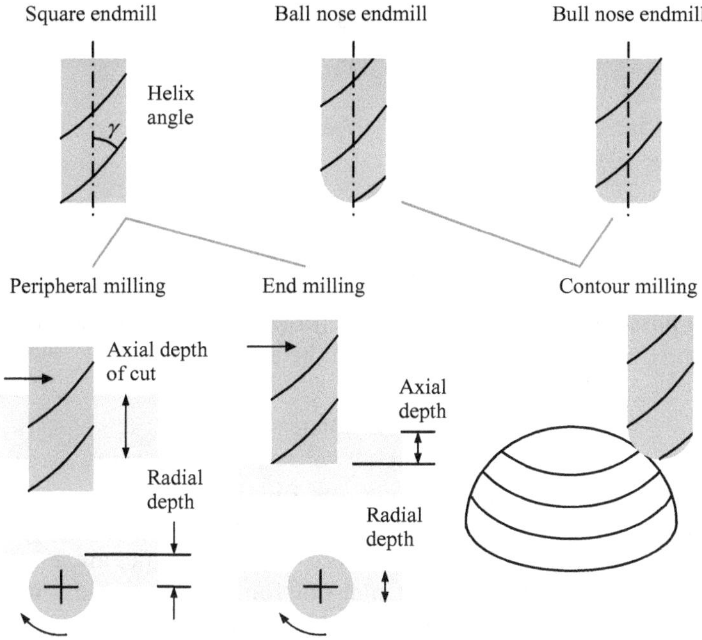

Fig. 4.1.2 Example milling applications and tool geometries.

traditionally use steel bodies with carbide or ceramic inserts. Coatings designed to improve tool life are also often applied in single or multiple layers. For either ground or solid body endmills, the cutting edge is not usually parallel to the tool's rotating axis (referred to as "straight" teeth). Rather, the edge is inclined so that the chip to be removed is spread over an increased length and the cutting edge pressure is reduced. For solid body cutters, the edge is typically ground with a helical profile around the tool periphery using a constant helix angle, γ. The chip width, $\tilde{b}$, is then related to the axial depth of cut, b, as:

$$\tilde{b} = \frac{b}{\cos(\gamma)} \tag{4.1.1}$$

This will have implications for the surface location error calculations described in Chapter 5. However, for now we will neglect the cutting edge's helical shape and assume straight cutter teeth ($\tilde{b} = b$ for $\gamma = 0$). This simplification enables us to obtain some insight into the milling operation and is reasonable if the axial depth of cut is small. A second assumption that we will apply is referred to as the "circular tool path" approximation. The actual path followed by any point on the cutter's periphery as the tool rotates during translation is

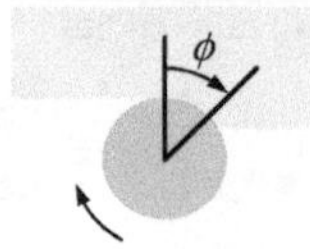

Fig. 4.1.3 Cutter angle definition

cycloidal in nature. However, because the linear advance is generally small relative to the product of the rotational speed and tool radius, the path may be approximated as a series of circles, each offset by the feed per tooth, f_t, so that the time dependent chip thickness can be expressed as:

$$h = f_t \sin(\phi) \tag{4.1.2}$$

where ϕ is the tool's rotational angle; see Fig. 4.1.3. The feed per tooth is described in terms of the linear feed, f, spindle speed, Ω, and number of teeth on the cutter, N_t, in Eq. 4.1.3. Typical units for these variables are mm/tooth for feed per tooth, rpm for spindle speed, mm/min for linear feed rate, and teeth/rev for number of teeth. An improved analytical model for chip thickness is provided in [1], but the circular tooth path approximation offers sufficient accuracy for our purposes.

$$f_t = \frac{f}{\Omega N_t} \tag{4.1.3}$$

As Eq. 4.1.2 shows, the chip thickness in milling varies periodically, even in the absence of tool or workpiece vibrations. It is zero when $\phi = 0$ and 180 deg and maximum (equal to f_t) when $\phi = 90$ deg. Figure 4.1.4 shows this variation for both conventional, or up, and climb, or down, peripheral milling operations. Note that the chip thickness increases during up milling and decreases in down milling. In both cases, it is zero for $180 < \phi < 360$ deg because no cutting occurs between these angles. The entry, or start, angle

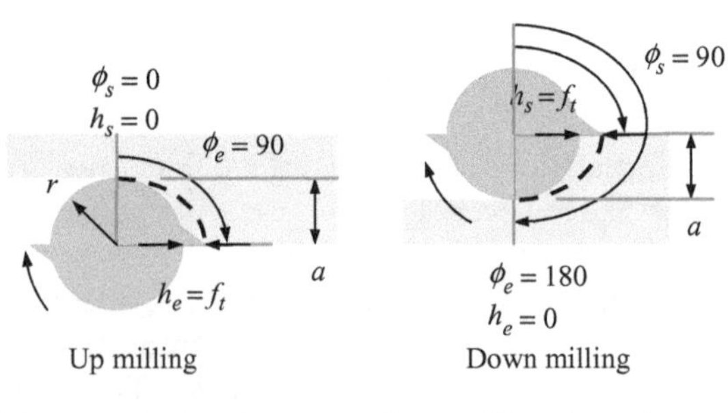

Fig. 4.1.4 Chip thickness variation for up and down milling ($a = r$)

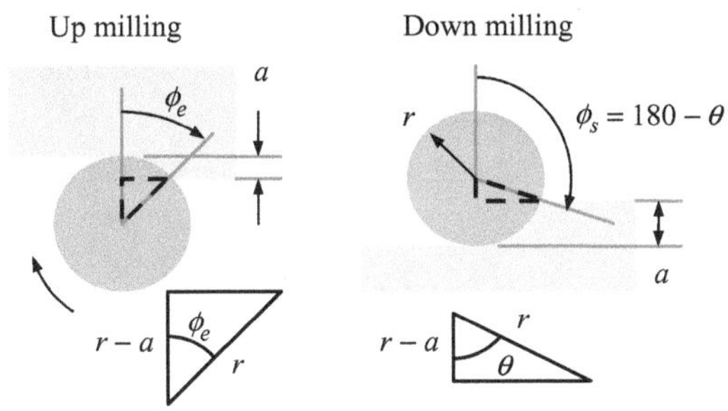

Fig. 4.1.5 Exit and start angle geometry for up and down milling

for up milling is $\phi_s = 0$, while the exit angle, ϕ_e, depends on the radial depth of cut, a, and tool radius, r:

$$\phi_e = \cos^{-1}\left(\frac{r-a}{r}\right) \tag{4.1.4}$$

In down milling, the exit angle is $\phi_e = 180$ deg. Similar to up milling, the start angle is written as a function of the radial depth and tool radius. See Eq. 4.1.5 and Fig. 4.1.5.

$$\phi_s = 180 - \theta = 180 - \cos^{-1}\left(\frac{r-a}{r}\right)\ \text{(deg)} \tag{4.1.5}$$

IN A NUTSHELL Whereas in turning operations the chip thickness and chip width are fixed, this is not the case in milling. In a straight slotting cut, the chip thickness encountered by each tooth varies continuously as that tooth enters and exits the cut. In pocket milling, the radial depth of cut may also change. Finally, in sculptured surface milling, the axial depth of cut may vary as well.

Example 4.1.1: Start and exit angles for up milling Consider the peripheral up milling cut shown in Fig. 4.1.6 where the radial depth of cut is 1.9 mm. For a 19 mm diameter (9.5 mm radius) cutter, we will refer to this as a 10% radial immersion cut. (Using this notation, a slotting cut with a radial depth of 19 mm

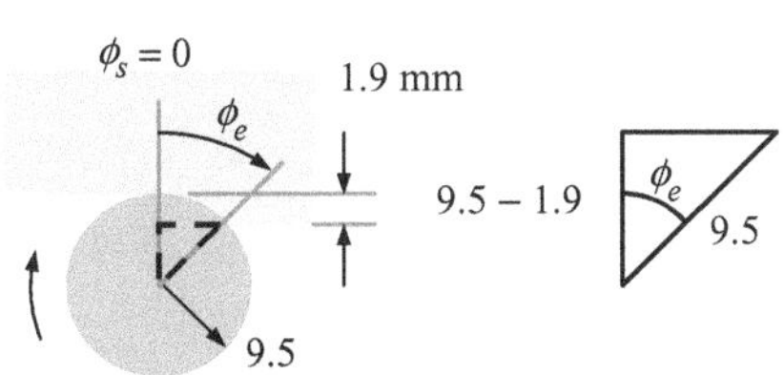

Fig. 4.1.6 Exit angle for 10% radial immersion up milling cut

would be described as 100% radial immersion for the 19 mm diameter cutter.) Because it is an up milling cut, the start angle is $\phi_s = 0$. The exit angle is:

$$\phi_e = \cos^{-1}\left(\frac{9.5 - 1.9}{9.5}\right) = \cos^{-1}(0.8) = 37 \text{ deg}$$

and the instantaneous chip thickness between the start and exit angles can be defined by Eq. 4.1.2.

As with turning, let's begin our discussion of cutting force by assuming a rigid tool and workpiece; that is, by considering the geometry alone. Unlike turning, however, the cutting force is not constant under these conditions. Rather, it is a function of the cutting angle. As we'll see, the cutting force expression is complicated by the chip thickness variation with cutter angle, the number of teeth simultaneously engaged in the cut at any instant, and the projection of the cutting force into a non-rotating coordinate frame. Similar to Eqs. 3.1.1–3.1.3, we can express the cutting force on any cutting edge as a function of the chip area and specific force:

$$F = K_s A = K_s bh. \tag{4.1.6}$$

The normal and tangential components can be written using Eqs. 4.1.7 and 4.1.8, as demonstrated in Fig. 4.1.7:

$$F_n = \cos(\beta)F = \cos(\beta)K_s bh = k_n bh \text{ and} \tag{4.1.7}$$

$$F_t = \sin(\beta)F = \sin(\beta)K_s bh = k_t bh. \tag{4.1.8}$$

This figure shows the cutting force exerted on a single tooth, where the force is described in a coordinate frame that rotates with the tool. For measurement purposes, however, it is generally more convenient to express the force in a fixed frame. For example, the workpiece may be mounted on a cutting force dynamometer and the x, y, and z direction force components recorded during milling (Fig. 4.1.8). To describe these forces analytically, we must project the normal and tangential components into the x and y directions using the cutter angle ϕ. Based on our assumption of straight cutter teeth, we may neglect the z direction component.

$$F_x = F_t \cos(\phi) + F_n \sin(\phi) \tag{4.1.9}$$

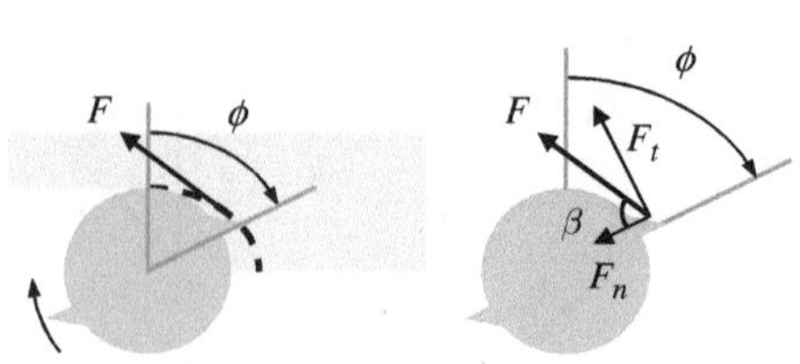

Fig. 4.1.7 Cutting force geometry for milling

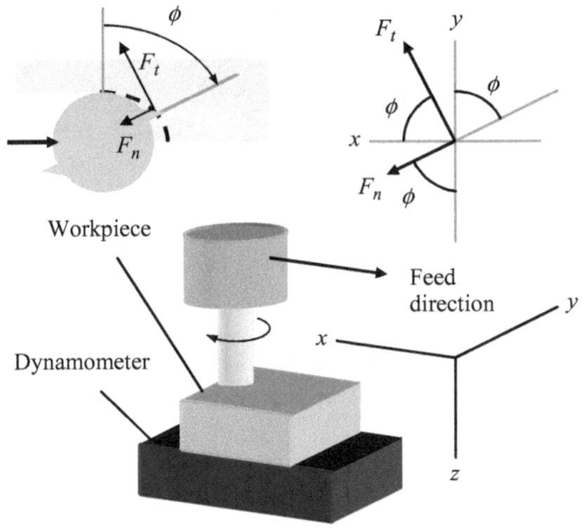

Fig. 4.1.8 Projection geometry for tangential and normal cutting force components into x and y directions.

$$F_y = F_t \sin(\phi) - F_n \cos(\phi) \tag{4.1.10}$$

We can now substitute for the tangential and normal force components in Eqs. 4.1.9 and 4.1.10. This yields:

$$F_x = k_t b f_t \sin(\phi)\cos(\phi) + k_n b f_t \sin(\phi)\sin(\phi) \text{ and} \tag{4.1.11}$$

$$F_y = k_t b f_t \sin(\phi)\sin(\phi) - k_n b f_t \sin(\phi)\cos(\phi), \tag{4.1.12}$$

where we have also replaced the instantaneous chip thickness with the expression provided in Eq. 4.1.2. These forces are clearly periodic with ϕ, but we must also consider the cut geometry in the final force calculations. See the following example.

IN A NUTSHELL As the tool rotates, the force on each tooth changes. In addition, the force orientation (direction) on that tooth varies during the tool rotation. Therefore, the forces generated by a single tooth, as seen by the workpiece or machine tool, vary periodically with the rotation of the tool. In addition, the total force changes as additional teeth enter or leave the cut.

Example 4.1.2: Cutting forces for rigid tool and workpiece Consider a 25% radial immersion peripheral up milling operation carried out using a tool with

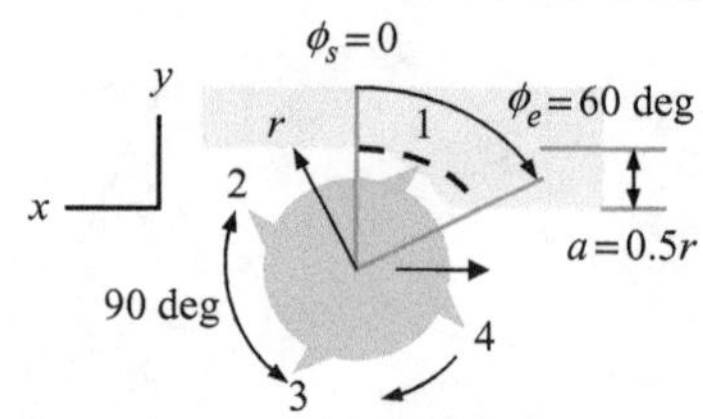

Fig. 4.1.9 25% radial immersion up milling cutting forces for Ex. 4.1.2

four teeth; see Fig. 4.1.9. The radial depth of cut is 25% of the cutter diameter or one half of its radius, $a = \frac{r}{2}$. Because it is an up milling operation, $\phi_s = 0$ and the exit angle is:

$$\phi_e = \cos^{-1}\left(\frac{r - \frac{r}{2}}{r}\right) = \cos^{-1}(0.5) = 60 \text{ deg}.$$

Therefore, each of the four teeth, equally spaced at 90 deg intervals around the periphery of the cutter, is engaged in the cut for only 60 deg. For this 60 deg range, the force components in the fixed x-y frame can be described using Eqs. 4.1.11 and 4.1.12. For the remaining 30 degrees, the forces are zero. The x and y direction cutting forces over one revolution are shown in Fig. 4.1.10, where the resultant force, $F = \sqrt{F_x^2 + F_y^2}$, is also included. For illustration purposes, the following parameters were selected: $k_t = 750$ N/mm^2 and $k_n = 250$ N/mm^2 (corresponds to $K_s = 791$ N/mm^2 and $\beta = 71.6$ deg – these values would be reasonable for an aluminum alloy), $b = 5$ mm, and $f_t = 0.1$ mm/tooth. We see that tooth 1 is engaged between 0 and 60 deg (assuming a start angle of $\phi = 0$ when tooth 1 is vertical). Because the chip thickness increases as the cutter angle

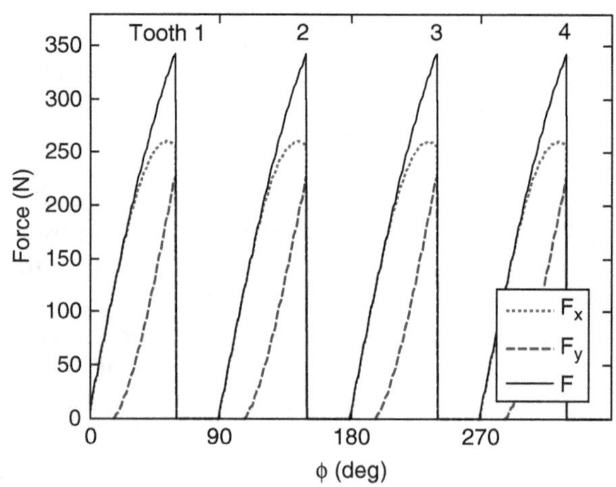

Fig. 4.1.10 25% radial immersion up milling cutting forces for Ex. 4.1.2

increases in up milling, the force levels grow. At 60 deg, the forces drop to zero until tooth 2 enters the cut at $\phi = 90$ deg. It is engaged until 150 deg (60 deg beyond 90 deg) when the force again drops to zero and so on for one full revolution (all four teeth). The MATLAB® program used to produce Fig. 4.1.10 is provided on the companion CD as p_4_1_2_1.m. A detailed description of the code is given in Section 4.4.

For comparison purposes, Fig. 4.1.11 shows the geometry for a 25% radial immersion down milling cut and Fig. 4.1.12 displays the corresponding single revolution cutting force profile. All parameters remain the same, except that the starting angle is:

$$\phi_s = 180 - \cos^{-1}\left(\frac{r - \frac{r}{2}}{r}\right) = 180 - \cos^{-1}(0.5) = 120\,\text{deg}$$

and the exit angle is $\phi_e = 180$ deg. It is observed that tooth 4 enters the cut first (again assuming a start angle of $\phi = 0$ when tooth 1 is vertical). This entry occurs after a 30 deg delay where no cutting occurs (i.e., the 90 deg lead of tooth 4

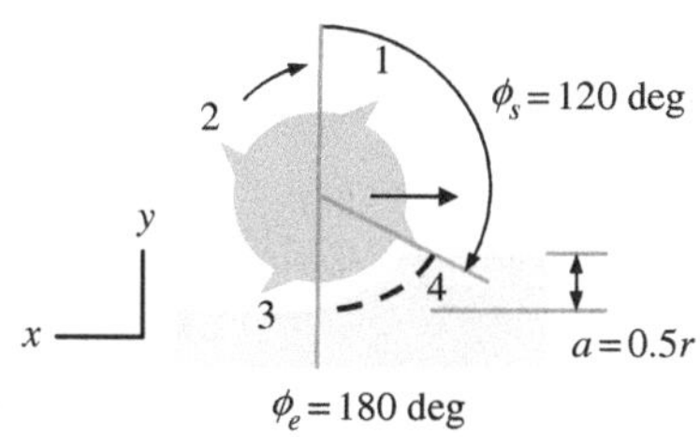

Fig. 4.1.11 25% radial immersion down milling geometry for Ex. 4.1.2

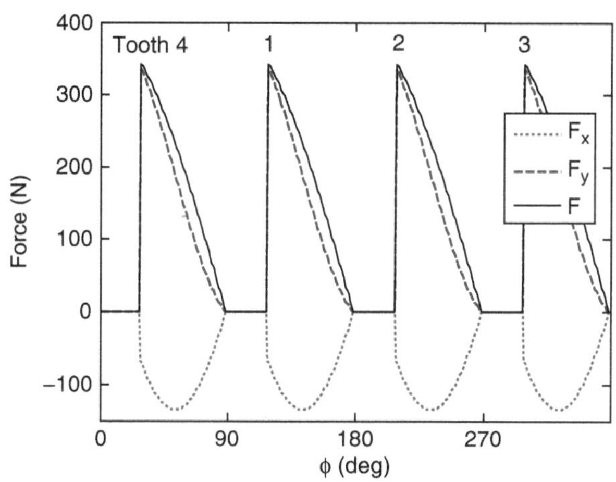

Fig. 4.1.12 25% radial immersion down milling cutting forces for Ex. 4.1.2

relative to tooth 1 plus 30 deg gives the 120 deg cut starting angle). The maximum force level is encountered at 30 deg and then decreases with the chip thickness to an angle of 90 deg; this trend of force reduction as the final surface is being created explains why down milling is often selected for finishing passes when surface finish is most critical. After $\phi = 90$ deg is reached, the force is again zero until tooth 1 enters the cut at 120 deg and the cycle is repeated. Note that the x direction force is now negative (acting to the right). This is in direct contrast to the positive up milling x force seen in Fig. 4.1.10. The MATLAB® program used to produce Fig. 4.1.12 is provided on the companion CD as p_4_1_2_2.m.

4.1.1 Tooth Passing Frequency

As a final point of emphasis for this example, let's discuss the frequency content of cutting force signals. As seen in Figs. 4.1.10 and 4.1.12, the forces during these partial immersion (up and down milling) cuts resemble trains of periodic impulses. To determine the resulting frequency spectrum, we first convert the abscissa to time (s), rather than tooth angle (deg). This conversion requires that the spindle speed, Ω, (rpm) is specified:

$$t = \frac{\phi \cdot 60}{\Omega \cdot 360} \text{(s)}. \tag{4.1.13}$$

Figures 4.1.13 (up milling) and 4.1.14 (down milling) show the resulting time vs. force plots for a spindle speed of 7500 rpm. It is seen that each tooth passage (90 deg angular interval for the cutter with four teeth) requires 0.002 s, or 2×10^{-3}s. The Fourier transform may now be applied to determine the

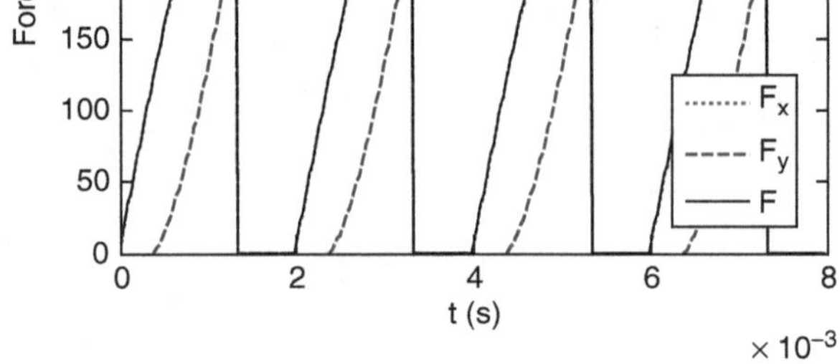

Fig. 4.1.13 Time-domain 25% radial immersion up milling cutting forces for Ex. 4.1.2

frequency content. Because the time-domain signal is a succession of impulse-like peaks, frequency content is observed not only at the tooth passing frequency:

$$f_{tooth} = \frac{\Omega \cdot N_t}{60} \text{ (Hz)}, \tag{4.1.14}$$

where the spindle speed is again expressed in rpm, but also at integer multiples of f_{tooth}. We refer to f_{tooth} as the fundamental tooth passing frequency or first harmonic, $2f_{tooth}$ as the second harmonic, and so on. The relative magnitudes of these harmonics depend on the "sharpness" of the force impulses; sharper peaks lead to increased magnitudes of higher order harmonics. Zero frequency, or DC, content is also typically observed. For example, the perfectly sinusoidal x direction force profile obtained for a two tooth cutter in a slotting cut exhibits content at only DC (because its average value is nonzero) and the fundamental tooth passing frequency (because it is a pure sinusoid and not impulsive in nature). Returning to Figs. 4.1.13 and 4.1.14, we saw that the tooth period was 0.002 s. This corresponds to a frequency of $\frac{1}{0.002} = 500$ Hz, which matches the result obtained from Eq. 4.1.14, $f_{tooth} = \frac{7500 \cdot 4}{60} = 500$ Hz. The magnitude vs. frequency plot for the down milling resultant force, F, is shown in Fig. 4.1.15. As expected, peaks are observed at 500 Hz, 1000 Hz, 1500 Hz, and higher order harmonics. This figure was generated using the MATLAB® program p_4_1_2_3.m included on the companion CD. The MATLAB® function `fft` was implemented to compute the fast Fourier transform of the force data.

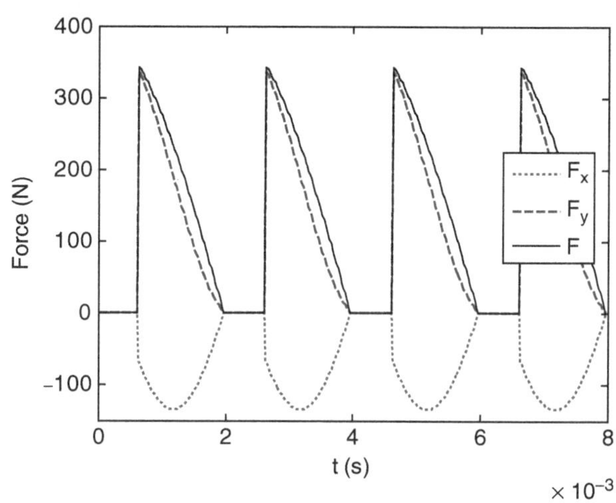

Fig. 4.1.14 Time-domain 25% radial immersion down milling cutting forces for Ex. 4.1.2

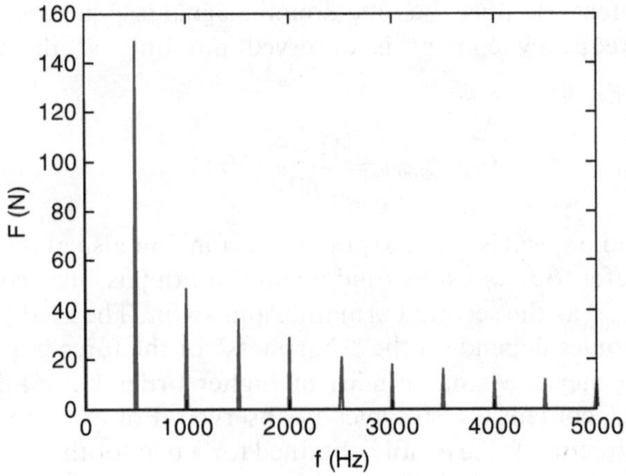

Fig. 4.1.15 25% radial immersion down milling frequency spectrum for resultant force in Ex. 4.1.2

IN A NUTSHELL The cutting force varies with the angle of the cutting edge. By including the spindle speed, we see that the cutting force varies in time and that the forces generated by the teeth are an external excitation that cause forced vibration. Changing the spindle speed modifies the excitation frequency. Because the force is periodic, but not purely sinusoidal, it can be thought of as a combination of many sinusoidal forces and these can be determined using the mathematical technique of the Fourier transform. The resulting plot shows the "frequency content" of the force. In many ways it is like looking at a graphic equalizer in a home stereo system[1] with the height of the peaks in the graph representing the force level at the corresponding frequency.

4.1.2 Multiple Teeth in the Cut

A natural extension of the previous example is to consider cases where there are more teeth on the cutter or the radial immersion is increased. In these instances, it is possible for more than one tooth to be engaged in the cut at a given time. Further, it is possible to alternate between, for example, one tooth cutting

[1] The "old" authors realize that graphic equalizer displays on stereos are not as common as they were in our teenage years. For that matter, they don't call them "boom boxes" any more, do they?

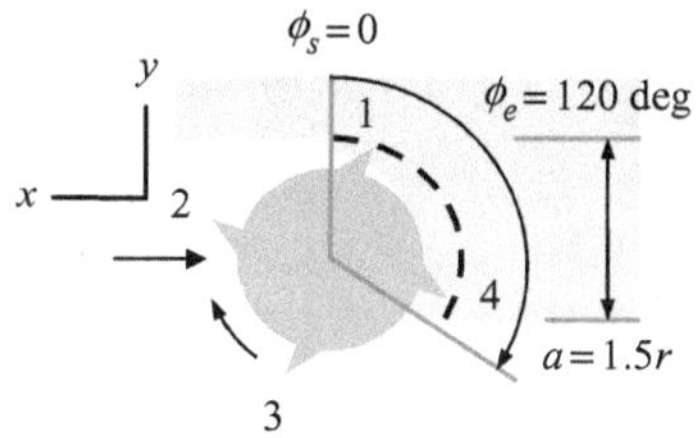

Fig. 4.1.16 75% radial immersion milling geometry

over some interval and two teeth in another. This situation is illustrated in Fig. 4.1.16, where the radial immersion is 75% and the cutter again has four teeth. First, we notice that the cut includes both up ($0 \leq \phi \leq 90$ deg) and down ($90 < \phi \leq 120$ deg) milling portions. Next, we see that when $0 \leq \phi \leq 30$ deg, both teeth 4 and 1 are cutting (assume $\phi = 0$ when tooth 1 is vertical). However, when $30 < \phi \leq 90$ deg, only tooth 1 is engaged. For $90 < \phi \leq 120$ deg, two teeth (1 and 2) are again cutting simultaneously. As ϕ continues to increase, the cycle continues and, unlike the previous example, the cutting force components never drop to zero. As seen in Fig. 4.1.17, during the intervals when two teeth are engaged simultaneously, the cutting forces are constant. This surprising result leads to constant cutting forces at all times for slotting operations with an even number of teeth, where $N_t > 2$. The MATLAB® program p_4_1_2_3.m can be modified to: 1) produce Fig. 4.1.17 by selecting `phie = 120`; and 2) demonstrate the constant force for slotting with $N_t = 4, 6, 8, \ldots$ by setting `phis = 0`, `phie = 180`, and `Nt` to the appropriate value. Because multiple teeth can be cutting at any instant, the force expressions in Eqs. 11 and 12 should be updated to include the necessary summations:

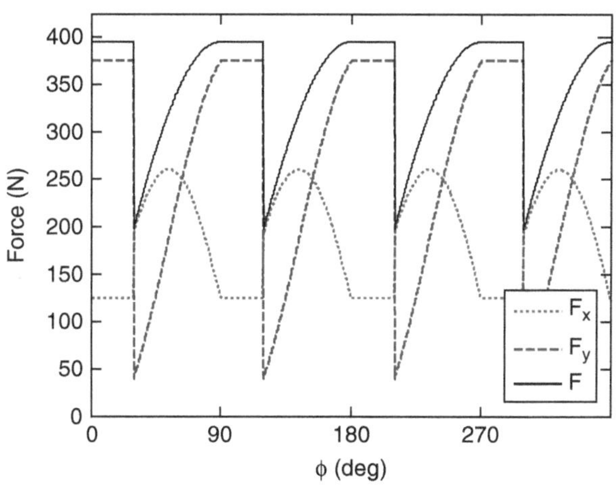

Fig. 4.1.17 75% radial immersion cutting forces

$$F_x = \sum_{j=1}^{N_t} k_t b f_t \sin(\phi_j)\cos(\phi_j) + k_n b f_t \sin(\phi_j)\sin(\phi_j) \text{ and} \tag{4.1.15}$$

$$F_y = \sum_{j=1}^{N_t} k_t b f_t \sin(\phi_j)\sin(\phi_j) - k_n b f_t \sin(\phi_j)\cos(\phi_j), \tag{4.1.16}$$

where ϕ_j is the angle of each tooth on the cutter. For a cutter with four teeth, as shown in Fig. 4.1.9 for example, if the angle for tooth 1 is $\phi_1 = 40$ deg, then the remaining angles are $\phi_2 = 310$ deg, $\phi_3 = 220$ deg, and $\phi_4 = 130$ deg. In this case, each successive tooth angle is decremented by 90 deg (the tooth pitch for equally spaced teeth is equal to 360 deg divided by the number of teeth), where $0 \le \phi_j < 360$. As we've seen, to calculate the final cutting force we must also verify that the jth tooth in question is between ϕ_s and ϕ_e prior to summing its contribution to the overall force value. Otherwise, the tooth is not engaged in the cut and zero force is added to the summation at that instant.

FOR INSTANCE If the radial depth of cut is very small, there are long periods of time when no tooth is cutting at all. In this case, the cutting force looks like a series of very short impacts. If the radial depth of cut is larger, then the number of teeth cutting simultaneously may be constant or vary depending on the cut geometry and the tooth spacing.

Prior to beginning our discussion of regenerative chatter, let's identify four other important terms related to milling performance.

1. The cutting speed, v, is the peripheral velocity of the cutter and is described by:

$$v = \frac{r\Omega \cdot 2\pi}{60} = \frac{d\Omega \cdot \pi}{60} \text{ (mm/s)}, \tag{4.1.17}$$

where Ω is given in rpm and the tool radius (or diameter) is stated in mm. This quantity is important because the cutting temperature generally increases with cutting speed. Since diffusive tool wear (typically characterized by diffusion of the tool material into the workpiece/chip) is temperature dependent, high temperatures at the tool-chip interface can lead to prohibitive wear and, therefore, place an upper bound on the allowable cutting speed. This, in turn, limits the top available spindle speed (for a given tool diameter) and can reduce the availability of the large stable zones observed to the right of stability lobe diagrams at higher spindle speeds. High temperatures are particularly problematic when the workpiece material exhibits low thermal conductivity. This causes the heat to remain at the tool-chip interface, rather than being conducted away into the workpiece, and accelerates diffusive wear. A well-known example is the difficulty in machining the titanium alloy 6Al-4V, with a thermal conductivity value approximately 20 times less than 7075 aluminum for instance, due to the associated excessive wear rates at high cutting speeds.

2. For any cut to be carried out, the spindle must possess the required torque to continue rotating the tool against the retarding tangential component of the cutting force. The instantaneous torque, T, is expressed as the sum of the products of the tool radius and tangential force for each tooth engaged in the cut. Similar to Eqs. 4.1.15 and 4.1.16, the torque can be written as:

$$T = \sum_{j=1}^{N_t} r \cdot F_{t,j} = r \sum_{j=1}^{N_t} F_{t,j} = r \sum_{j=1}^{N_t} k_t b f_t \sin(\phi_j) = r k_t b f_t \sum_{j=1}^{N_t} \sin(\phi_j), \qquad (4.1.18)$$

where the same ϕ_j definition holds and typical units for torque are N-m. For most spindles, the available torque is spindle speed dependent and is provided as a "torque curve" with the spindle specifications. Because the rotating spindle has inertia, it acts like a flywheel. For that reason it is often the average torque, rather than the instantaneous torque, that matters. The cutting operation can briefly sustain a torque in excess of the torque curve (thereby slightly slowing the rotation of the spindle) if the average torque is sufficiently low.

3. Similar to torque, the available spindle power can limit the potential cut geometry (i.e., if the power is exceeded, the spindle will stall). The power is determined from the product of torque and rotational speed. If the spindle speed is expressed in rpm and torque in N-m, the power in W is:

$$P = \frac{T\Omega \cdot 2\pi}{60}. \qquad (4.1.19)$$

Again, the spindle power is a function of spindle speed and is expressed via a "power curve" supplied by the spindle manufacturer. Like torque, it is often the average power that counts. Interestingly, the average power is proportional to the average metal removal rate.

4. In order to reduce machining time and, subsequently, cost, it is often desired to increase the mean material removal rate, MRR, or average volume of material removed per unit time.

$$MRR = abf = abf_t N_t \Omega \qquad (4.1.20)$$

4.2 Regenerative Chatter in Milling

If we remove the assumption of a rigid tool, then the cutting force could cause deflections of the tool[2]. Additionally, if the tool is vibrating as it removes material, we would expect these vibrations to be "imprinted" on the workpiece

---

[2] We will assume a rigid workpiece in our analysis.

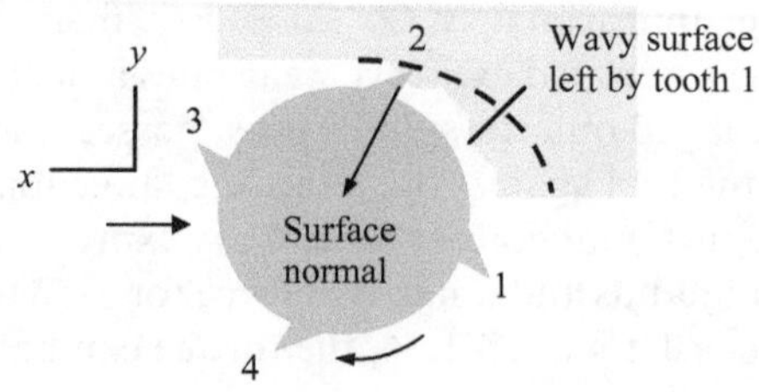

Fig. 4.2.1 Regeneration in milling. Cutter deflections during the passage of tooth 1 are copied onto the workpiece surface and are encountered by tooth 2. This varies the chip thickness and cutting force which, in turn, affects the resulting tool deflections

surface as a wavy profile just as we discussed for turning. The difference in milling is that the time-delayed surface regeneration step occurs from tooth to tooth, rather than from revolution to revolution as in turning. Figure 4.2.1 shows that the wavy surface left behind by tooth 1 is removed by tooth 2 and so on. This again provides a feedback mechanism because the instantaneous chip thickness depends on both the current vibration and the surface left by the previous tooth (one tooth period, or time from one tooth to the next, earlier). The variable chip thickness governs the cutting force which, in turn, affects subsequent tool vibrations. The result is the possibility for instability, or chatter, just as we saw with turning operations. We will again consider vibrations normal to the cut surface in order to determine the chip thickness. However, this normal direction (through the cutting edge toward the tool center) constantly varies as the cutter rotates; see Fig. 4.2.1.

Analogous to turning, the relative phasing between the surface waviness from one tooth to the next determines the level of force variation and, together with the chip width, whether the operation is stable or unstable. Figures 4.2.2 and 4.2.3 show two possibilities. In Fig. 4.2.2, the wavy surface and current vibration between two subsequent teeth are in phase. Therefore, even though the tool is vibrating during material removal, the chip thickness variation (along the instantaneous surface normal) is similar to what would be obtained from the cycloidal tool path alone. This tends to produce forced vibration only and

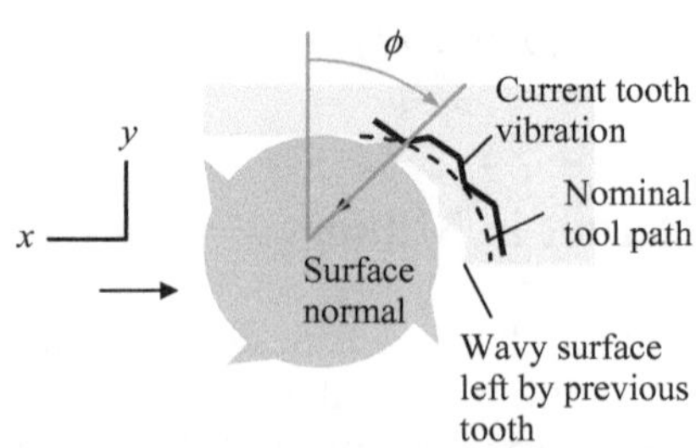

Fig. 4.2.2 Condition when the vibrations from one tooth to the next are in phase. This yields chip thickness variation that seen for the cycloidal tool path only

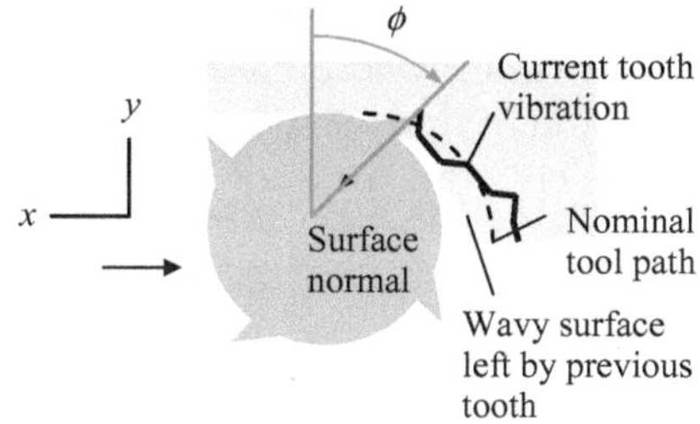

Fig. 4.2.3 Out of phase condition where the tool vibrations from one tooth to the next lead to significant chip thickness variation

stable cutting. Again, because the tool vibrates at its natural frequency (we'll assume a single degree of freedom system for now), it is intuitive that matching the tooth passing frequency to the tool's natural frequency will lead to this preferred "in phase" situation. We recognize, however, that driving the system at resonance does lead to larger vibration levels, although much less than what would be obtained during chatter. We will explore this issue further in Chapter 5. Figure 4.2.3 shows a less favorable phase relationship where there is significant variation in the chip thickness (beyond the periodic variation due to the cycloidal tool path). This can lead to self-excited vibrations and unstable cutting, depending on the chip width, due to the force variations and subsequent tool deflections.

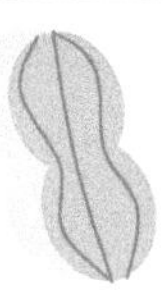

IN A NUTSHELL Although the geometry is more complicated, milling stability is conceptually similar to turning. The varying cutting force causes vibration of the tool, which generates a wavy surface. The wavy surface causes a variable chip thickness which, in turn, yields a variable cutting force. Depending on the cutting conditions and the alignment of the waves, the vibration can grow (chatter) or diminish (stable) with subsequent tooth passes.

As mentioned, the chip thickness is measured along the surface normal just as for turning, but the surface normal is a function of the cutter angle ϕ. The projection of the instantaneous chip thickness along the surface normal under the circular tool path assumption is characterized by Eq. 4.1.2 in the absence of tool vibrations. However, if we now allow the tool to undergo vibrations in the x and y directions due to the application of the cutting force to the non-rigid tool, the chip thickness equation must be modified. First, we can project the x and y vibrations onto the surface normal, n, according to:

$$n = x\sin(\phi) - y\cos(\phi). \tag{4.2.1}$$

Note that the positive direction for n is out of the cut. Because ϕ varies with time, $\phi = \Omega \cdot 2\pi \frac{180}{\pi} \frac{1}{60} t = 6 \cdot \Omega t \text{deg}$ where Ω is given in rpm and t in s, the direction of n is also a function of time. Similar to the chip thickness in turning, Eq. 3.2.1, the instantaneous chip thickness for milling can then be written as:

$$h(t) = f_t \sin(\phi) + n(t - \tau) - n(t), \tag{4.2.2}$$

where $f_t \sin(\phi)$ replaces the "mean chip thickness" in Eq. 3.2.1 and τ is the tooth period in seconds. The $n(t - \tau)$ term represents the vibration contributions along the surface normal by the previous tooth and $\tau = \frac{60}{\Omega N_t}$ (this is the reciprocal of Eq. 4.1.14), where Ω is again given in rpm. The geometry of Eq. 4.2.2 is shown in Fig. 4.2.4. As we saw in turning, larger (more positive) vibrations by the previous tooth increase the chip thickness because less material is removed than commanded. Larger vibrations by the current tooth lead to decreased chip thickness by the same logic. If the vibrations from one tooth to the next are equal, Eq. 4.2.2 reduces to Eq. 4.1.2. To complete our analogy to turning, let's "unwrap" the milled surface and show the current and previous vibrations along the "planar" instantaneous surface normal. We can observe the similarity of Fig. 4.2.5 for milling to Fig. 4.2.4 for turning. The inherent chip thickness variation due to the cycloidal tool path is schematically represented as a wedge (an up milling cut is presumed where the chip thickness grows as the cutter rotates). In Fig. 4.2.5, recall that v is the cutting speed; its direction is shown.

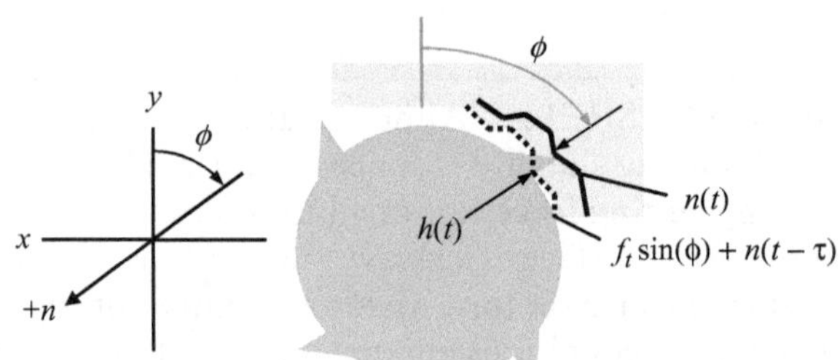

Fig. 4.2.4 Geometry of instantaneous chip thickness calculation for milling

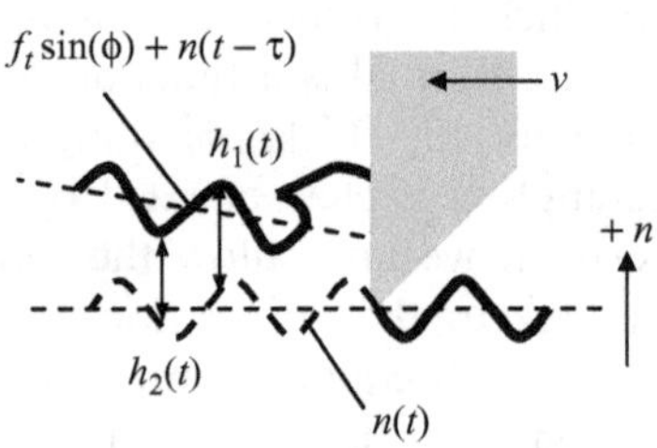

Fig. 4.2.5 Unwrapped view of milled surface for comparison to chip thickness variation in turning

4.3 Stability Lobe Diagrams

4.3.1 Average Tooth Angle Approach

Tlusty modified his turning analysis, presented in Section 3.3, to accommodate the milling process [2–4]. A primary obstacle to defining an analytical solution to milling stability (aside from the inherent time delay) is the time dependence of the cutting force direction. Tlusty solved this problem by assuming an "average" angle of the tooth in the cut and, therefore, an average force direction. This created an autonomous, or time invariant, system. He then made use of directional orientation factors to first project this force into the x and y mode directions and, second, project these results onto the surface normal. Just as for turning, depending on the feedback system "gain", represented by the chip width, b, and the spindle speed, Ω, the milling operation is either stable or it exhibits chatter. Let's first discuss the required equations and then determine the directional orientation factors for various radial immersions.

$$b_{\lim} = \frac{-1}{2K_s \mathrm{Re}[FRF_{orient}]N_t^*}. \tag{4.3.1}$$

$$\frac{f_c}{\Omega N_t} = N + \frac{\varepsilon}{2\pi} \tag{4.3.2}$$

$$\varepsilon = 2\pi - 2\tan^{-1}\left(\frac{\mathrm{Re}[FRF_{orient}]}{\mathrm{Im}[FRF_{orient}]}\right) \tag{4.3.3}$$

Due to the assumption of a fixed force direction, only minor modifications are necessary to adapt the turning stability equations to the milling case. In Eq. 4.3.1, a new variable, N_t^*, appears in the denominator. This represents the average number of teeth in the cut. As discussed in Section 4.1, it is possible for multiple teeth to be engaged simultaneously and, further, for the number of teeth in the cut to vary over one revolution. This average value is calculated by dividing the difference between the exit and starting angles by the angular tooth spacing (we'll assume equal, or proportional, teeth spacing for now). See Eq. 4.3.4, where the start and exit angles are given in deg. We also see that Eq. 4.3.2 differs from Eq. 3.3.2 by a single term. Specifically, N_t appears in the denominator of the left hand side. Because the product ΩN_t represents the tooth passing frequency, the left hand side gives the ratio of the chatter frequency to the forcing frequency. (Units compatibility requires that Ω is given in rev/s if f_c is expressed in Hz.) This is analogous to Eq. 3.3.2, where the ratio is $\frac{f_c}{\Omega}$ because the forcing frequency is simply the rotating speed in turning.

$$N_t^* = \frac{\phi_e - \phi_s}{\frac{360}{N_t}}. \tag{4.3.4}$$

The milling stability lobe diagram is generated by: 1) determining the oriented FRF and identifying the valid chatter frequency range(s), i.e., where the real part is negative; 2) solving for ε; the phase between the current and previous tooth vibrations, over the valid frequency range(s); 3) finding the average number of teeth in the cut for the selected radial immersion; 4) calculating b_{lim} over the valid frequency range(s); 5) selecting an N value (integer number of waves between teeth) and calculating the associated spindle speeds over the valid frequency range(s); and 6) plotting Ω vs. b_{lim} (both are a function of the same frequency vector, so the first spindle speed value corresponds to the first limiting chip width and so on) for each N value ($N = 0, 1, 2, \ldots$). Just as with turning, any (Ω, b_{lim}) pair that appears above the collective boundary indicates unstable behavior, while any pair below the boundary indicates stable cutting. Notice that because the N value denotes the integer number of waves between teeth, rather than the integer number of waves per revolution as in turning, the expanded stable zones appear at much lowers spindle speeds. For this reason, the use of the stability lobe diagram to find high performance stable zones in milling is a much more enticing prospect than it is in turning.

4.3.2 Oriented FRF

The oriented FRF is calculated by summing the products of the directional orientation factors and corresponding FRFs for the x and y directions; see Eq. 4.3.5. We'll show the directional orientation factor computations through examples.

$$FRF_{orient} = \mu_x FRF_x + \mu_y FRF_y \tag{4.3.5}$$

Example 4.3.1: Directional orientation factors for slotting For a 100% radial immersion (slotting) cut, the cut start angle is 0 deg and the exit angle is 180 deg (based on the circular tool path approximation). The average angle of a tooth in the cut is therefore:

$$\phi_{ave} = \frac{\phi_s + \phi_e}{2} = \frac{0 + 180}{2} = 90 \text{ deg.}$$

The surface normal is drawn at ϕ_{ave} with the positive direction out of the cut and the cutting force, F, is oriented at the force angle, β, (opposing the direction of rotation) with respect to the surface normal. Figure 4.3.1 shows this geometry.

Two steps are required to determine the x and y direction directional orientation factors. First, the force is projected onto the mode direction. Second, this result is projected onto the surface normal. These two steps are repeated once for each direction. For μ_x, projection onto the x direction gives $F_x = F\cos(\beta)$. The projection of this result onto the "average" surface normal is:

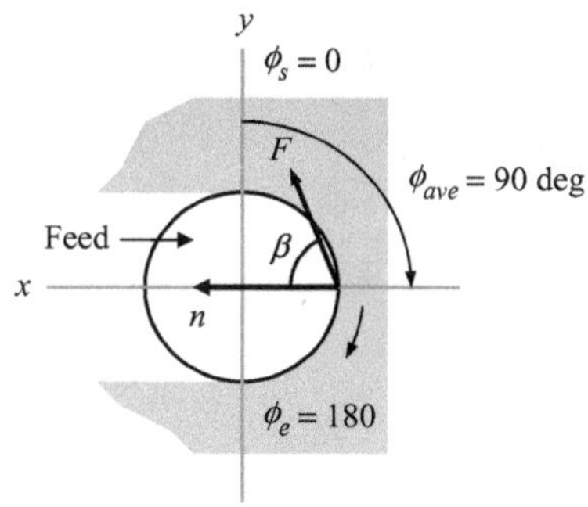

Fig. 4.3.1 Geometry for determining slotting directional orientation factors

$$F_n = F_x \cos(0) = F\cos(\beta)\cos(0) = F\cos(\beta)$$

because the surface normal is parallel to the mode direction in this instance. The directional orientation factor is then $\mu_x = \cos(\beta)$. For μ_y, projection onto the y direction gives $F_y = F\cos(90 - \beta) = F\sin(\beta)$, where β is expressed in deg. The projection of this result onto the "average" surface normal is $F_n = F_y\cos(90) = F\sin(\beta)\cos(90) = 0$ because the surface normal is perpendicular to the mode direction in this instance. The directional orientation factor is then $\mu_y = 0$. According to Eq. 4.3.5, the oriented FRF is $FRF_{orient} = \cos(\beta)FRF_x + 0 \cdot FRF_y$, which suggests that compliance in the y direction has no influence on the milling stability.

Example 4.3.2: Directional orientation factors for down milling Consider the 50% radial immersion down milling cut pictured in Fig. 4.3.2. To find μ_x, we first project F onto the x direction to obtain $F_x = F\cos(45 + \beta)$, where β is given in deg. The projection of this result onto the average surface normal, which occurs at the angle $\phi_{ave} = \frac{90+180}{2} = 135$ deg, is:

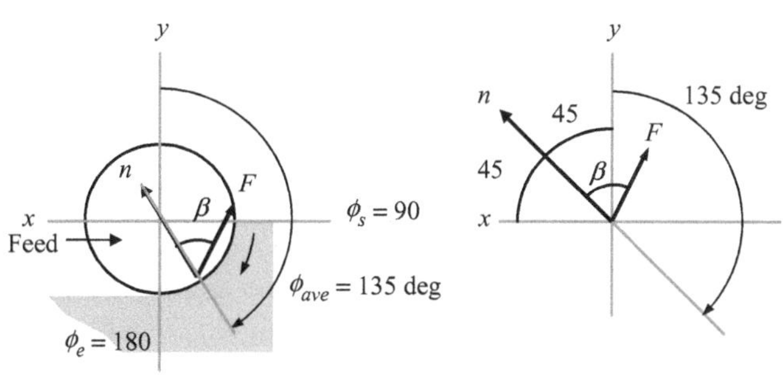

Fig. 4.3.2 Geometry for determining 50% radial immersion down milling directional orientation factors

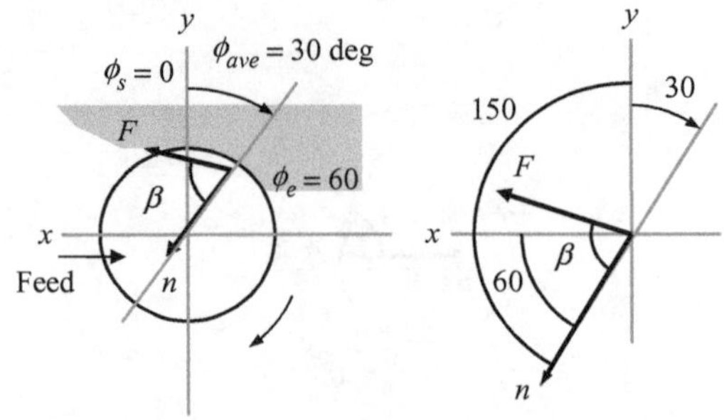

Fig. 4.3.3 Geometry for determining 25% radial immersion up milling directional orientation factors

$$F_n = F_x\cos(45) = F\cos(45 + \beta)\cos(45)$$

For μ_y, projection onto the y direction gives $F_y = F\cos(\beta - 45)$. The projection of this result onto the average surface normal is $F_n = F_y\cos(45) = F\cos(\beta - 45)\cos(45)$. The directional orientation factors are $\mu_x = \cos(45 + \beta)\cos(45)$ and $\mu_y = \cos(\beta - 45)\cos(45)$.

Example 4.3.3: Directional orientation factors for up milling A 25% radial immersion up milling cut in depicted in Fig. 4.3.3. As shown in Eq. 4.1.4, the exit angle is $\phi_e = \cos^{-1}\left(\frac{r-\frac{r}{2}}{r}\right) = \cos^{-1}(0.5) = 60$ deg. The start angle for any up milling cut is zero. To find μ_x, we first project F onto the x direction to obtain $F_x = F\cos(\beta - 60)$, where β is given in deg. The projection of this result onto the average surface normal, which occurs at the angle $\phi_{ave} = \frac{0+60}{2} = 30$ deg, is $F_n = F_x\cos(60) = F\cos(\beta - 60)\cos(60)$. For μ_y, projection onto the y direction gives $F_y = F\cos(150 - \beta)$. The projection of this result onto the average surface normal is $F_n = F_y\cos(150) = F\cos(150 - \beta)\cos(150)$. The directional orientation factors are $\mu_x = \cos(\beta - 60)\cos(60)$ and $\mu_y = \cos(150 - \beta)\cos(150)$.

Let's now examine stability lobes for milling. We'll see that the stability limit is again a mapping of the real part of the oriented FRF onto the (Ω, b_{lim}) diagram. In the following slotting example, we'll demonstrate analytical calculations for the best and worst spindle speeds and $b_{lim,crit}$ and show the resulting stability lobes. We'll then complete an example for 20% radial immersion down milling.

Example 4.3.4: Slotting stability lobe calculations For this example, the x and y direction dynamics are symmetric and can be described by $f_n = 500$ Hz, $k = 8 \times 10^6$ N/m $= 8 \times 10^3$ N/mm, and $\zeta = 0.02$. As shown in Ex. 4.3.1, the directional orientation factors are $\mu_x = \cos(\beta)$ and $\mu_y = 0$ for slotting and the oriented FRF is $FRF_{orient} = \cos(\beta)FRF_x$. We'll assume an aluminum alloy machined with a four tooth, 19 mm diameter square endmill that together exhibit a specific force of 750 N/mm^2 and a force angle of 68 deg.

For this single degree of freedom example, the best spindle speed is defined for a phase angle of $\varepsilon = 2\pi$ rad $= 360$ deg (see Fig. 3.3.6) just as we observed for turning. The corresponding chatter frequency is $f_c = f_n$. Equation 4.3.2 shows that the best spindle speed for the $N = 0$ lobe is:

$$\Omega_1 = \frac{f_c}{N_t} \cdot \frac{1}{N + \frac{\varepsilon}{2\pi}} = \frac{f_n}{4} \cdot \frac{1}{0 + \frac{2\pi}{2\pi}} = \frac{f_n}{4} = \frac{500}{4} = 125 \text{ rps}$$

or $125 \cdot 60 = 7500$ rpm. This provides a reasonable target spindle speed for milling. Note that for turning (with $N_t = 1$, effectively), the best spindle speed for $N = 0$ is 500 rps or 30000 rpm. Also, the tooth passing frequency for this spindle speed is:

$$f_{tooth} = \frac{\Omega \cdot N_t}{60} = \frac{7500 \cdot 4}{60} = 500 \text{ Hz}$$

Point 1, where f_{tooth} is set equal to f_n, is identified in Fig. 4.3.4. The theoretical limiting chip width for point 1 is:

$$b_1 = \frac{-1}{2K_s \text{Re}[FRF_{orient}]N_t^*} = \frac{-1}{2K_s \cdot 0 \cdot N_t^*} = \infty.$$

However, just as with turning, the $N = 1$ lobe truncates the $N = 0$ lobe and gives a finite chip width where they cross. The average number of teeth in the cut is $N_t^* = \frac{180-0}{\frac{360}{4}} = 2$ for this slotting example.

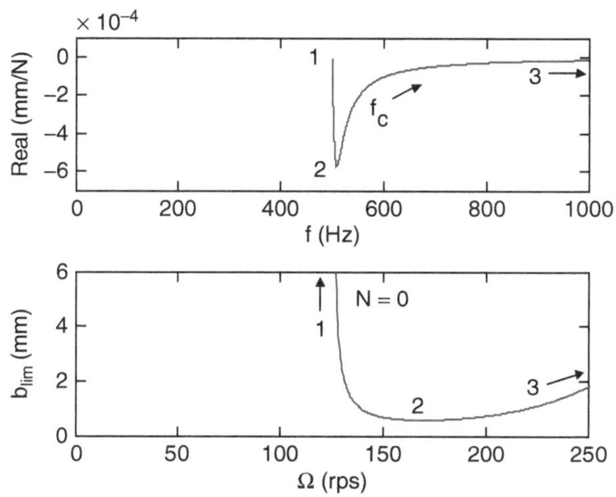

Fig. 4.3.4 Mapping of the oriented FRF onto the (Ω, b_{lim}) diagram for Ex. 4.3.4

The worst spindle speed is obtained when ε is $\frac{3\pi}{2}$ rad or 270 deg (see Fig. 3.3.4). The chatter frequency for this case (i.e., the frequency for $\min(\text{Re}[FRF_{orient}]$ or $\min(\text{Re}[\mu_x FRF_x]))$ is $f_c = f_n(1+\zeta)$. Substitution in Eq. 4.3.2 gives:

$$\Omega_2 = \frac{f_c}{N_t} \cdot \frac{1}{N + \frac{\varepsilon}{2\pi}} = \frac{f_n(1+\zeta)}{4} \cdot \frac{1}{0 + \frac{\frac{3\pi}{2}}{2\pi}} = \frac{f_n(1+\zeta)}{4} \frac{1}{0 + \frac{3}{4}} = \frac{500(1+0.02)}{3} = 170 \text{ rps}$$

or 10200 rpm. Point 2 is also shown in Fig. 4.3.4. The corresponding minimum limiting chip width, $b_{\text{lim,crit}}$, is determined by rewriting Eq. 4.3.1, $b_{\text{lim,crit}} = \frac{-1}{2K_s \min(\text{Re}[\mu_x FRF_x])N_t^*}$. For a single degree of freedom system, the minimum real part can be approximated by:

$$\min(\text{Re}[FRF]) = \frac{-1}{4k\zeta(1+\zeta)}. \tag{4.3.6}$$

We can therefore write a simple expression for the point 2 chip width (in this single degree of freedom, slotting example):

$$b_{\text{lim,crit}} = \frac{4k\zeta(1+\zeta)}{2K_s \mu_x N_t^*} = \frac{4 \cdot 8 \times 10^3 \cdot 0.02 \cdot (1+0.02)}{2 \cdot 750 \cdot \cos(68) \cdot 2} = 0.58 \text{ mm}$$

Point 3, where $f_c \to \infty$, is similar to point 1 in Fig. 4.3.4. As shown previously in Fig. 3.3.8, even though the tooth to tooth phase relationship is unfavorable ($\varepsilon \to 180$ deg, or exactly out of phase), the response amplitude approaches zero, $\text{Re}(FRF) \to 0$, so the vibration level is small. Substitution in Eq. 4.3.1 yields an infinite chip thickness, $b_3 = \frac{-1}{2K_s \cdot 0 \cdot N_t^*} = \infty$. However, the left end of the $N = 0$ curve, where $f_c = 0$, serves to limit the right hand side of the $N = 1$ curve, where $f_c \to \infty$. Figure 4.3.5 shows the $N = 0$ to 3 lobes for this example. The

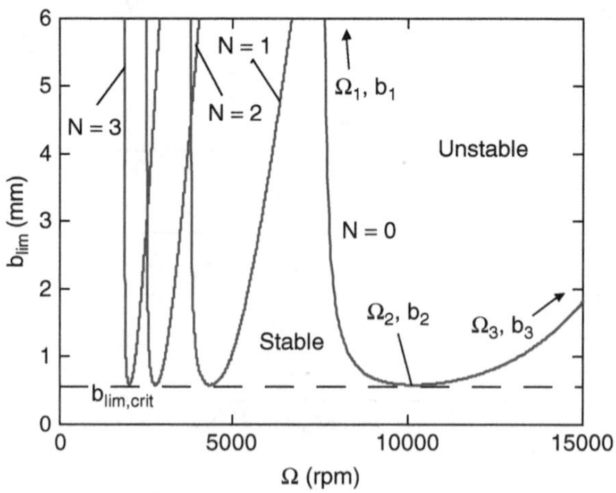

Fig. 4.3.5 Stability lobe diagram for Ex. 4.3.4 ($N = 0$ to 3)

MATLAB® program used to produce Fig. 4.3.5 is provided on the companion CD as p_4_3_4_1.m.

To conclude this example, let's calculate the mean material removal rate if we selected (Ω_2, b_2) with a feed per tooth of 0.15 mm/tooth as the operating parameters. As shown in Eq. 4.1.20, the mean MRR for milling is computed from the product of the axial depth of cut, radial depth (equal to the tool diameter for slotting), feed per tooth, number of teeth, and spindle speed:

$$MRR = b_{\text{lim,crit}} \cdot d \cdot f_t \cdot N_t \cdot \Omega_2 = 0.58 \cdot 19 \cdot 0.15 \cdot 4 \cdot 10200 = 67442\,\text{mm}^3/\text{min}.$$

To visualize this removal rate, let's divide by 60×10^3 to determine the cubic centimeters per second, $MRR = 1.1240\,\text{cm}^3/\text{sec}$. This means that the equivalent of approximately a 1 cm x 1 cm x 1 cm cube of aluminum is removed each second under these conditions.

Example 4.3.5: 20% radial immersion down milling stability lobe calculations Figure 4.3.6 shows the geometry for a 20% radial immersion down milling cut. The start angle is $\phi_s = 180 - \cos^{-1}(0.6) = 126.9$ deg and the exit angle is $\phi_e = 180$ deg. The average angle of a tooth in the cut is $\phi_{ave} = \frac{126.9+180}{2} = 153.4$ deg. The x (feed) direction dynamics are given by $f_{nx} = 900$ Hz, $k_x = 9 \times 10^6$ N/m $= 9 \times 10^3$ N/mm, and $\zeta_x = 0.02$. The y direction dynamics are $f_{ny} = 950$ Hz, $k_y = 1 \times 10^7$ N/m $= 1 \times 10^4$ N/mm, and $\zeta_y = 0.01$. The workpiece material is a low carbon steel alloy and it is machined with a three tooth, 19 mm diameter square endmill. Together they give a specific force of $K_s = 2250$ N/mm^2 and a force angle of $\beta = 75$ deg.

As shown in Fig. 4.3.6, the directional orientation factor for the x direction is determined from the two projections:

$$F_x = F\cos(63.4 + \beta) \text{and} F_n = F_x \cos(63.4) = F\cos(63.4 + \beta)\cos(63.4),$$

where β is expressed in deg. Therefore, $\mu_x = \cos(63.4 + \beta)\cos(63.4) = -0.335$. For the y direction, the projections are:

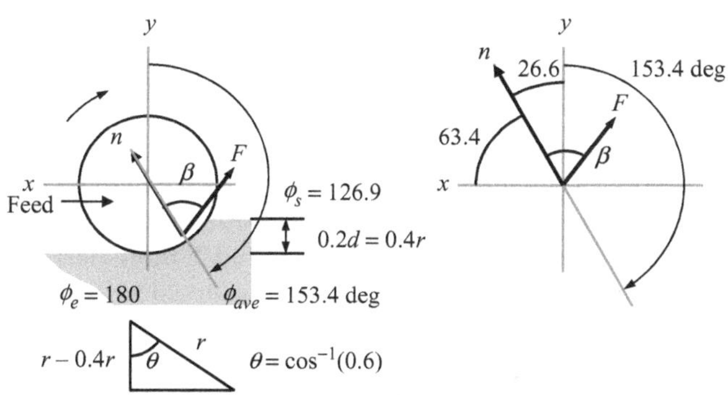

Fig. 4.3.6 Geometry for Ex. 4.3.5 20% radial immersion down milling cut

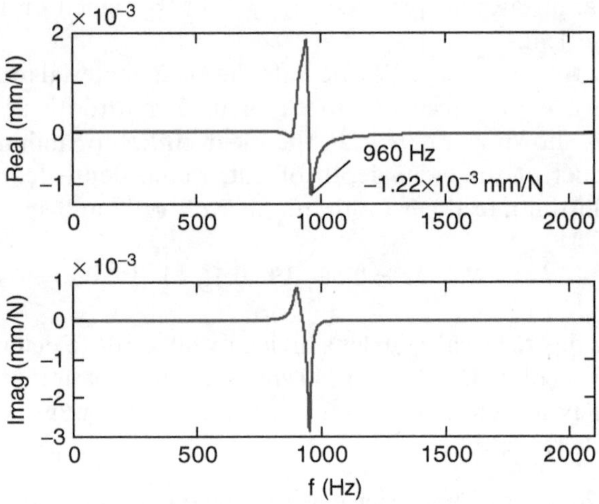

Fig. 4.3.7 Oriented FRF for Ex. 4.3.5

$$F_y = F\cos(\beta - 26.6) \text{ and } F_n = F_y\cos(26.6) = F\cos(\beta - 26.6)\cos(26.6)$$

and $\mu_y = \cos(\beta - 26.6)\cos(26.6) = 0.594$. Additionally, the average number of teeth in the cut is: $N_t^* = \frac{180-126.9}{\frac{360}{3}} = 0.443$.

The oriented FRF is shown in Fig. 4.3.7. The unusual shape of the real and imaginary parts is due to the negative μ_x value. The minimum real part is observed at 960 Hz with a value of -1.22×10^{-3} mm/N. The corresponding worst spindle speed (for the $N = 0$ lobe) and $b_{\text{lim,crit}}$ are approximated by:

$$\Omega_{worst} = \frac{960}{3} \cdot \frac{1}{0 + \frac{\frac{3\pi}{2}}{2\pi}} = \frac{960}{3}\frac{4}{3} = 427 \text{ rps} = 25600 \text{ rpm and}$$

$$b_{\text{lim,crit}} = \frac{-1}{2 \cdot 2250(-1.22 \times 10^{-3})0.443} = 0.41 \text{ mm}$$

The real part of the oriented FRF for the valid chatter frequency ranges ($\text{Re}[FRF_{orient}] < 0$) and the corresponding $N = 0$ stability lobe are shown in Fig. 4.3.8. Two valid ranges are seen; the first, lower frequency range is much stiffer than the second. The "competing lobe" effect of this stiffer section is observed in Fig. 4.3.9, where the $N = 0$ to 4 lobes are plotted. In this case, the stability boundary is nearly completely established by the much more flexible section. The "best speeds" equation from Section 3.3 can also be updated for milling:

$$\Omega_{best} = \frac{f_n \cdot 60}{(N+1) \cdot N_t} \text{(rpm)}. \tag{4.3.7}$$

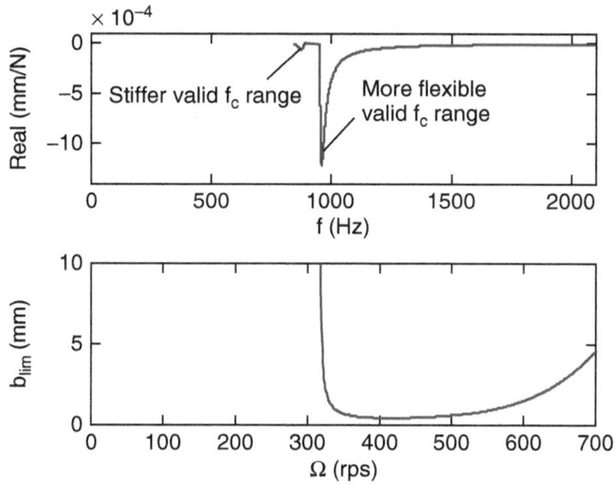

Fig. 4.3.8 (Top) Valid chatter frequency ranges for the real part of the oriented FRF. (Bottom) $N = 0$ stability lobe

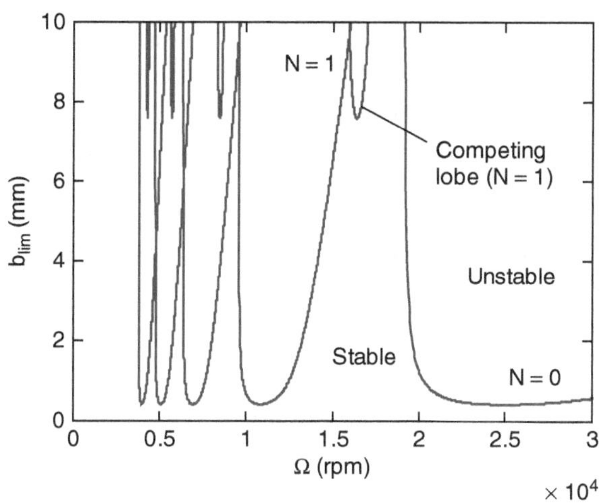

Fig. 4.3.9 Example 4.3.5 stability lobe diagram ($N = 0$ to 4)

If we choose f_n to be the zero crossing for the more flexible real part valid chatter frequency range (at 951 Hz from Figs. 4.3.7 and 4.3.8), then the associated best speeds are 19020 rpm, 9510 rpm, 6340 rpm, and so on for the three tooth endmill. These speeds match those seen in Fig. 4.3.9 for the right hand side of the stable zones moving from right to left on the diagram. For example, 19020 rpm corresponds to the stable gap created by the intersection of the $N = 0$ and $N = 1$ lobes. The MATLAB® program used to produce these figures is provided on the companion CD as p_4_3_5_1.m.

4.3.3 Fourier Series Approach

In [5], Altintas and Budak use an alternate technique to transform the dynamic milling equations into a time invariant, but radial immersion dependent system. Similar to Tlusty, they approximate the time dependent cutting forces by an average value, but employ a different approach to identify this mean. Rather than using the average angle of the tooth in the cut, the time varying coefficients of the dynamic milling equations, which depend on the angular orientation of the tool as it rotates through the cut, are expanded into a Fourier series and then the series is truncated to include only the average component. We'll first present a brief review of this approach and then rearrange the analytical stability equations into the form expected by the MATLAB® `eig` function.

The development of the time dependent cutting force equations closely follows the previous analysis. The primary difference is that the positive x direction is taken to be in the direction of the feed (following the convention in [5]); compare Fig. 4.3.10 (Fourier series) to Fig. 4.1.9 (average tooth angle). The projection of the x and y vibrations onto the surface normal (referred to as the radial direction in [5]) is now:

$$n = -x\sin(\phi) - y\cos(\phi) \tag{4.3.8}$$

and the instantaneous chip thickness equation for tooth j becomes:

$$h(\phi_j) = \left(f_t \sin(\phi_j) + n_{j-1} - n_j\right) \cdot g(\phi_j), \tag{4.3.9}$$

where the switching function, $g(\phi_j)$, is equal to one when the jth tooth is engaged in the cut (i.e., between the cut start and exit angles) and zero otherwise; see Eq. 4.3.10. Also, n_{j-1} indicates the normal direction vibration from the previous tooth and n_j the current tooth vibration.

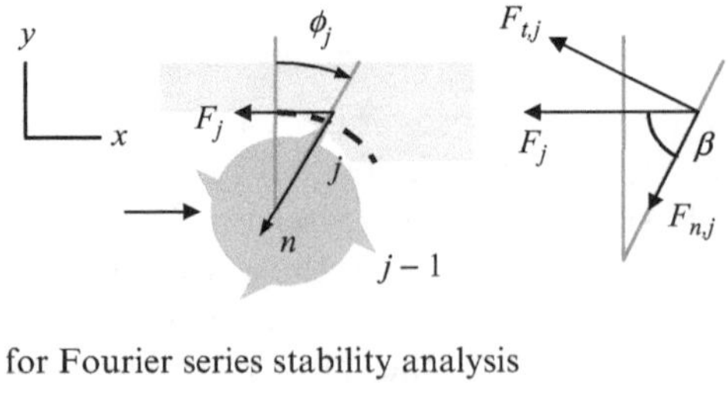

Fig. 4.3.10 Notation for Fourier series stability analysis

$$g(\phi_j) = \begin{Bmatrix} 1, & \text{when} & \phi_s \leq \phi_j \leq \phi_e \\ 0, & \text{when} & \phi_j < \phi_s, \phi_j > \phi_e \end{Bmatrix} \tag{4.3.10}$$

If the variable x and y contributions are substituted into Eq. 4.3.9 and the nominal component $f_t \sin(\phi_j)$ is neglected (because we are interested in linear stability), the cutter angle dependent chip thickness in the normal direction may be written as shown in Eq. 4.3.11.

$$\begin{aligned} h(\phi_j) &= ((-x_{j-1}\sin(\phi_j) - y_{j-1}\cos(\phi_j)) - (-x_j \sin(\phi_j) - y_j \cos(\phi_j))) \cdot g(\phi_j) \\ h(\phi_j) &= ((x_j - x_{j-1})\sin(\phi_j) + (y_j - y_{j-1})\cos(\phi_j)) \cdot g(\phi_j) \\ h(\phi_j) &= (\Delta x \cdot \sin(\phi_j) + \Delta y \cdot \cos(\phi_j)) \cdot g(\phi_j) \end{aligned} \tag{4.3.11}$$

The tangential and normal (radial) force components are then written as a function of the chip thickness:

$$F_{t,j} = K_t b h(\phi_j) \text{ and } F_{n,j} = K_n F_{t,j} = K_n K_t b h(\phi_j), \tag{4.3.12}$$

where K_t is equal to the previously defined tangential specific force k_t. The relationship between (K_t, K_n) and K_s is defined using Fig. 4.3.10:

$$F_j = K_s b h(\phi_j) \quad F_j = \sqrt{F_{t,j}^2 + F_{n,j}^2} = \sqrt{K_t^2 + K_n^2 K_t^2}\, b h(\phi_j) = K_t\sqrt{1 + K_n^2}\, b h(\phi_j), \tag{4.3.13}$$

so that $K_s = K_t\sqrt{1 + K_n^2}$. Also, $\tan(\beta) = \frac{F_{t,j}}{F_{n,j}} = \frac{K_t b h(\phi_j)}{K_n K_t b h(\phi_j)} = \frac{1}{K_n}$. The projections of the tangential and normal force components onto the fixed (x and y) coordinate frame are:

$$F_{x,j} = -F_{t,j}\cos(\phi_j) - F_{n,j}\sin(\phi_j) \text{ and } F_{y,j} = F_{t,j}\sin(\phi_j) - F_{n,j}\cos(\phi_j). \tag{4.3.14}$$

Substitution into the $F_{x,j}$ component gives Eq. 4.3.15.

$$\begin{aligned} F_{x,j} &= \begin{pmatrix} -(K_t b(\Delta x \cdot \sin(\phi_j) + \Delta y \cdot \cos(\phi_j))g(\phi_j))\cos(\phi_j) - \\ (K_n K_t b(\Delta x \cdot \sin(\phi_j) + \Delta y \cdot \cos(\phi_j))g(\phi_j))\sin(\phi_j) \end{pmatrix} \\ F_{x,j} &= -K_t b g(\phi_j) \begin{pmatrix} (\Delta x \cdot \sin(\phi_j) + \Delta y \cdot \cos(\phi_j))\cos(\phi_j) + \\ K_n(\Delta x \cdot \sin(\phi_j) + \Delta y \cdot \cos(\phi_j))\sin(\phi_j) \end{pmatrix} \\ F_{x,j} &= -K_t b g(\phi_j) \begin{pmatrix} \Delta x(\sin(\phi_j)\cos(\phi_j) + K_n \sin^2(\phi_j)) + \\ \Delta y(\cos^2(\phi_j) + K_n \sin(\phi_j)\cos(\phi_j)) \end{pmatrix} \end{aligned} \tag{4.3.15}$$

Similarly, the $F_{y,j}$ component is expressed as shown in Fig. 4.3.16.

$$F_{y,j} = \begin{pmatrix} (K_t b(\Delta x \cdot \sin(\phi_j) + \Delta y \cdot \cos(\phi_j))g(\phi_j)) \sin(\phi_j) - \\ (K_n K_t b(\Delta x \cdot \sin(\phi_j) + \Delta y \cdot \cos(\phi_j))g(\phi_j)) \cos(\phi_j) \end{pmatrix}$$

$$F_{y,j} = K_t b g(\phi_j) \begin{pmatrix} (\Delta x \cdot \sin(\phi_j) + \Delta y \cdot \cos(\phi_j)) \sin(\phi_j) - \\ K_n(\Delta x \cdot \sin(\phi_j) + \Delta y \cdot \cos(\phi_j)) \cos(\phi_j) \end{pmatrix} \tag{4.3.16}$$

$$F_{y,j} = K_t b g(\phi_j) \begin{pmatrix} \Delta x(\sin^2(\phi_j) - K_n \cos(\phi_j) \sin(\phi_j)) + \\ \Delta y(\cos(\phi_j) \sin(\phi_j) - K_n \cos^2(\phi_j)) \end{pmatrix}$$

Substituting the trigonometric identities $\sin(\phi)\cos(\phi) = \frac{\sin(2\phi)}{2}$, $\sin^2(\phi) = \frac{1-\cos(2\phi)}{2}$, and $\cos^2(\phi) = \frac{1+\cos(2\phi)}{2}$ into Eqs. 4.3.15 and 4.3.16 yields:

$$F_{x,j} = -\frac{1}{2} K_t b g(\phi_j) \begin{pmatrix} \Delta x(\sin(2\phi_j) + K_n(1 - \cos(2\phi_j))) + \\ \Delta y(1 + \cos(2\phi_j) + K_n \sin(2\phi_j)) \end{pmatrix} \quad \text{and} \tag{4.3.17}$$

$$F_{y,j} = \frac{1}{2} K_t b g(\phi_j) \begin{pmatrix} \Delta x((1 - \cos(2\phi_j)) - K_n \sin(2\phi_j)) + \\ \Delta y(\sin(2\phi_j) - K_n(1 + \cos(2\phi_j))) \end{pmatrix}. \tag{4.3.18}$$

The force expressions are completed by including the summation over all teeth so that the contributions of (potentially) multiple teeth in the cut are considered.

$$F_x = \sum_{j=1}^{N_t} F_{x,j} \text{ and } F_y = \sum_{j=1}^{N_t} F_{y,j} \tag{4.3.19}$$

The x and y direction force expressions are now arranged in matrix form to obtain:

$$\begin{pmatrix} F_x \\ F_y \end{pmatrix} = \frac{1}{2} b K_t \begin{bmatrix} a_{xx} & a_{xy} \\ a_{yx} & a_{yy} \end{bmatrix} \begin{pmatrix} \Delta x \\ \Delta y \end{pmatrix} = \frac{1}{2} b K_t [A](\Delta). \tag{4.3.20}$$

where the individual A matrix entries, referred to in [5] as the “time varying directional dynamic force coefficients”, are:

$$a_{xx} = \sum_{j=1}^{N_t} -g(\phi_j)\left(\sin(2\phi_j) + K_n(1 - \cos(2\phi_j))\right),$$

$$a_{xy} = \sum_{j=1}^{N_t} -g(\phi_j)\left(1 + \cos(2\phi_j) + K_n \sin(2\phi_j)\right),$$

$$a_{yx} = \sum_{j=1}^{N_t} g(\phi_j)\left((1 - \cos(2\phi_j)) - K_n \sin(2\phi_j)\right), \text{and}$$

$$a_{yy} = \sum_{j=1}^{N_t} g(\phi_j)\left(\sin(2\phi_j) - K_n(1 + \cos(2\phi_j))\right).$$

Here we note that, because these expressions are periodic with the tooth pitch, $\phi_p = \frac{2\pi}{N_t}$ (rad), they are also periodic in time over the tooth period, $\tau = \frac{60}{\Omega N_t}$ (s), where Ω is expressed in rpm. Therefore, although the A matrix is written as a function of ϕ in Eq. 4.3.20, it is equivalently a function of time, $t = \frac{\phi}{\Omega}\frac{60}{2\pi}$ (s), where Ω is given in rpm and ϕ in rad. In order to remove the time dependence of the A matrix, now written explicitly as $[A(t)]$, it is first expanded into a Fourier series:

$$[A(t)] = \sum_{r=-\infty}^{\infty} [A_r]e^{ir\omega_{tooth}t}, \tag{4.3.21}$$

where $\omega_{tooth} = 2\pi \cdot f_{tooth}$ (rad/s) and the Fourier coefficients are $[A_r] = \frac{1}{N_t\tau}\int_0^{N_t\tau} [A(t)]e^{-ir\omega_{tooth}t}dt$, and then only the $r = 0$ term is retained. Although $[A(t)]$ is best represented using multiple harmonics of ω_{tooth}, setting $r = 0$ leads to the desired time invariance because the exponential term becomes unity ($e^0 = 1$). The tradeoff is that the accuracy degrades as the radial immersion is reduced (the same is true for the average tooth angle approximation). This is particularly true when the radial immersion is only a few percent and the cutting force becomes more "impulse-like" [6–9]. Equation 4.3.22 shows the $r = 0$ coefficient, $[A_0]$, written explicitly as a function of ϕ (as in Eq. 4.3.20). By modifying the integration limits to be ϕ_s and ϕ_e (all angles in rad), the switching function is effectively removed since it is equal to one within these limits and zero otherwise.

$$[A_0] = \frac{1}{2\pi}\int_0^{2\pi} [A(\phi)]d\phi = \frac{1}{2\pi}\int_{\phi_s}^{\phi_e} [A(\phi)]d\phi = \frac{N_t}{2\pi}\begin{bmatrix} \alpha_{xx} & \alpha_{xy} \\ \alpha_{yx} & \alpha_{yy} \end{bmatrix} \tag{4.3.22}$$

The individual α terms, similar to the directional orientation factors defined in the average tooth angle approach, are provided in Eq. 4.3.23, where the integrals must be evaluated between ϕ_s and ϕ_e to obtain the final scalar values. Note that the N_t term is introduced in the final Eq. 4.3.22 expression due to the summations ($\sum_{j=1}^{N_t}\ldots$) in the a terms from Eq. 4.3.20.

$$\begin{aligned}
\alpha_{xx} &= \frac{1}{2}\left(\cos(2\phi) - 2K_n\phi + K_n\sin(2\phi)\right)\Big|_{\phi_s}^{\phi_e} \\
\alpha_{xy} &= \frac{1}{2}\left(-\sin(2\phi) - 2\phi + K_n\cos(2\phi)\right)\Big|_{\phi_s}^{\phi_e} \\
\alpha_{yx} &= \frac{1}{2}\left(-\sin(2\phi) + 2\phi + K_n\cos(2\phi)\right)\Big|_{\phi_s}^{\phi_e} \\
\alpha_{yy} &= \frac{1}{2}\left(-\cos(2\phi) - 2K_n\phi - K_n\sin(2\phi)\right)\Big|_{\phi_s}^{\phi_e}
\end{aligned} \tag{4.3.23}$$

The stability analysis begins by equating the frequency-domain vibrations in the x and y directions (the j subscripts indicate the current tooth) with the product of the frequency response function matrix and the frequency-domain representation of the cutting forces.

$$\begin{pmatrix} X_j \\ Y_j \end{pmatrix} = \begin{bmatrix} FRF_{xx} & FRF_{xy} \\ FRF_{yx} & FRF_{yy} \end{bmatrix} \begin{pmatrix} F_x \\ F_y \end{pmatrix} e^{i\omega_c t} = \begin{bmatrix} FRF_{xx} & 0 \\ 0 & FRF_{yy} \end{bmatrix} \begin{pmatrix} F_x \\ F_y \end{pmatrix} e^{i\omega_c t} \tag{4.3.24}$$

In Eq. 4.3.24, the cross FRFs (FRF_{xy}, representing measurement in the x direction with excitation in the y direction, and FRF_{yx}) are set equal to zero because the x and y directions are orthogonal and zero cross-talk between the two directions is assumed. Also, the ω_c notation indicates chatter frequencies (rad/s). The vibrations one tooth period earlier (denoted by the j–1 subscripts) are similarly written as:

$$\begin{pmatrix} X_{j-1} \\ Y_{j-1} \end{pmatrix} = \begin{pmatrix} X_j \\ Y_j \end{pmatrix} e^{-i\omega_c \tau} = \begin{bmatrix} FRF_{xx} & 0 \\ 0 & FRF_{yy} \end{bmatrix} \begin{pmatrix} F_x \\ F_y \end{pmatrix} e^{i\omega_c t} \cdot e^{-i\omega_c \tau}, \tag{4.3.25}$$

where $e^{-i\omega_c \tau}$ represents the delay of one tooth period. Taking the difference between Eqs. 4.3.24 and 4.3.25 gives:

$$\begin{pmatrix} X_j \\ Y_j \end{pmatrix} - \begin{pmatrix} X_{j-1} \\ Y_{j-1} \end{pmatrix} = \left(1 - e^{-i\omega_c \tau}\right) \begin{bmatrix} FRF_{xx} & 0 \\ 0 & FRF_{yy} \end{bmatrix} \begin{pmatrix} F_x \\ F_y \end{pmatrix} e^{i\omega_c t}. \tag{4.3.26}$$

The left hand side of Eq. 4.3.26 can be written as $\begin{pmatrix} X_j \\ Y_j \end{pmatrix} - \begin{pmatrix} X_{j-1} \\ Y_{j-1} \end{pmatrix} = \begin{pmatrix} \Delta X \\ \Delta Y \end{pmatrix}$, which enables substitution into the frequency-domain representation of Eq. 4.3.20, the "dynamic milling equation" according to [5].

$$\begin{pmatrix} F_x \\ F_y \end{pmatrix} e^{i\omega_c t} = \frac{1}{2} b K_t [A_0] \left(1 - e^{i\omega_c T}\right) \begin{bmatrix} FRF_{xx} & 0 \\ 0 & FRF_{yy} \end{bmatrix} \begin{pmatrix} F_x \\ F_y \end{pmatrix} e^{i\omega_c t} \tag{4.3.27}$$

Rearranging Eq. 4.3.27 gives

$\begin{pmatrix} F_x \\ F_y \end{pmatrix} e^{i\omega_c t} \left([I] - \frac{1}{2} b K_t (1 - e^{i\omega_c T}) [A_0] \begin{bmatrix} FRF_{xx} & 0 \\ 0 & FRF_{yy} \end{bmatrix} \right) = 0$, which has a non-trivial solution only if:

$$\det\left([I] - \frac{1}{2} b K_t \left(1 - e^{i\omega_c T}\right) [A_0] \begin{bmatrix} FRF_{xx} & 0 \\ 0 & FRF_{yy} \end{bmatrix} \right) = 0, \tag{4.3.28}$$

where $[I] = \begin{bmatrix} 1 & 0 \\ 0 & 1 \end{bmatrix}$ is the 2×2 identity matrix. This is the characteristic equation of the closed loop dynamic milling system [5]. The product $[A_0][FRF]$ from Eq. 4.3.28 is analogous to the oriented FRF from the average tooth angle approach. Expanding gives Eq. 4.3.29.

$$\begin{aligned} [A_0][FRF] &= \frac{N_t}{2\pi} \begin{bmatrix} \alpha_{xx} & \alpha_{xy} \\ \alpha_{yx} & \alpha_{yy} \end{bmatrix} \begin{bmatrix} FRF_{xx} & 0 \\ 0 & FRF_{yy} \end{bmatrix} = \\ \frac{N_t}{2\pi} &\begin{bmatrix} \alpha_{xx} FRF_{xx} & \alpha_{xy} FRF_{yy} \\ \alpha_{yx} FRF_{xx} & \alpha_{yy} FRF_{yy} \end{bmatrix} = \frac{N_t}{2\pi} [FRF_{or}] \end{aligned} \tag{4.3.29}$$

A new variable, Λ, is now defined as:

$$\Lambda = \frac{N_t}{2\pi} \left(-\frac{1}{2} b K_t \left(1 - e^{-i\omega_c T}\right) \right) = -\frac{N_t}{4\pi} b K_t \left(1 - e^{-i\omega_c T}\right) \tag{4.3.30}$$

so that the characteristic equation can be rewritten as shown in Eq. 4.3.31, which includes the new oriented FRF, FRF_{or}, from Eq. 4.3.29.

$$\det([I] + \Lambda[FRF_{or}]) = 0 \tag{4.3.31}$$

Computing the determinant, $\det \begin{bmatrix} 1 + \Lambda \alpha_{xx} FRF_{xx} & \Lambda \alpha_{xy} FRF_{yy} \\ \Lambda \alpha_{yx} FRF_{xx} & 1 + \Lambda \alpha_{yy} FRF_{yy} \end{bmatrix} = 0$, gives an expression which is quadratic in Λ.

$$(1+\Lambda\alpha_{xx}FRF_{xx})(1+\Lambda\alpha_{yy}FRF_{yy})-\Lambda^2(\alpha_{xy}FRF_{yy})(\alpha_{yx}FRF_{xx})=0$$
$$\Lambda^2(\alpha_{xx}\alpha_{yy}FRF_{xx}FRF_{yy}-\alpha_{xy}\alpha_{yx}FRF_{xx}FRF_{yy})+\Lambda(\alpha_{xx}FRF_{xx}+\alpha_{yy}FRF_{yy})+1=0$$

This frequency dependent equation is rewritten as $a_0\Lambda^2+a_1\Lambda+1=0$, where $a_0=FRF_{xx}FRF_{yy}(\alpha_{xx}\alpha_{yy}-\alpha_{xy}\alpha_{yx})$ and $a_1=\alpha_{xx}FRF_{xx}+\alpha_{yy}FRF_{yy}$. The two roots are the system eigenvalues, Λ_1 and Λ_2. These complex eigenvalues can be determined using the quadratic equation:

$$\Lambda_{1,2}=\frac{-a_1\pm\sqrt{a_1^2-4a_0}}{2a_0}=-\frac{1}{2a_0}\left(a_1\mp\sqrt{a_1^2-4a_0}\right).$$

Note that these eigenvalues are a function of ω_c due to the dependence on the x and y direction FRFs. At each potential chatter frequency, the minimum b_{lim} value (from the two available eigenvalues) is selected to establish the limiting chip width. The relationship between b_{lim} and Ω is detailed next.

Equation 4.3.30 is now rewritten by recognizing the complex nature of the eigenvalues, $\Lambda=\Lambda_{\text{Re}}+i\Lambda_{\text{Im}}$, and substituting the Euler identity $e^{-i\omega_c\tau}=\cos(\omega_c\tau)-i\sin(\omega_c\tau)$:

$$\Lambda_{\text{Re}}+i\Lambda_{\text{Im}}=-\frac{N_t}{4\pi}bK_t(1-\cos(\omega_c\tau)+i\sin(\omega_c\tau)). \tag{4.3.32}$$

Equation 4.3.32 is solved for b to obtain the limiting chip width, b_{lim}:

$$b_{\text{lim}}=-(\Lambda_{\text{Re}}+i\Lambda_{\text{Im}})\frac{4\pi}{N_tK_t}\frac{1}{(1-\cos(\omega_c\tau)+i\sin(\omega_c\tau))}. \tag{4.3.33}$$

Rationalizing Eq. 4.3.33 gives Eq. 4.3.34.

$$b_{\text{lim}}=-(\Lambda_{\text{Re}}+i\Lambda_{\text{Im}})\frac{4\pi}{N_tK_t}\frac{(1-\cos(\omega_c\tau)-i\sin(\omega_c\tau))}{\left((1-\cos(\omega_c\tau))^2+(\sin(\omega_c\tau))^2\right)}$$

$$b_{\text{lim}}=-\frac{4\pi}{N_tK_t}\frac{(\Lambda_{\text{Re}}+i\Lambda_{\text{Im}})(1-\cos(\omega_c\tau)-i\sin(\omega_c\tau))}{(2-2\cos(\omega_c\tau))} \tag{4.3.34}$$

$$b_{\text{lim}}=-\frac{2\pi}{N_tK_t}\left(\begin{array}{c}\frac{(\Lambda_{\text{Re}}(1-\cos(\omega_c\tau))+\Lambda_{\text{Im}}\sin(\omega_c\tau))}{(1-\cos(\omega_c\tau))}+\\ i\frac{(\Lambda_{\text{Im}}(1-\cos(\omega_c\tau))-\Lambda_{\text{Re}}\sin(\omega_c\tau))}{(1-\cos(\omega_c\tau))}\end{array}\right)$$

Next, it is recognized that, because b_{lim} must be real valued, the imaginary part of the final expression in Eq. 4.3.34 must be equal to zero, giving $\Lambda_{\text{Im}}(1-\cos(\omega_c\tau))-\Lambda_{\text{Re}}\sin(\omega_c\tau)=0$. From this statement, the new variable κ is defined.

$$\kappa = \frac{\Lambda_{\text{Im}}}{\Lambda_{\text{Re}}} = \frac{\sin(\omega_c \tau)}{1 - \cos(\omega_c \tau)} \tag{4.3.35}$$

Equation 4.3.34 is now rewritten to obtain the final, frequency dependent expression for the stability limit.

$$\begin{aligned}
b_{\text{lim}} &= -\frac{2\pi}{N_t K_t}\left(\frac{(\Lambda_{\text{Re}}(1 - \cos(\omega_c \tau)) + \Lambda_{\text{Im}} \sin(\omega_c \tau))}{(1 - \cos(\omega_c \tau))}\right) \\
b_{\text{lim}} &= -\frac{2\pi}{N_t K_t}\left(\Lambda_{\text{Re}} + \frac{\Lambda_{\text{Im}} \sin(\omega_c \tau)}{(1 - \cos(\omega_c \tau))}\right) \\
b_{\text{lim}} &= -\frac{2\pi}{N_t K_t}\Lambda_{\text{Re}}\left(1 + \frac{\Lambda_{\text{Im}}}{\Lambda_{\text{Re}}}\frac{\sin(\omega_c \tau)}{(1 - \cos(\omega_c \tau))}\right) = -\frac{2\pi}{N_t K_t}\Lambda_{\text{Re}}(1 + \kappa^2)
\end{aligned} \tag{4.3.36}$$

The corresponding frequency dependent spindle speeds are determined by first writing the phase shift in the surface undulations between subsequent tooth passages (similar to the average tooth angle derivation), $\varepsilon = \pi - 2\psi$ (rad), where $\psi = \tan^{-1}(\kappa)$ (rad). The tooth passing periods are next expressed as $\tau = \frac{1}{\omega_c}(\varepsilon + j \cdot 2\pi)$ (s), where $j = 0, 1, 2\ldots$ refers to the integer number of waves between teeth. Incrementing j leads to the individual lobes; it serves in the same capacity as N in Eq. 4.3.2. Finally, the spindle speeds are obtained from:

$$\Omega = \frac{60}{N_t \tau} \text{(rpm)}. \tag{4.3.37}$$

Because our computing platform of choice is MATLAB®, we will rewrite the eigenvalue problem as stated in Eq. 4.3.31, $\det([I] + \Lambda[FRF_{or}]) = 0$, in the more traditional format required by the `eig` function. This formulation is $\det([FRF_{or}] - \lambda[I]) = 0$, where the new eigenvalue expression is $\lambda = \frac{4\pi}{N_t} \frac{1}{bK_t(1 - e^{-i\omega_c \tau})}$. The corresponding revised stability limit is:

$$\tilde{b}_{\text{lim}} = \frac{2\pi}{N_t K_t (\lambda_{\text{Re}}^2 + \lambda_{\text{Im}}^2)} \lambda_{\text{Re}}(1 + \tilde{\kappa}^2) \tag{4.3.38}$$

where $\tilde{\kappa} = \frac{\lambda_{\text{Im}}}{\lambda_{\text{Re}}}$. The derivation that leads to Eq. 4.3.38 is included in Appendix B. The two complex eigenvalues, λ_1 and λ_2, are determined using the MATLAB® function call `eig(FRF_or)` in p_4_3_6_1.m, where `FRF_or` is the oriented FRF defined in Eq. 4.3.29. The spindle speed equations are also modified to be $\tilde{\psi} = \tan^{-1}(\tilde{\kappa})$ (rad), $\tilde{\varepsilon} = \pi - 2\tilde{\psi}$ (rad), $\tilde{\tau} = \frac{1}{\omega_c}(\tilde{\varepsilon} + j \cdot 2\pi)(\text{s})$ (s), and

$$\tilde{\Omega} = \frac{60}{N_t \tilde{\tau}} \text{(rpm)}. \tag{4.3.39}$$

The stability lobe diagram is obtained by plotting $\tilde{\Omega}$ versus the two $\tilde{b}_{\text{lim}}$ values for each chatter frequency (the minimum at each spindle speed is selected to define the stability boundary). Note that in this analysis the valid chatter frequencies are not limited to those corresponding to negative real values of the oriented FRF. Instead, the full frequency range of FRF_{or} is applied.

IN A NUTSHELL There are two related, but slightly different, methods for dealing with the varying orientation of the cutting force as the teeth pass through the cutting zone. Both approaches are simplifications of reality and have cutting conditions and machine dynamic characteristics where they predict experimental results more or less accurately.

Example 4.3.6: Comparison with Ex. 4.3.5 stability results In this example, the conditions are the same as for Ex. 4.3.5. Specifically, a 20% radial immersion down milling cut with a start angle of $\phi_s = 126.9$ deg and exit angle of $\phi_e = 180$ deg is analyzed. The x (feed) direction dynamics are $f_{nx} = 900$ Hz, $k_x = 9 \times 10^6$ N/m, and $\zeta_x = 0.02$. The y direction dynamics are $f_{ny} = 950$ Hz, $k_y = 1 \times 10^7$ N/m, and $\zeta_y = 0.01$. The workpiece material is a low carbon steel alloy and it is machined with a three tooth, 19 mm diameter square endmill. For a specific force value of $K_s = 2250$ N/mm^2 and a force angle of $\beta = 75$ deg, the corresponding cutting force coefficients are $K_n = \frac{1}{\tan(\beta)} = 0.268$ and $K_t = \frac{K_s}{\sqrt{1+K_n^2}} = 2173$ N/mm^2.

Based on the selected start and exit angles and K_n value, the α values from Eq. 4.3.23 required to calculate FRF_{or} are $\alpha_{xx} = 0.5198$, $\alpha_{xy} = -1.2356$, $\alpha_{yx} = 0.6180$, and $\alpha_{yy} = -1.0165$. The stability lobe diagram obtained from the Fourier series analysis is provided in Fig. 4.3.11. Two limits are seen; the dotted line is obtained from λ_1 and the solid line from λ_2. It is observed that the composite stability limit (defined collectively from the minimum of the λ_1 and λ_2 b_{lim} values at each spindle speed) is very similar to the average tooth angle limit shown in Fig. 4.3.9. For example, $b_{\text{lim,crit}} = 0.50$ mm from Fig. 4.3.11 compares favorably with the 0.41 mm value from Fig. 4.3.9 and the stability boundary peaks and troughs occur at the same spindle speeds. As noted previously, the MATLAB® program used to produce Fig. 4.3.11 is provided on the companion CD as p_4_3_6_1.m.

Example 4.3.7: Comparison with Ex. 4.3.4 stability results In contrast to the close agreement between the two approaches seen in Ex. 4.3.6, the methods diverge somewhat in the case of slotting. Although the best speeds agree, the allowable chip widths differ. As in Ex. 4.3.4, the symmetric x and y direction dynamics are $f_n = 500$ Hz, $k = 8 \times 10^6$ N/m, and $\zeta = 0.02$. An aluminum alloy is machined with a four tooth square endmill and the cutting force coefficients are $K_t = 695$ N/mm^2 and $K_n = 0.404$, corresponding to $K_s = 750$ N/mm^2 and

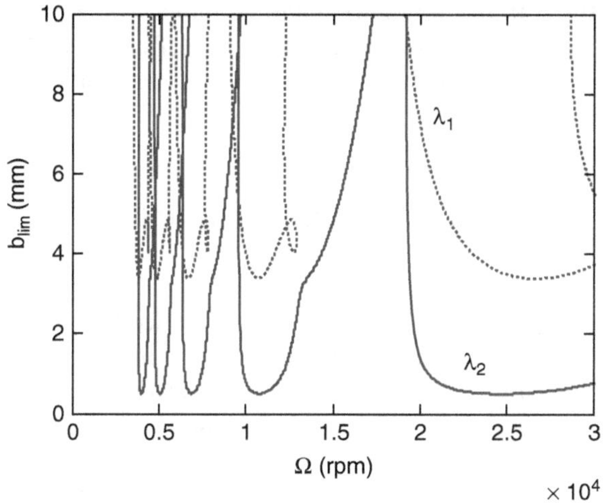

Fig. 4.3.11 Example 4.36 Stability lobe diagram (j = 0 to 4)

β = 68 deg. For $\phi_s = 0$ and $\phi_e = 180$ deg, the α values are $\alpha_{xx} = -1.2693$, $\alpha_{xy} = -3.1416$, $\alpha_{yx} = 3.1416$, and $\alpha_{yy} = -1.2693$.

The Fourier series approach stability limit is shown as the solid line in Fig. 4.3.12. Only the λ_1 boundary is seen at the selected scale (the λ_2 boundary occurs at higher chip widths in this case). For comparison convenience, the result from the average tooth angle approach (previously shown in Fig. 4.3.5) is also included as the dotted line. In this case, the $b_{\mathrm{lim,crit}}$ values are 0.22 mm (Fourier series) and 0.58 mm (average tooth angle). Additionally, the stable

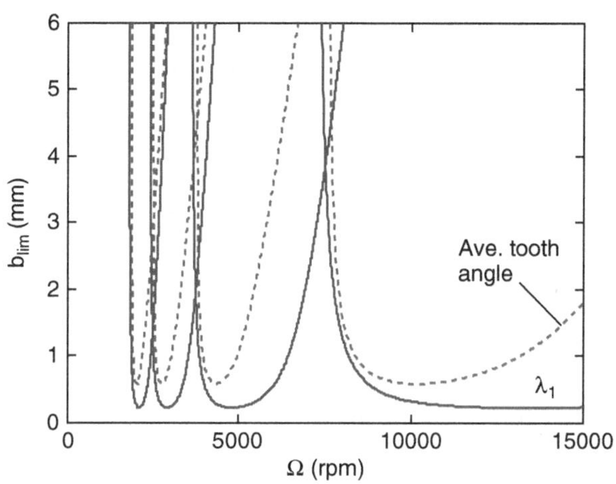

Fig. 4.3.12 Example 4.3.7 stability lobe diagram. The solid line shows the Fourier series stability boundary(j = 0 to 3), while the dotted line gives the average tooth angle result (N = 0 to 3)

zone (beneath the stability boundary) has a smaller area for the Fourier series calculations. We will explore this disagreement further in Section 4.4. The MATLAB® program used to produce Fig. 4.3.12 is included as p_4_3_7_1.m on the companion CD.

4.4 Milling Time-Domain Simulation with Straight Teeth

In this section, a time-domain simulation for the milling problem is detailed. It is based on the 'Regenerative Force, Dynamic Deflection Model' described by Smith and Tlusty [10]. As with the turning analysis in Chapter 3, the analytical stability lobe diagrams detailed in Section 4.3 provide a "global" picture of the stability behavior, but do not provide information regarding the "local" cutting force or tool vibrations. The time-domain simulation, on the other hand, gives this "local" force and vibration information for the selected cutting conditions. The simulation again applies numerical integration to solve the time delayed differential equations of motion and includes the nonlinearity that occurs if the tooth leaves the cut. Assumptions include straight cutter teeth, the circular tool path, and a square endmill geometry. Similar to turning, the simulation proceeds as follows: 1) the instantaneous chip thickness is determined using the vibration of the current and previous teeth at the selected tooth angle; 2) the cutting force is calculated; 3) the force is used to find the new displacements; and 4) the tooth angle is incremented and the process is repeated. The simulation model geometry is the same as was presented in Fig. 4.1.8. We use modal parameters to describe the dynamics of the non-rigid tool in the x and y directions, where multiple degrees of freedom in each direction can be accommodated as described in Section 3.5.

IN A NUTSHELL Nothing is a better representation of reality than reality (actual cutting). However, milling simulation removes many of the simplifying assumptions required by the analytical derivations. As in turning, time-domain simulation provides detailed information about a specific case, but loses the global information provided by the analytical solutions. The production of a stability lobe diagram requires many executions of the simulation.

4.4.1 Chip Thickness Calculation

The instantaneous chip thickness depends on the nominal, tooth angle dependent chip thickness (Eq. 4.1.2), the current normal direction vibration, and the vibration of the previous tooth at the same angle. As described in Section 4.2, the chip thickness can be expressed as $h(t) = f_t \sin(\phi) + n(t - \tau) - n(t)$, where τ

is the tooth period, $\tau = \frac{60}{\Omega N_t}$ (s) and Ω is given in rpm. The vibrations in the direction of the surface normal for the current tooth depend on the x and y vibrations as well as the tooth angle, according to $n = x\sin(\phi) - y\cos(\phi)$. Figure 4.2.5 presented an "unwrapped" view of the milled surface that demonstrates the regeneration mechanism in the chip thickness equation. Note that the positive x direction again opposes the feed direction. For this simulation, we will neglect the possibility that the current surface may depend on more than the prior and current tooth vibrations (as shown for turning in Fig. 3.5.1). Also, because we have assumed straight teeth, we will ignore vibrations in the z direction.

As discussed in the stability lobe development, the milling equations may be equivalently considered to be a function of time or angle. For the simulation, our strategy is to divide the angle of the cut into a discrete number of steps. At each small time step, dt, we increment the cutter angle by the corresponding small angle, $d\phi$. This approach enables convenient computation of the chip thickness for each simulation step because: 1) we have predefined the possible teeth orientations; and 2) we can store the surface created by the previous tooth at each angle. The cutter rotation by the $d\phi$ increment is depicted in Fig. 4.4.1, where the size of the increment depends on the selection of the number of steps per revolution (`steps _rev` in p_4_4_1_1.m on the companion CD), specifically, $d\phi = \frac{360}{steps\_rev}$ (deg). The corresponding time step is $dt = \frac{60}{steps\_rev \cdot \Omega}$ (s), where Ω is the spindle speed in rpm. (Note that the teeth number labeling convention differs from that shown in Sections 4.1 and 4.2. Previously, the teeth were numbered according to their order of entry into the cut for convenience of description, while they are now labeled according to increasing angle in a clockwise manner for "bookkeeping" purposes within the simulation.) A vector of angles is defined to represent the potential orientations of the teeth as the cutter is rotated through one revolution of the circular tool path, $\phi = [0, d\phi, 2d\phi, 3d\phi, \ldots, (steps\_rev - 1)d\phi]$. The locations of the teeth within the cut are then defined by referencing entries in this vector (`phi` in p_4_4_1_1.m). For

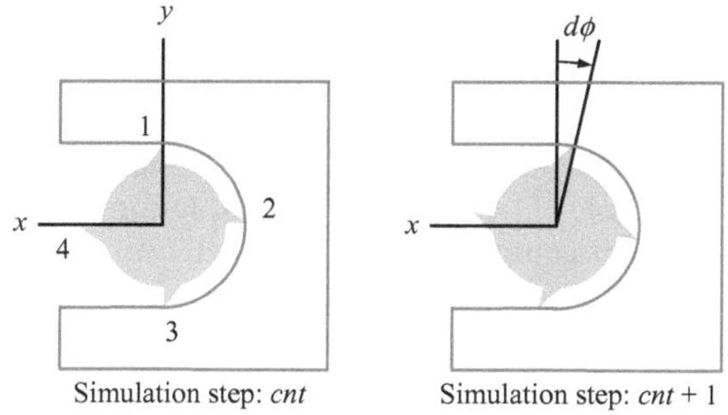

Fig. 4.4.1 Cutter rotation by $d\phi$ from one simulation step to the next (a slotting cut is depicted)

equal teeth spacing, this means that the teeth are located every $\frac{steps\_rev}{N_t}$. A vector of tooth orientations, that indicates particular entries in the ϕ vector, not actual angles, is then defined as: $teeth = [1, \frac{steps\_rev}{N_t} + 1, 2\frac{steps\_rev}{N_t} + 1, \ldots, (N_t - 1)\frac{steps\_rev}{N_t} + 1]$. Incrementing the teeth angles by $d\phi$ is then a simple matter of adding one to each entry in the *teeth* vector. Finally, individual tooth angles can be identified by the MATLAB® statement: `phi(teeth(cnt))` indicates the tooth in question. Therefore, a requirement for using the $\frac{steps\_rev}{N_t}$ ratio to indicate locations within the ϕ vector is that it is an integer greater than zero (necessary for MATLAB® indices). This is explored in Ex. 4.4.1.

Example 4.4.1: Simulation tooth angle definition We'll consider the case where 650 steps per revolution has been selected for a cutter with four teeth. The spindle speed is 10000 rpm. Our first requirement is that $\frac{steps\_rev}{N_t}$ is an integer so that the `phi` vector can be referenced using the `teeth` entries. In this case, $\frac{steps\_rev}{N_t} = \frac{650}{4} = 162.5$ steps/tooth. To correct the non-integer situation, the MATLAB® function `round` can be implemented to determine the nearest integer ratio:

```
temp = round(steps_rev/Nt);
steps_rev = temp*Nt;
```

as shown in p_4_4_1_1.m. The result is that the new ratio, saved to the `temp` variable, is 163 and the updated number of steps per revolution is 652. The angular increment is $d\phi = \frac{360}{652} = 0.5521$ deg and the time increment for numerical integration is $dt = \frac{60}{652.10000} = 9.2025 \times 10^{-6}$ s. The vector of possible tooth angles (in MATLAB® syntax) is `phi = [0 0.5521 1.1043 1.6564 ... 359.4479]` deg and the *teeth* vector can be initially described by `teeth = [1 164 327 490]`. For this cutter orientation, tooth 1 is positioned at `phi (teeth(1)) = phi(1) = 0`. Similarly, the angles for teeth 2 through 4 are `phi(teeth(2)) = phi(164) = 90` deg, `phi(teeth(3)) = phi(327) = 180` deg, and `phi(teeth(4)) = phi(490) = 270` deg, respectively. In the next time increment, 9.2025×10^{-6} s later, the *teeth* vector values are each incremented by one to be `teeth = [2 165 328 491]`. The teeth angles are `phi(teeth(1)) = phi(2) = 0.5521` deg, `phi(teeth (2)) = phi(165) = 90.5521` deg, `phi(teeth(3)) = phi(328) = 180.5521` deg, and `phi(teeth (4)) = phi(491) = 270.5521` deg.

The next step in determining the (discrete) chip thickness at each simulation step is to address the $n(t - \tau)$ and $n(t)$ terms in the continuous $h(t) = f_t \sin(\phi) + n(t - \tau) - n(t)$ chip thickness equation. The $n(t)$ term is simply the current normal direction vibration level determined using:

```
n = (x*sin(phi(teeth(cnt3))*pi/180) - y*cos(phi
(teeth(cnt3))*pi/180));
```

in p_4_4_1_1.m, where `x` and `y` are the current x and y direction vibrations, the `sin` and `cos` functions require arguments in rad, and `cnt3` is a simulation index that counts through the individual teeth to sum the force contributions.

To accommodate the $n(t-\tau)$ term, the normal direction vibration for the previous tooth at the current angular orientation is required. To organize these values, a new variable *surf*[3] is defined. Values in this vector are indexed in the same way as `phi` (according to entries in the *teeth* vector), i.e., `surf (teeth(cnt3))`. The *surf* entries are updated after the current chip thickness calculation so that each time the value is queried at a particular orientation, the entry for that tooth angle is referencing the result from the last time a tooth was positioned at the current angle. There are two scenarios for updating the `surf` value. First, if the current tooth is cutting (it is between ϕ_s and ϕ_e and the tangential cutting force is greater than or equal to zero), then the value is set to the current vibration using:

```
surf(teeth(cnt3)) = n;
```

in p_4_4_1_1.m. Recall that this value will not be referenced until the next tooth is at the current angle. Second, if the current tooth is bounded by the start and exit angles, but the tangential force is less than zero, then the tooth has vibrated out of the cut and the updating command is:

```
surf(teeth(cnt3)) = surf(teeth(cnt3)) + ft*sin(phi
(teeth(cnt3))*pi/180);
```

so that the previous value is appended by the nominal feed per tooth that was not removed. If the current angle is not bounded by the cut, then no updating is necessary. Finally, the instantaneous chip thickness equation is:

```
h = ft*sin(phi(teeth(cnt3))*pi/180) + surf(teeth
(cnt3)) - n;
```

in p_4_4_1_1.m.

4.4.2 Force Calculation

Once the chip thickness is computed, the tangential component of the force in the current time step is determined using Eq. 4.1.8, where the chip width, b, is equal to the axial depth of cut for straight teeth. As with turning, the calculated chip thickness is negative if the current tool vibration in the normal direction is larger than the surface location (equal to the sum of the nominal chip thickness, $f_t \sin\phi$, and the vibration of the previous tooth at the same angle). The tangential cutting force is set to zero in this situation that no cutting is occurring. The normal force is then computed and these results are projected into the x and y directions as shown in Eqs. 4.1.11 and 4.1.12.

[3] This variable is named to indicate the machined surface, not surfing along Florida's Atlantic coast.

4.4.3 Displacement Calculation

Considering a single degree of freedom in the x and y directions, the equations of motion are:

$$m_x\ddot{x} + c_x\dot{x} + k_x x = F_{x_1} \text{ and } m_y\ddot{y} + c_y\dot{y} + k_y y = F_{y_1}. \tag{4.4.1}$$

Rewriting Eq. 4.4.1 yields expressions for the x and y direction accelerations in the current time step:

$$\ddot{x} = \frac{F_x - c_x\dot{x} + k_x x}{m_x} \text{ and } \ddot{y} = \frac{F_y - c_y\dot{y} + k_y y}{m_y}. \tag{4.4.2}$$

where the velocities, $\dot{x}$ and $\dot{y}$, and positions, x and y, from the previous time step are used (initial values are zero to begin the simulation). The new velocities and positions are then determined by numerical (Euler) integration:

$$\dot{x} = \dot{x} + \ddot{x} \cdot dt \qquad \dot{y} = \dot{y} + \ddot{y} \cdot dt \tag{4.4.3}$$

$$x = x + \dot{x} \cdot dt \qquad y = y + \dot{y} \cdot dt, \tag{4.4.4}$$

where the velocities on the right hand side of the equal signs in Eq. 4.4.3 are retained from the previous time step. The new velocities are then applied to determine the new displacements in Eq. 4.4.4. Again, the displacements on the right hand side of Eq. 4.4.4 are those from the previous time step. As mentioned previously, multiple degrees of freedom in each direction can be accommodated by summing the individual modal contributions as described in Section 3.5. Again, considering the dynamic characteristics of the machine in a modal sense facilitates the computation.

4.4.4 Simulation Summary and Implementation

The milling simulation provided in p_4_4_1_1.m completes three basic activities at each time step. First, the cutter is rotated by $d\phi$ by adding one to each entry in the *teeth* vector. Second, within a `for` loop (indexed by `cnt3`) that sums over all the cutter teeth, it is first verified that the tooth in question is bounded by the start and exit cut angles. If so, the chip thickness is determined and the cutting force is calculated (including the nonlinearity if a tooth leaves the cut due to excessive vibration). If not, the force is set to zero. Third, the displacement is determined by numerical integration. To exercise the simulation, comparisons between the stability limits described in Ex. 4.3.7 and the simulated forces and displacements are made in Ex. 4.4.2.

Example 4.4.2: Comparison with Ex. 4.3.7 using time-domain simulation To review, the x and y direction dynamics are symmetric with $f_n = 500$ Hz,

$k = 8 \times 10^6$ N/m, and $\zeta = 0.02$. An aluminum alloy is machined with a four tooth square endmill and the cutting force coefficients are $k_t = 695$ N/mm^2 and $k_n = 281$ N/mm^2, corresponding to $K_s = 750$ N/mm^2 and $\beta = 68$ deg. For the slotting cut, $\phi_s = 0$ and $\phi_e = 180$ deg. Additionally, there are 652 steps per revolution, the feed per tooth is 0.15 mm/tooth, and 20 revolutions are simulated, where the number of revolutions (*rev*) is related to the total number of simulation steps (*steps*) by $rev = \frac{steps \cdot dt \cdot \Omega}{60}$. Four different cases are considered: 1) $\Omega = 7500$ rpm and $b = 3$ mm; 2) $\Omega = 7500$ rpm and $b = 5$ mm; 3) $\Omega = 5000$ rpm and $b = 0.1$ mm; and 4) $\Omega = 5000$ rpm and $b = 0.5$ mm. These operating points are identified in Fig. 4.4.2, which includes the average tooth angle and Fourier series stability limits.

According to the two analytical approaches, cases 1 and 3 should provide stable operating conditions. The time-domain simulation corroborates these predictions. Figures 4.4.3 and 4.4.4 show the x and y direction forces and displacements, respectively, for case 1. It is seen that, once the initial transients attenuate after approximately 0.07s, the expected constant force for a four tooth cutter in a slotting cut is obtained. Similar results are observed for case 3 in Figs. 4.4.5 and 4.4.6.

For cases 2 and 4, however, the Fourier series approach predicts that the cuts will be unstable, while the average tooth angle approximation shows these should be stable conditions. Figures 4.4.7 and 4.4.8 display the x and y direction results for case 2. Chatter is observed with the forces and displacement increasing over time. Additionally, Fig. 4.4.9 shows the resultant force and nonlinearity when the force drops to zero as the tooth vibrates out of the cut (in the vicinity of 0.1 s for example). Similar results are seen for case 4 in Figs. 4.4.10

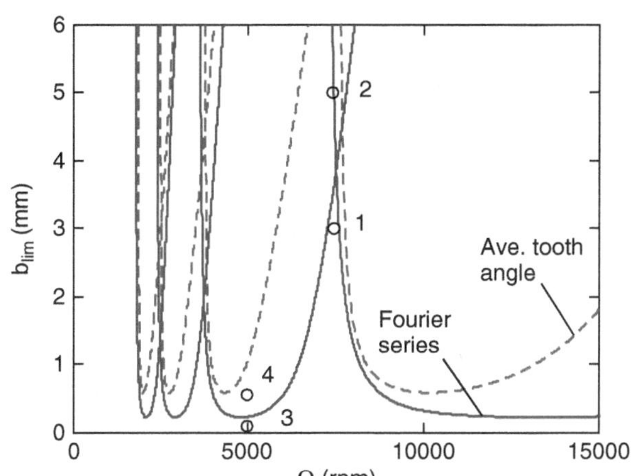

Fig. 4.4.2 Operating points for comparison between analytical stability boundaries from Ex. 4.3.7 and time-domain simulation. 1) $\Omega = 7500$ rpm and $b = 3$ mm; 2) $\Omega = 7500$ rpm and $b = 5$ mm; 3) $\Omega = 5000$ rpm and $b = 0.1$ mm; and 4) $\Omega = 5000$ rpm and $b = 0.5$ mm

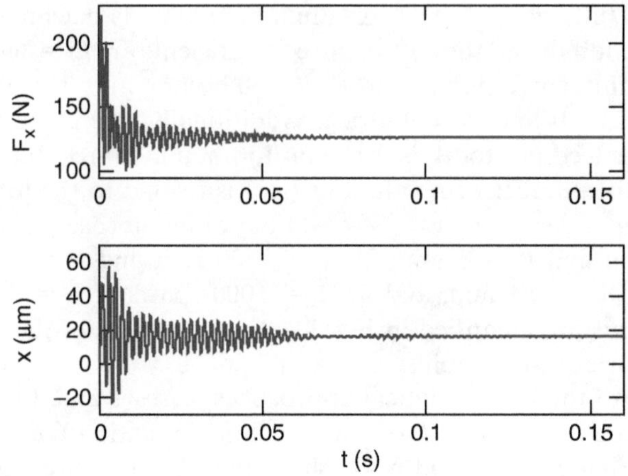

Fig. 4.4.3 Example 4.4.2 stable case 1 (Ω = 7500 rpm and b = 3 mm) simulation results for x direction force (top) and displacement (bottom)

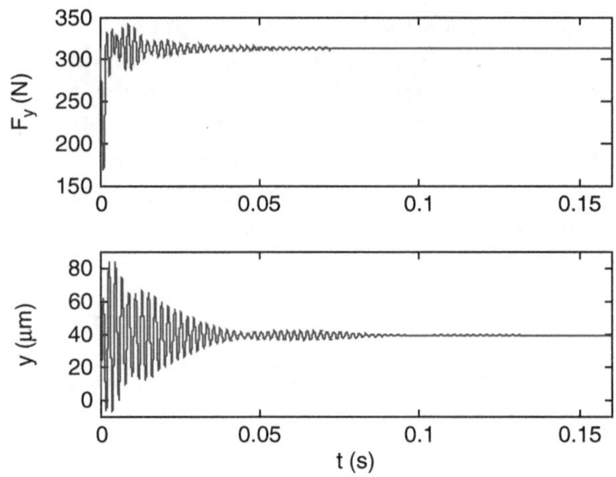

Fig. 4.4.4 Example 4.4.2 stable case 1 (Ω = 7500 rpm and b = 3 mm) simulation results for y direction force (top) and displacement (bottom)

through 4.4.12. Here the vibration and force levels grow quickly until the nonlinearity is reached and then the levels persist. The disagreement between the average tooth angle stability boundary and time-domain simulation for cases 2 and 4 is due to the orthogonality between the average surface normal and y direction in slotting. This causes the directional orientation factor for the y direction, μ_y, to be zero (see Ex. 4.3.4) and the contribution of the dynamics in the y direction to be neglected. The simulation used to produce Figs. 4.4.3 through 4.4.12 is included as p_4_4_1_1.m on the companion CD.

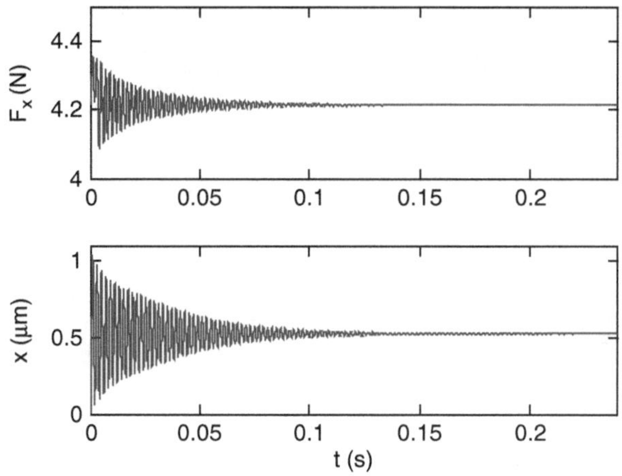

Fig. 4.4.5 Example 4.4.2 stable case 3 ($\Omega = 5000$ rpm and $b = 0.1$ mm) simulation results for x direction force (top) and displacement (bottom)

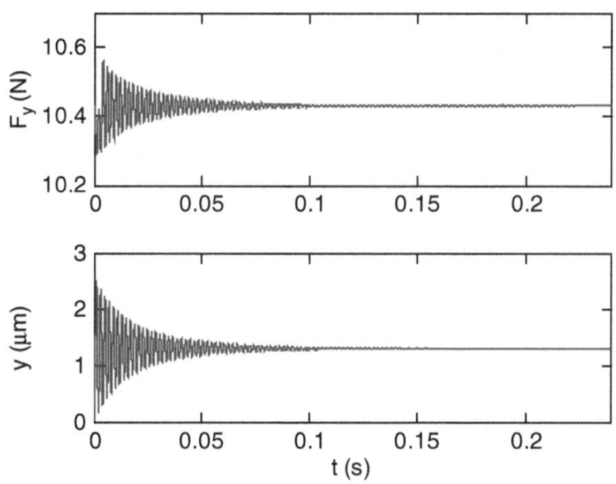

Fig. 4.4.6 Example 4.4.2 stable case 3 ($\Omega = 5000$ rpm and $b = 0.1$ mm) simulation results for y direction force (top) and displacement (bottom)

Example 4.4.3: Comparison with Ex. 4.3.6 using time-domain simulation The milling conditions are 20% radial immersion down milling cut with a start angle of $\phi_s = 126.9$ deg and exit angle of $\phi_e = 180$ deg. The x direction dynamics are $f_{nx} = 900$ Hz, $k_x = 9 \times 10^6$ N/m, and $\zeta_x = 0.02$. The y direction dynamics are $f_{ny} = 950$ Hz, $k_y = 1 \times 10^7$ N/m, and $\zeta_y = 0.01$. The workpiece material is a low carbon steel alloy and it is machined with a three tooth, 19 mm diameter square endmill and feed per tooth of 0.2 mm/tooth. For a specific force value of $K_s = 2250$ N/mm^2 and a force angle of $\beta = 75$ deg, the corresponding cutting

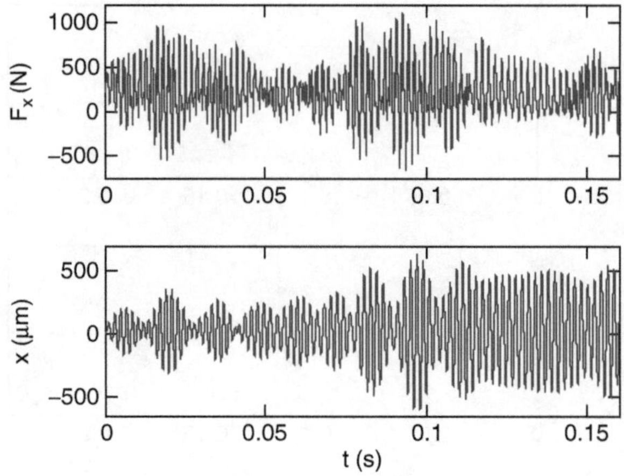

Fig. 4.4.7 Example 4.4.2 unstable case 2 (Ω = 7500 rpm and b = 5 mm) simulation results for x direction force (top) and displacement (bottom)

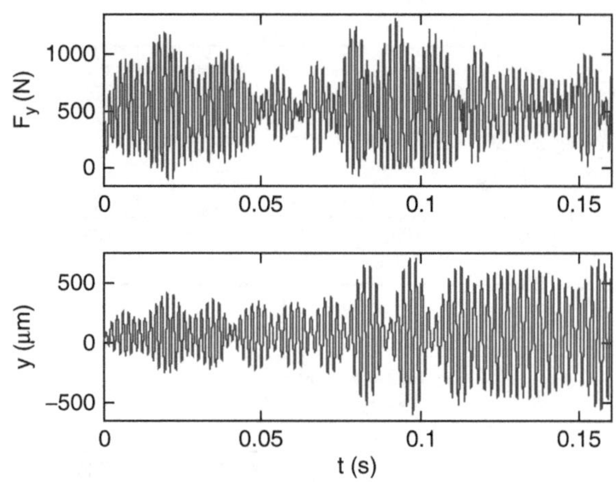

Fig. 4.4.8 Example 4.4.2 unstable case 2 (Ω = 7500 rpm and b = 5 mm) simulation results for y direction force (top) and displacement (bottom)

force coefficients are $k_t = 2173$ N/mm^2 and $k_n = 582$ N/mm^2. The number of steps per revolution is 801 and 40 revolutions are simulated.

Based on the analytical stability lobe results in Figs. 4.3.9 and 4.3.11, let's select two (Ω, b) combinations for numerical simulation. Case 1 with Ω = 17000 rpm and b = 4 mm should be stable according to the stability lobes. Case 2 with Ω = 13000 rpm and b = 4 mm should be unstable. Figure 4.4.13 displays the y direction force and vibration for case 1. It takes approximately half of the first 40 revolutions for the transients to attenuate, but the final result is stable cutting. The force and vibrations results between 0.13 s and 0.14 s are

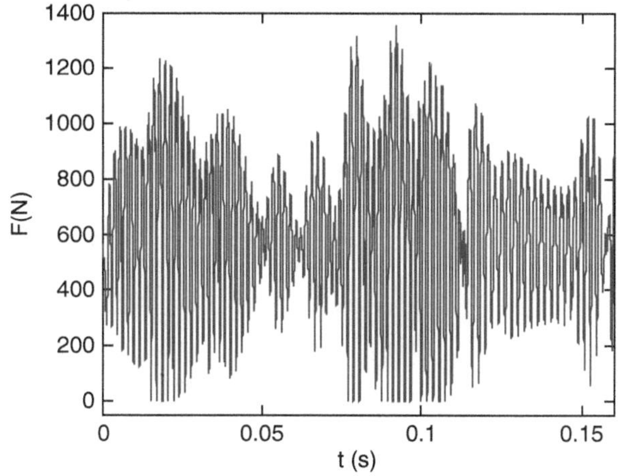

Fig. 4.4.9 Example 4.4.2 unstable case 2 (Ω = 7500 rpm and b = 5 mm) simulation results for resultant force

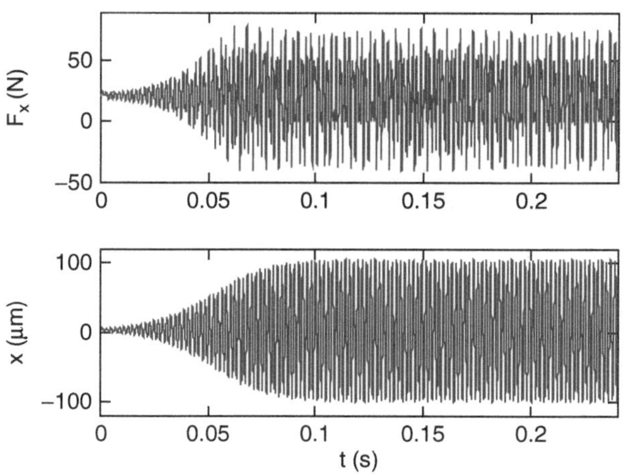

Fig. 4.4.10 Example 4.4.2 unstable case 4 (Ω = 5000 rpm and b = 0.5 mm) simulation results for x direction force (top) and displacement (bottom)

displayed in Fig. 4.4.14. We see the expected force profile for the down milling cut. Note that only one tooth is engaged at any time so the force drops to zero during the delay between one tooth leaving the cut and the next tooth entering the cut (we refer to this as interrupted cutting). Also, the force is largest at the cut entry when the chip thickness is at is maximum value for down milling and drops to zero as the chip thickness decreases to zero, according to $f_t \sin(\phi)$, at the cut exit. The resulting tool vibrations are sinusoidal in nature.

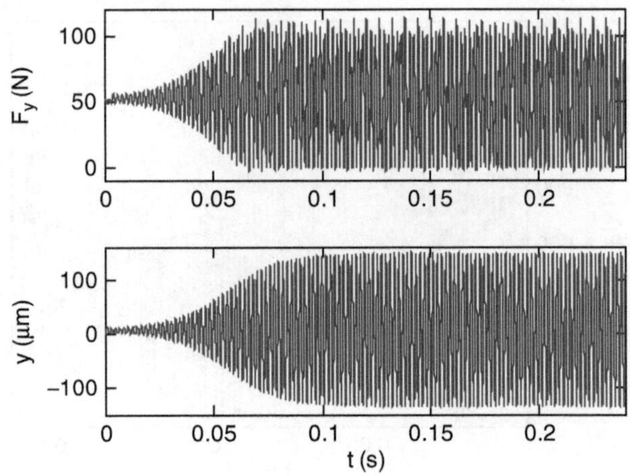

Fig. 4.4.11 Example 4.4.2 unstable case 4 (Ω = 5000 rpm and b = 0.5 mm) simulation results for y direction force (top) and displacement (bottom)

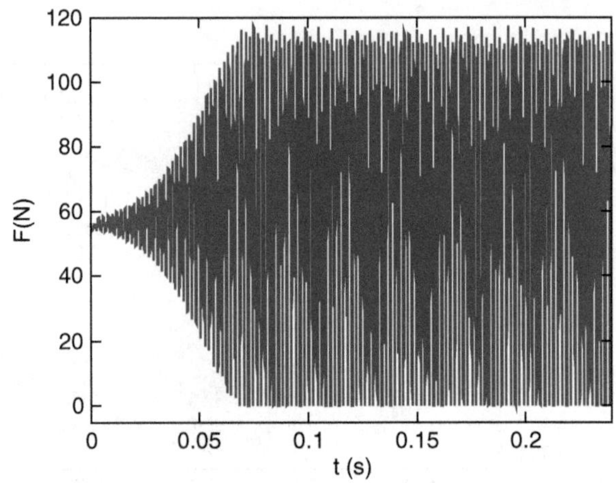

Fig. 4.4.12 Example 4.4.2 unstable case 4 (Ω = 5000 rpm and b = 0.5 mm) simulation results for resultant force

The y direction force and vibration for case 2 are provided in Fig. 4.4.15. The behavior differs substantially from Figs. 4.4.14 and 4.4.15. For the previous stable cut (case 1), the vibration exhibited only a forced vibration response to the cutting force. For the unstable case 2, the vibration now occurs at both the chatter frequency (near the tool's natural frequency) and the forcing frequencies (the tooth passing frequency and its harmonics as demonstrated in Fig. 4.1.15). We will examine the frequency content of stable and unstable milling signals further in Chapter 6. Figure 4.4.16 shows the portion of the force and vibration

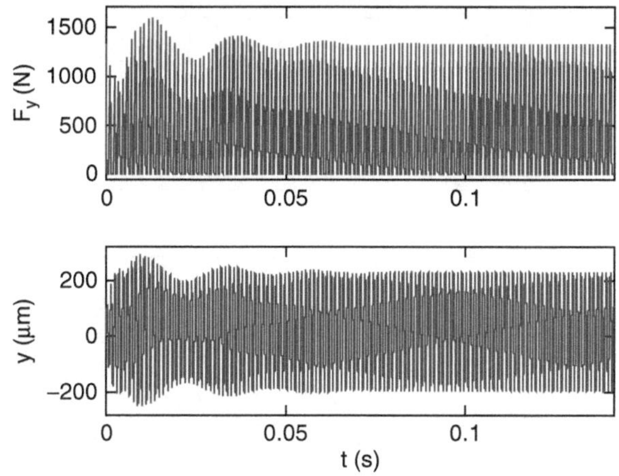

Fig. 4.4.13 Example 4.4.3 stable case 1 (Ω = 17000 rpm and b = 4 mm) simulation results for y direction force (top) and displacement (bottom)

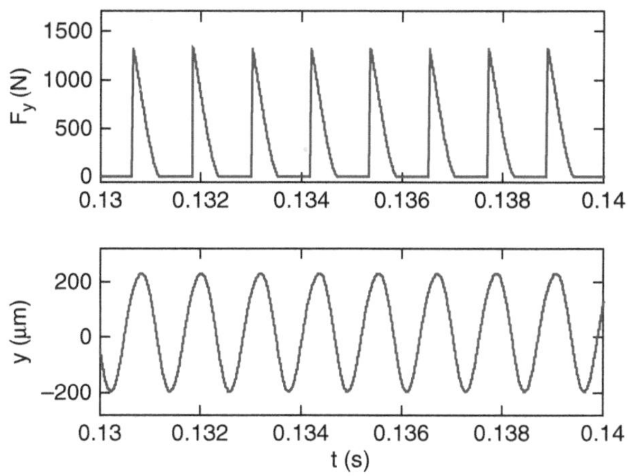

Fig. 4.4.14 Portion of results from Fig. 4.4.13 between 0.13 s and 0.14 s

between 0.13 and 0.16 s. The behavior is clearly different than the stable case shown in Fig. 4.4.14. The simulation used to produce the figures for this example is included as p_4_4_3_1.m on the companion CD.

4.5 Milling Time-Domain Simulation with Helical Teeth

While the simulation described in Section 4.4 is capable of predicting forces and displacement in milling, the assumption of straight cutter teeth is rarely applicable in practice. As discussed in Section 4.1, the cutting edges are typically

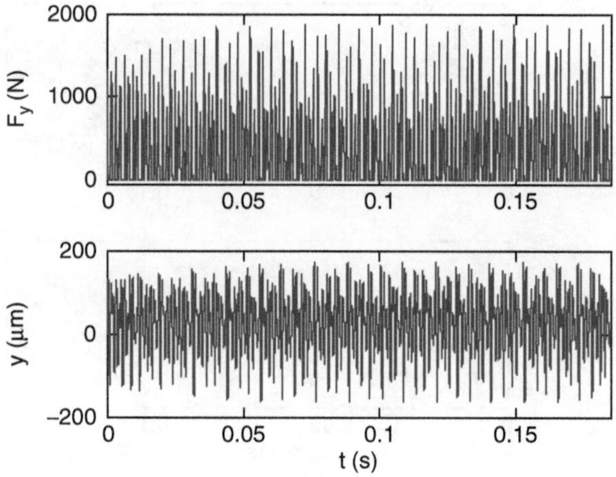

Fig. 4.4.15 Example 4.4.3 unstable case 2 ($\Omega = 13000$ rpm and $b = 4$ mm) simulation results for y direction force (top) and displacement (bottom)

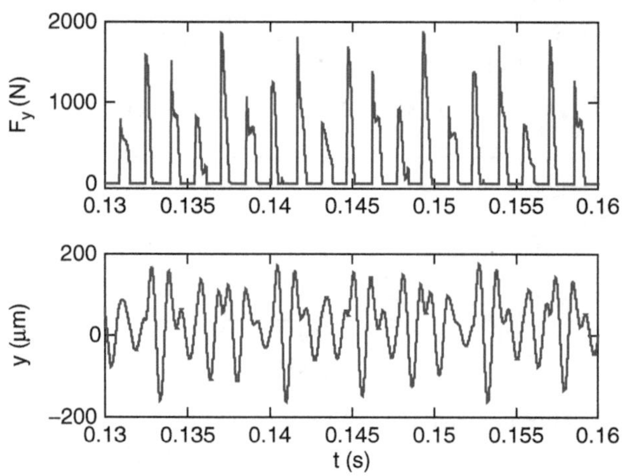

Fig. 4.4.16 Portion of results from Fig. 4.4.15 between 0.13 s and 0.16 s

inclined at the helix angle, γ, so that the chip to be removed is spread over an increased length and the cutting edge pressure is reduced. The result of the helical cutting edge geometry is that the full length of the cutting edge does not enter (or exit) the cut at the same instant. Instead, there is an increasing delay of the cut entry (and exit) when moving from the free end of the cutter toward the spindle. The situation is depicted in Fig. 4.5.1 for the helical square endmill geometry considered in this section.

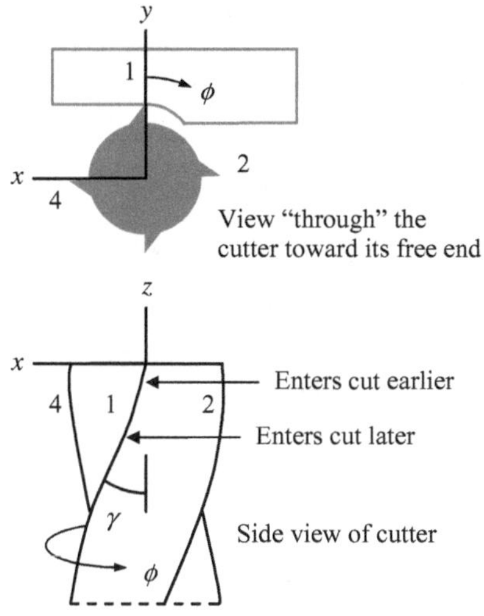

Fig. 4.5.1 The helical endmill geometry causes the cutting edge to enter and exit the cut at later instants in time when moving from its free end toward the spindle

We can visualize the angular delay, χ, of the helical cutting edge, which increases with distance from the cutter free end, by "unrolling" the periphery of the cylindrical endmill as described in [11]. The helical teeth now appear as straight lines and angles, ϕ (in rad), become distances, $r\phi$, where r is the endmill radius. Note that the velocity of every point on the tooth edge is the cutting speed, v (Eq. 4.1.17), with a direction perpendicular to the z axis. The unrolled view of the endmill pictured in Fig. 4.5.1 is presented in Fig. 4.5.2. The delay distance, $r\chi$, is zero at $z = 0$ (endmill free end) because we have referenced the cutter rotation angle, ϕ, to this axial location. However, at $z = -b$ the delay distance is now:

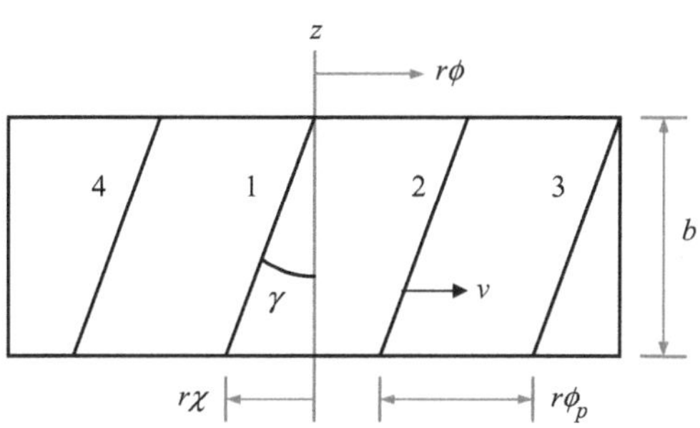

Fig. 4.5.2 Unrolled view of helical endmill geometry. The helical teeth appear as straight lines and angles are seen as distances

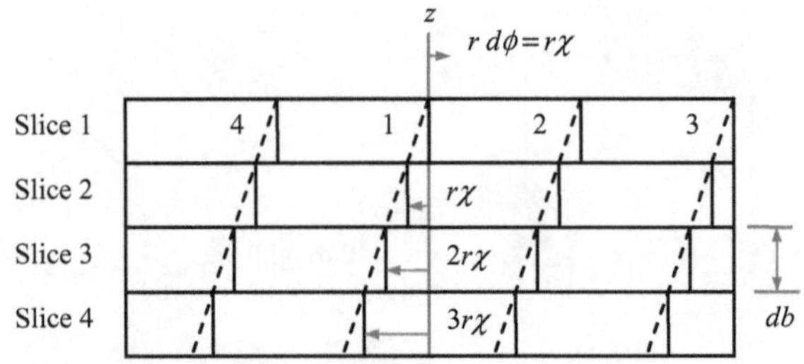

Fig. 4.5.3 Discretized version of unrolled helical endmill geometry. The axial depth is sectioned into multiple slices with thickness *db*. Each slice is treated as a straight tooth cutter with the delay distance $r\chi = r\,d\phi$ relative to the adjacent slice nearer the free end of the endmill

$$r\chi = b\tan(\gamma). \tag{4.5.1}$$

In order to discretize this relationship and retain the $d\phi$ angular increments in our time-domain simulation, we section the tool into a number of slices (or disks) perpendicular to the z axis. Each slice is treated as an individual straight tooth endmill, where the thickness of each slice is a small fraction, db, of the axial depth of cut, b. This approximation is shown in Fig. 4.5.3. Note that each slice incorporates the distance delay $r\chi = db\tan(\gamma)$ relative to the prior slice (nearer the cutter free end), which becomes the angular delay between slices of:

$$\chi = \frac{db\tan(\gamma)}{r} = \frac{2db\tan(\gamma)}{d}\,(\text{rad}) \tag{4.5.2}$$

for the rotating endmill. In Eq. 4.5.2, d is the endmill diameter. In order to ensure that the angles for each axial slice match the `phi` vector entries in the time-domain simulation, we require that the delay angle between slices is $\chi = d\phi$. This places a constraint on the db value. By substituting $d\phi$ for χ in Eq. 4.5.2 and rearranging, we see that we must select it such that:

$$\text{db} = \frac{d \cdot d\phi}{2\tan(\gamma)}. \tag{4.5.3.}$$

The integer number of axial slices for the simulation is then determined using the MATLAB® expression: `steps_axial = round(b/db)`. Naturally, the accuracy is improved if $db << b$, which requires that $d\phi$ is small (i.e., the `steps_ rev` value is high).

The primary modifications to the time-domain simulation described in Section 4.4 are that: 1) an additional `for` loop is added to count through the individual axial slices (`cnt4` in p_4_5_1_1.m included on the companion CD); and 2) the `surf` vector is now replaced by an array, where each row in the array includes the `surf` information for a single axial slice. This array is also named `surf` in p_4_5_1_1.m. It is organized similar to Fig. 4.5.3, where the first row, `surf(1, 1:steps_rev)`, contains the data for the axial slice nearest the free end of the cutter ($z = 0$) and there are `steps_axial` total rows. Within the `cnt4` loop, the angle of the current axial slice, `phia`, is determined by decrementing the angle for the tooth in question, referenced to the free end of the cutter and indexed by `cnt3` in the next outer `for` loop, by: `phi_counter = teeth(cnt3) – (cnt4-1)`. Because we selected `db` using Eq. 4.5.3 to ensure that the angular delay corresponds to `dphi` between slices, the current axial slice angle is then: `phia = phi(phi_counter)`.

A secondary modification in p_4_5_1_1.m, relative to p_4_4_1_1.m and p_4_4_3_1.m, is that a vector notation for the modal parameters, velocities, and displacements is added. This new format enables the convenient inclusion of multiple modes in the models of the system dynamics for the x and y directions. For each mode in the x direction, for example, a separate numerical integration is completed to update the modal velocity, `dp(cnt5)`, and displacement, `p(cnt5)`, where `cnt5` is the mode number index. Because the response in local (physical) coordinates is the sum of the modal contributions, the individual modal displacements are simply added to obtain the x direction displacement. The y direction approach is the same, except the modal velocities and displacement are contained in the `dq` and `q` vectors, respectively. Strictly speaking, due to the helical geometry, we should also consider the axial (z direction) forces and potential deflections. However, for most endmilling applications, the z direction dynamic stiffness is much higher than the x or y direction stiffness values, so it is common to consider the z direction to be rigid.

Finally, because these simulations tend to take longer than the straight tooth cases, a progress bar has been added (for the impatient among us). The MATLAB® function `waitbar` was used to accomplish this task[4].

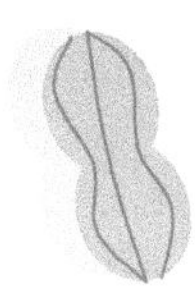

IN A NUTSHELL Because the teeth are helical, the portion of each tooth that is engaged in the cut varies as the tool rotates. The changing deflection of the tool is imprinted on the machined surface parallel to the tools' axis of rotation. It is difficult to include this complexity in the analytical formulations, but relatively straightforward to include in the time-domain simulation. For this and many other reasons, time-domain simulation is a better representation of the cutting process than either of the analytical formulations described earlier. As with all of engineering, however, there is an inherent trade-off to be made. In this case, it is that increased accuracy is computationally more intensive.

Example 4.5.1: Comparison of forces between straight and helical teeth In this example, we compare the cutting forces produced by straight and helical teeth with all other conditions being equal. We model a 30% radial immersion up milling cut with a zero start angle and exit angle of 66.4 deg. There are two

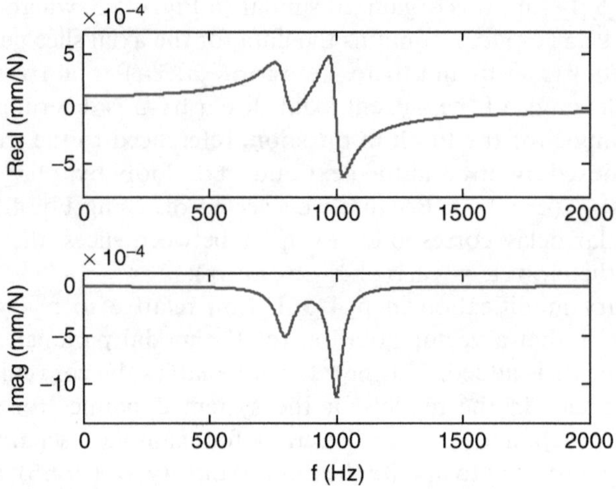

Fig. 4.5.4 Two mode FRF for Ex. 4.5.1 (*x* and *y* direction responses are equal)

identical modes in both the *x* and *y* directions. These are expressed in modal coordinates as: $f_{n1} = 800$ Hz, $k_{q1} = 2 \times 10^7$ N/m, and $\zeta_{q1} = 0.05$; and $f_{n2} =$ 1000 Hz, $k_{q2} = 1.5 \times 10^7$ N/m, and $\zeta_{q2} = 0.03$. The FRF in local coordinates for these modal parameters is displayed in Fig. 4.5.4. The workpiece material is an aluminum alloy and it is machined with a four tooth, 19 mm diameter square endmill using a feed per tooth of 0.15 mm/tooth. For a specific force value of $K_s =$ 600 N/mm^2 and force angle of $\beta = 60$ deg, the corresponding cutting force coefficients are $k_t = 520$ N/mm^2 and $k_n = 300$ N/mm^2 or, equivalently, $K_t =$ 520 N/mm^2 and $K_n = 0.577$.

We wish to consider stable cutting conditions for the time-domain force simulations. To aid in the selection of stable (Ω, b) combinations, we first generate the stability lobe diagrams using the average tooth angle and Fourier series approaches. The directional orientation factors for the average tooth angle approach are $\mu_x = \cos(\beta - 56.8)\cos(56.8)$ and $\mu_y = \cos(146.8 - \beta)\cos(146.8)$ as described in Ex. 4.3.3, where the average tooth angle in the cut is now 33.2 deg. The stability lobe diagrams for the two analytical methods are shown in Figs. 4.5.5 and 4.5.6. We observe that the stability limits are similar and a preferred speed is near 15000 rpm. This agrees with the best spindle speed selection rule given in Eq. 4.3.7 even though this is a two degree of freedom system. The MATLAB® program p_4_5_1_2.m (available on the companion CD) was used to generate these two figures.

$$\Omega_{best} = \frac{f_n \cdot 60}{(N+1) \cdot N_t} = \frac{1000 \cdot 60}{(0+1) \cdot 4} = 15000 \text{ rpm}$$

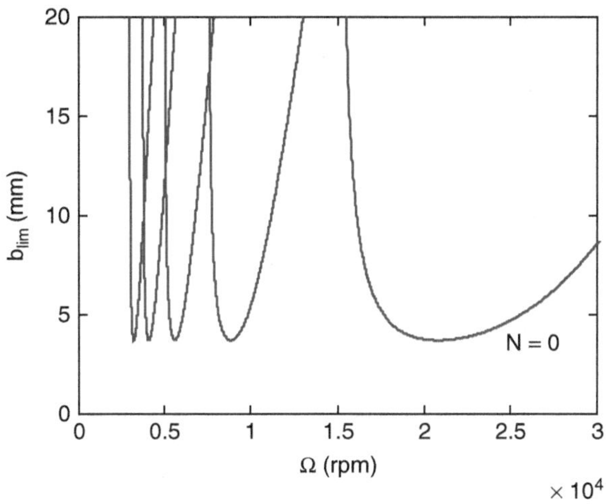

Fig. 4.5.5 Stability lobe diagram for Ex. 4.5.1 using average tooth angle approach ($N = 0$ to 4)

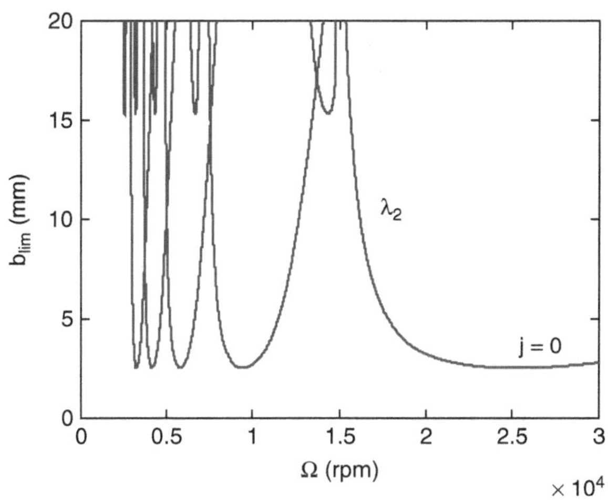

Fig. 4.5.6 Stability lobe diagram for Ex. 4.5.1 using Fourier series approach ($j = 0$ to 4)

For the time-domain simulations, let's choose the number of steps per revolution to be 800 and the number of revolutions to be 30. The spindle speed is 15000 rpm and two axial depths of cut are considered. Figures 4.5.7 and 4.5.8 show the resultant cutting forces (in the x-y plane) for a small time portion of the simulation result with $b = 5$ mm and helix angles of zero and 45 deg, respectively. A comparison of the two figures shows clear differences. First, the maximum cutter force is lower for the helical teeth endmill. Second, the force grows to its maximum value and then abruptly drops to zero at the cut

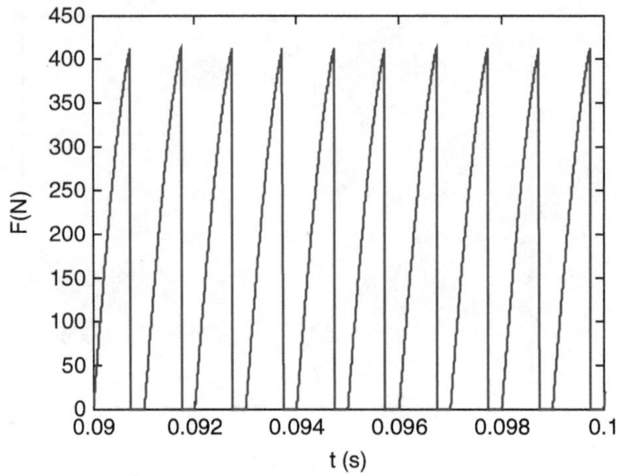

Fig. 4.5.7 Resultant cutting force versus time for zero helix angle endmill with Ω = 15000 rpm and b = 5 mm

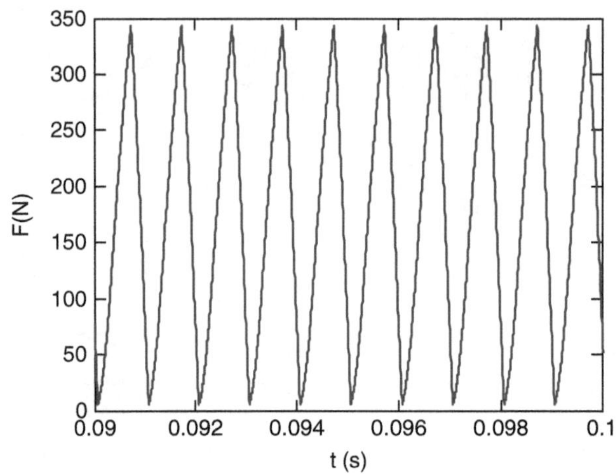

Fig. 4.5.8 Resultant cutting force versus time for 45 deg helix angle endmill with Ω = 15000 rpm and b = 5mm

exit angle for the straight tooth endmill. For the helical teeth cutter, on the other hand, the force grows and then decreases in a more saw tooth pattern and does not quite reach zero. We can understand this behavior using Eq. 4.5.2 and Fig. 4.5.2. The lag angle between the free end of the cutter and the helical edge at b = 5 mm is $\chi = \frac{2b\tan(\gamma)}{d} = \frac{2 \cdot 5\tan(45)}{19} = 30.2$ deg. This means that the cutter rotates 30.2 deg between the time that the free end of the helical tooth enters the cut (assume that ϕ = 0 here) and the time that the helical edge at b = 5 mm is engaged. At $\phi = 66.4$ deg, the free end cutting edge begins to exit the cut, but the helical portion nearer the spindle remains engaged. The force therefore does not drop immediately to zero. In fact, it does not drop to zero at all because, by the time the $b = 5$ mm helical portion exits the cut (30.2 deg later), the next tooth has entered the cut (i.e., $30.2 + 66.4 = 96.6$ deg is greater than the tooth pitch of 90 deg).

As noted in [11, Example 9.12], this "wrapping" behavior of the helical cutting edge can be exploited to achieve a constant cutting force at particular axial depths of cut. These b values are obtained when the lag angle is equal to the pitch angle. In this case, the same cutting edge length is engaged regardless of the cutter angle or radial depth of cut. The situation is depicted in Fig. 4.5.9, where the left gray box covers the cutting edge length for tooth 1. At a later instant in time, the right gray box represents the same lag distance for the helical endmill. Although tooth 1 is no longer engaged, we see that the same cutting edge length is covered but it is now made up of portions of teeth 2 and 3. This constant cutting edge length leads to the same force value for the two instants. The b value at which this behavior is observed is determined by rewriting Eq. 4.5.3 to be: $b = \frac{d \cdot \phi_p}{2\tan(\gamma)}$. For this example (45 deg helix with four teeth), the axial depth for constant cutting force is:

$$b = \frac{d \cdot \phi_p}{2\tan(\gamma)} = \frac{19 \cdot 90\frac{\pi}{180}}{2\tan(45)} = 14.9\,\text{mm}$$

Multiples of this axial depth will yield the same constant force behavior provided the cutting conditions remain stable (and sufficient flute length is available), although this is unlikely for typical tool-workpiece combinations.

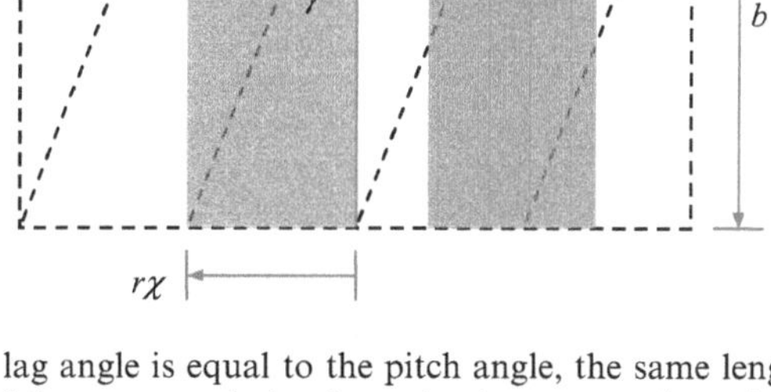

Fig. 4.5.9 when the lag angle is equal to the pitch angle, the same length of cutting edge is engaged regardless of the cutter angle (as shown by the two gray boxes). The force is constant for this axial depth

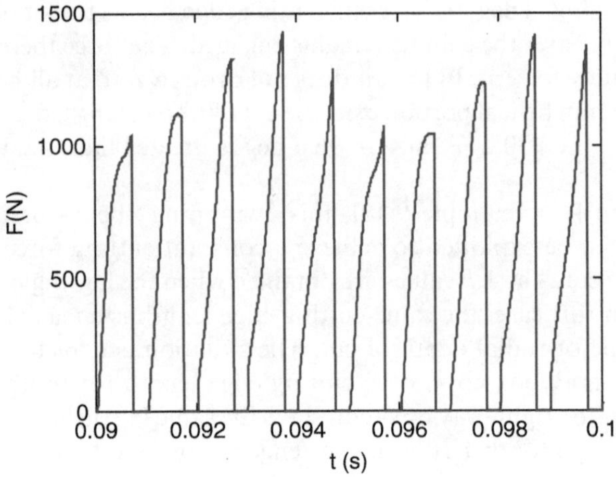

Fig. 4.5.10 Resultant cutting force versus time for 45 deg helix angle endmill with Ω = 15000 rpm and b = 14.9 mm

Figures 4.5.10 and 4.5.11 show the resultant forces for Ω = 15000 rpm and b = 14.9 mm with helix angles of zero and 45 deg, respectively. It is seen that the cut is bordering on instability in Fig. 4.5.10; the force is disturbed from its nominal values by regeneration. In Fig. 4.5.11, however, the force is nearly constant with a small oscillation that is retained from the initial transients. Note that the same force scale is used on both plots. Clearly, the maximum force is much lower for the helical cutter. Figures 4.5.8 through 4.5.11 were produced using program p_4_5_1_1.m.

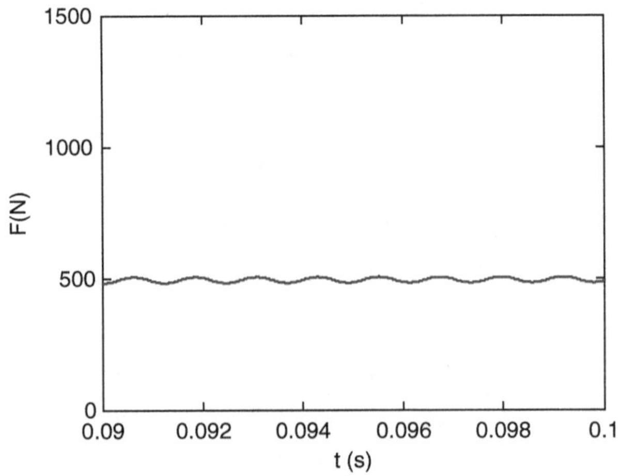

Fig. 4.5.11 Resultant cutting force versus time for 45 deg helix angle endmill with Ω = 15000 rpm and b = 14.9 mm

4.6 Ball Milling Time-Domain Simulation with Helical Teeth

In this section we extend the helical teeth milling simulation for square endmills from Section 4.5 to incorporate the spherical geometry of ball endmills. The circular tool path assumption is again applied. Additionally, the tool is sectioned into slices along its axis, as before, and the tool axis is perpendicular to the feed direction (in practice, the ball endmill axis is often inclined with respect to the feed direction to avoid cutting with the zero velocity point at the ball apex, but we will not treat this case). The ball endmilling simulation geometry and variables are displayed in Fig. 4.6.1, where we now compute the z direction force component although we maintain our assumption of rigid z direction dynamics. As shown, the width of the axial slices along the tool axis is dz. This value is again selected to ensure that the instantaneous cutter angle is coincident with one of the predefined `phi` angles as in Section 4.5. See Eq. 4.6.1, where γ is the global helix angle (the local helix on the spherical end varies with the z location as described in [12]).

$$dz = \frac{d \cdot d\phi}{2\tan(\gamma)} \tag{4.6.1}$$

The new variable, κ', is identified in Fig. 4.6.1 (the prime is used to differentiate it from the variable κ used in the Fourier series stability analysis). This represents the angle between the tool axis and the ball surface normal for the current axial slice and is calculated using Eq. 4.6.2, where j is the axial slice in question (`cnt4` in p_4_6_1_1.m) and r is the radius. It is used to first project the chip thickness from the x-y plane onto the tool surface normal direction and, second, project the tangential, normal, and axial force components onto the x-y-z coordinate directions (together with the tooth angle) [12].

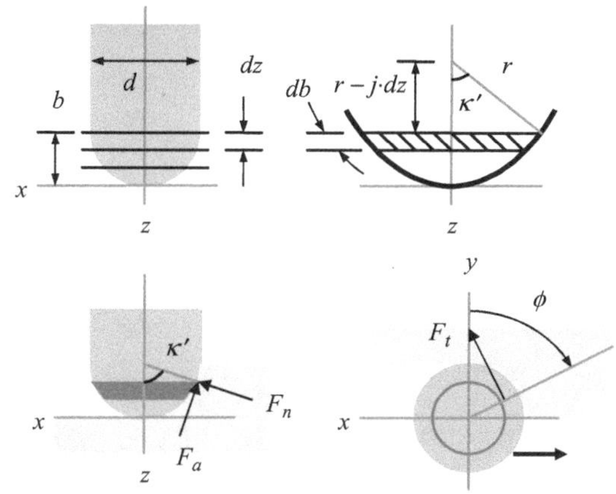

Fig. 4.6.1 Ball endmilling simulation geometry. The normal force component, F_n, is oriented along the ball surface normal

$$\kappa' = \cos^{-1}\left(1 - \frac{j \cdot dz}{r}\right) = \cos^{-1}\left(1 - \frac{2j \cdot dz}{d}\right) \tag{4.6.2}$$

As with the helical square endmill simulation, the instantaneous chip thickness in the x-y plane is determined from the: 1) nominal chip thickness, which depends on the feed per tooth and angle of the current tooth for the selected slice, `phia`; 2) vibrations in the x and y directions from the previous tooth, collected in the `surf` array (defined in the same way as described in Section 4.5); and 3) the x and y vibrations of the current tooth projected in the radial direction (i.e., the direction from the current tooth toward the tool axis in the x-y plane – this was the normal direction, `n`, for the helical square endmill simulation p_4_5_1_1.m). However, the x-y plane thickness must then be projected onto the ball surface normal direction using κ' (`kappa_p`). The relevant lines from the MATLAB® program p_4_6_1_1.m are provided here.

```
n = x* sin(phia*pi/180) - y* cos(phia*pi/180);
h = (ft* sin(phia*pi/180) + surf(cnt4, phi_counter)
    - n)* sin(kappa_p);
```

In addition to calculating the chip thickness, the chip width for each slice, *db*, must also be determined. While this was simply the width of each axial slice for the square endmill geometry, it is now the arc length of the current slice when cutting with the ball surface. The situation is depicted in Fig. 4.6.2, where the arc length is the product of the radius and the angle θ. See Eq. 4.6.3. The angle θ is, in turn, the difference between κ' and θ', $\theta = \kappa' - \theta'$. The angle θ' is defined in Eq. 4.6.4.

$$db = r\theta \tag{4.6.3}$$

$$\theta' = \cos^{-1}\left(\frac{r - dz(j-1)}{r}\right) = \cos^{-1}\left(1 - \frac{dz(j-1)}{r}\right) = \cos^{-1}\left(1 - \frac{2dz(j-1)}{d}\right) \tag{4.6.4}$$

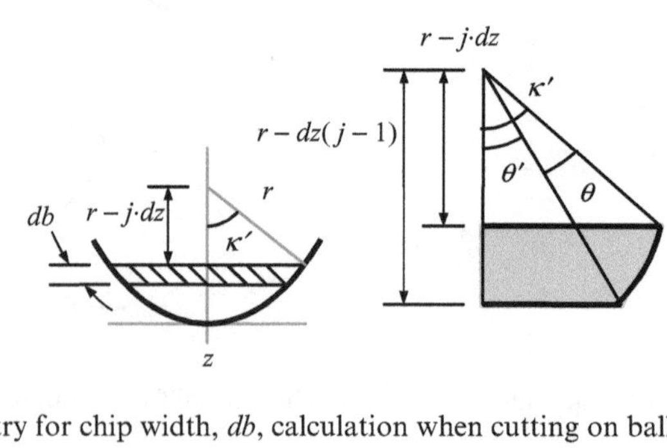

Fig. 4.6.2 Geometry for chip width, *db*, calculation when cutting on ball surface

Once the chip thickness and width are determined, the cutting force components in the tangential, normal, and axial directions are determined for each axial slice. See Eqs. 4.6.5 through 4.6.7 (referring to the F_a component as axial is actually something of a misnomer since it is directed along the tool axis only when κ' is 90 deg). Notice that a new cutting force coefficient, k_a, is defined in Eq. 4.6.7. It fills the same role as k_t and k_n, to relate the force to chip area, and is expressed in units of N/mm^2, or equivalent.

$$F_t = k_t h db \tag{4.6.5}$$

$$F_n = k_n h db \tag{4.6.6}$$

$$F_a = k_a h db \tag{4.6.7}$$

As shown in Fig. 4.6.1, the projection of these components on the *x*-*y*-*z* fixed coordinate frame depends on both the instantaneous cutter angle, ϕ, and the ball surface normal direction, κ', for the selected tooth and slice. The force relationships are provided in matrix form in Eq. 4.6.8, where ϕ_a is the tooth angle `phia` from the MATLAB® program p_4_6_1_1.m.

$$\begin{Bmatrix} F_x \\ F_y \\ F_z \end{Bmatrix} = \begin{bmatrix} \cos(\phi_a) & \sin(\phi_a)\sin(\kappa') & -\sin(\phi_a)\cos(\kappa') \\ \sin(\phi_a) & -\cos(\phi_a)\sin(\kappa') & \cos(\phi_a)\cos(\kappa') \\ 0 & -\cos(\kappa') & -\sin(\kappa') \end{bmatrix} \begin{Bmatrix} F_t \\ F_n \\ F_a \end{Bmatrix} \tag{4.6.8}$$

The only remaining issue that must be addressed in the simulation is treating the case where the commanded axial depth is greater than the ball radius. In this situation κ' is set to 90 deg and Eq. 4.6.8 collapses to Eq. 4.6.9. The *x* and *y* force projections are now identical to the helical square endmill simulation and the *z* component is equal to the axial force (with the appropriate sign convention applied).

$$\begin{Bmatrix} F_x \\ F_y \\ F_z \end{Bmatrix} = \begin{bmatrix} \cos(\phi_a) & \sin(\phi_a) & 0 \\ \sin(\phi_a) & -\cos(\phi_a) & 0 \\ 0 & 0 & -1 \end{bmatrix} \begin{Bmatrix} F_t \\ F_n \\ F_a \end{Bmatrix} \tag{4.6.9}$$

Example 4.6.1: Comparison of forces between square and ball endmills Here we compare the cutting forces produced by helical square and ball endmills. As in Ex. 4.5.1, we consider a 30% radial immersion up milling cut with a zero start angle and exit angle of 66.4 deg with an axial depth of 5 mm (which is less than the tools' radii of 9.5 mm). The dynamic responses in the two directions are the same as was previously presented. The workpiece material is an aluminum alloy and it is assumed that the cutting force coefficients are the same for both four tooth endmills. For a specific force value of K_s = 600 N/mm^2 and force angle of β = 60 deg, the corresponding cutting force coefficients are k_t = 520 N/mm^2

and $k_n = 300\ \mathrm{N/mm^2}$ and the axial coefficient, k_a, is taken to be equal to k_n. The feed per tooth is 0.15 mm/tooth.

For the simulations, let's use 1200 steps per revolution and complete 10 full revolutions. Results for the cutting forces in the x, y, and z directions under these conditions are displayed in Figs. 4.6.3 through 4.6.5, respectively (the time span includes data that follows the attenuation of the initial transients). Differences are observed in all three directions. This is due to the variation in the ball surface normal angle and the corresponding

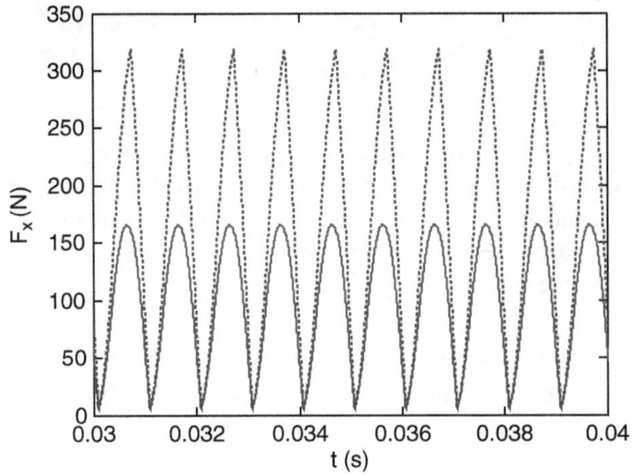

Fig. 4.6.3 Comparison of x direction cutting force for ball (solid line) and square (dotted line) helical endmills

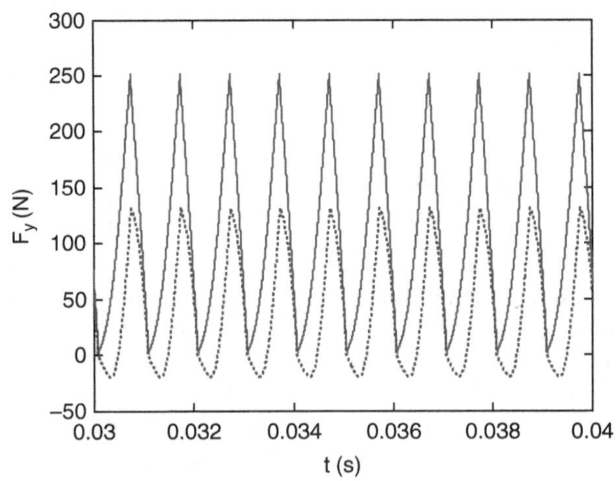

Fig. 4.6.4 Comparison of y direction cutting force for ball (solid line) and square (dotted line) helical endmills

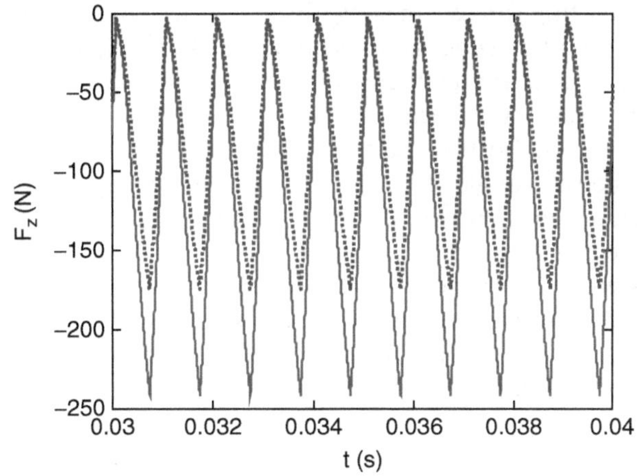

Fig. 4.6.5 Comparison of z direction cutting force for ball (solid line) and square (dotted line) helical endmills

projections of the normal and axial components according to Eq. 4.6.8. Naturally, the resultant force, $F = F_x^2 + F_y^2 + F_z^2$, is the same for both endmills. The MATLAB® programs p_4_6_1_1_.m (helical ball endmilling) and p_4_6_1_2_.m (helical square endmilling) may be used to obtain Figs. 4.6.3 through 4.6.5.

4.7 Experimental Cutting Force Coefficients

To conclude this chapter, let's discuss experimental determination of the cutting force coefficients for the milling force model (a square endmill geometry is again presumed). As shown in Fig. 4.1.8, these tests are carried out by prescribing a known feed per tooth and axial depth and measuring the x (feed), y, and z (axial) direction cutting force components in the dynamometer's fixed coordinate frame. Prior to detailing the numerical and experimental techniques used to obtain the coefficient values, however, let's revisit the force model and corresponding coefficients.

4.7.1 Updated Force Model

Equations 4.1.6 through 4.1.10 and 4.3.12 are all based on the assumption that the resultant cutting force, F in Fig. 4.1.7, is directly proportional to the chip thickness (and axial depth) and independent of other operating parameters. Although this is a reasonable assumption for the development of stability lobe diagrams used to guide the selection of preferred spindle speeds, comparison with measured cutting forces can show discrepancies with this approximation. First, because the cutting edge radius is nonzero, as the chip thickness is reduced to values near the edge radius there is increased rubbing[5] between the cutting edge and work surface which leads to a larger cutting force than the $F = K_s bh$ model would predict. With respect to chip formation in orthogonal cutting (Section 3.1), the rake angle, or inclination of the cutting edge relative to the surface normal, depends on both the chip thickness and edge radius in this case and can become negative if the ratio is unity or smaller, although the tool may have been designed with a positive rake angle at the macroscopic level. This rubbing phenomenon can be incorporated in the force model by adding a constant, or DC, force component that scales with the axial depth only, such that the force is nonzero even as the chip thickness approaches zero. This is referred to as the "edge effect" in [13] and threshold value in [14]. Additionally, the force may not linearly increase with chip thickness, but could follow a power law such as $F = K_s bh^c$, where c is a positive constant less than one [14]. Second, due to the variation in strain rate, temperature, and subsequent chip formation behavior, decreases in cutting force can be observed for higher cutting speeds [15]. This imposes a spindle speed dependence on the cutting force coefficients. Both of these issues (rubbing effect and cutting speed dependence) highlight the notion that these coefficients should not be considered material properties of the workpiece, but rather as approximate descriptions of the process behavior which depend on the tool-workpiece combination and, to a lesser extent, the operating conditions. Additional investigations of cutting force modeling may be reviewed in [16–26].

In order to improve the accuracy of the cutting force model without introducing significant complexity, let's augment Eqs. 4.1.7 and 4.1.8 to incorporate edge effects. Equations 4.7.1 and 4.7.2 now each include two coefficients: one is associated with "cutting" (or shearing) and includes the chip thickness dependence; and the other is the "rubbing" (or plowing) term, which is independent of chip thickness (denoted by the "e" subscript extension). We note that the DC rubbing terms, $k_{ne}b$ and $k_{te}b$, would be neglected in our analytical linear stability analyses anyway, so this does not invalidate our previous stability lobe diagram derivations. We have also added the axial force term in Eq. 4.7.3, which was neglected in our straight tooth analysis.

$$F_n = k_n bh + k_{ne} b \tag{4.7.1}$$

$$F_t = k_t bh + k_{te} b \tag{4.7.2}$$

$$F_a = k_a bh + k_{ae} b \tag{4.7.3}$$

[5] Plowing (alternately ploughing), or plastic deformation without material removal, can also occur for significant interference between the tool relief surface and workpiece.

We determine the six coefficients via linear regression using the average cutting forces measured by the dynamometer over a range of feed per tooth values. Projection of the normal and tangential components into the x, y, and z directions for the square endmill geometry, as shown in Fig. 4.1.8, gives Eqs. 4.7.4 through 4.7.6.

$$F_x = k_t b f_t \sin(\phi)\cos(\phi) + k_{te} b \cos(\phi) + k_n b f_t \sin^2(\phi) + k_{ne} b \sin(\phi) \quad (4.7.4)$$

$$F_y = k_t b f_t \sin^2(\phi) + k_{te} b \sin(\phi) - k_n b f_t \sin(\phi)\cos(\phi) - k_{ne} b \cos(\phi) \quad (4.7.5)$$

$$F_z = -k_a b f_t \sin(\phi) - k_{ae} b \quad (4.7.6)$$

If we apply the double angle identities: $\sin(\phi)\cos(\phi) = \frac{\sin(2\phi)}{2}$ and $\sin^2(\phi) = \frac{1-\cos(2\phi)}{2}$, we obtain Eqs. 4.7.7 and 4.7.8. The z direction force equation is unchanged.

$$F_x = k_t b f_t \frac{\sin(2\phi)}{2} + k_{te} b \cos(\phi) + k_n b f_t \frac{(1-\cos(2\phi))}{2} + k_{ne} b \sin(\phi) \quad (4.7.7)$$

$$F_y = k_t b f_t \frac{(1-\cos(2\phi))}{2} + k_{te} b \sin(\phi) - k_n b f_t \frac{\sin(2\phi)}{2} - k_{ne} b \cos(\phi) \quad (4.7.8)$$

To determine the mean cutting force per revolution, we must first augment the previous equations with the summation that accounts for all teeth on the cutter and the switching function that is nonzero only when the tooth angle is bounded by the cut start and exit angles.

$$F_x = \sum_{j=1}^{N_t} \left(k_t b f_t \frac{\sin(2\phi_j)}{2} + k_{te} b \cos(\phi_j) + k_n b f_t \frac{(1-\cos(2\phi_j))}{2} + k_{ne} b \sin(\phi_j) \right) g(\phi_j) \quad (4.7.9)$$

$$F_y = \sum_{j=1}^{N_t} \left(k_t b f_t \frac{(1-\cos(2\phi_j))}{2} + k_{te} b \sin(\phi_j) - k_n b f_t \frac{\sin(2\phi_j)}{2} - k_{ne} b \cos(\phi_j) \right) g(\phi_j) \quad (4.7.10)$$

$$F_z = \sum_{j=1}^{N_t} (-k_a b f_t \sin(\phi_j) - k_{ae} b) g(\phi_j) \quad (4.7.11)$$

The mean force per revolution in the x direction, for example, is then determined by $\bar{F}_x = \frac{1}{2\pi} \int_{\phi_s}^{\phi_e} F_x d\phi$. Because the integration limits are set between the start and exit angles, the switching function is always equal to one and is effectively removed from the integral. Also, the summation is incorporated by the multiplication of the integral by N_t as shown in Eq. 4.7.12. This equation is rewritten in Eq. 4.7.13.

$$\bar{F}_x = \frac{N_t}{2\pi}\int_{\phi_s}^{\phi_e}\left(k_t b f_t \frac{\sin(2\phi)}{2} + k_{te} b\cos(\phi) + k_n b f_t \frac{(1-\cos(2\phi))}{2} + k_{ne} b\sin(\phi)\right)d\phi \quad (4.7.12)$$

$$\bar{F}_x = \frac{N_t b}{4\pi}\int_{\phi_s}^{\phi_e}(k_t f_t\sin(2\phi) + 2k_{te}\cos(\phi) + k_n f_t - k_n f_t\cos(2\phi) + 2k_{ne}\sin(\phi))d\phi \quad (4.7.13)$$

By application of $\int\sin(ax)dx = -\frac{1}{a}\cos(ax)$ and $\int\cos(ax)dx = \frac{1}{a}\sin(ax)$, the integral in Eq. 4.7.13 is determined to be:

$$\bar{F}_x = \frac{N_t b}{4\pi}\left[-\frac{k_t}{2}f_t\cos(2\phi) + 2k_{te}\sin(\phi) + k_n f_t\phi - \frac{k_n}{2}f_t\sin(2\phi) - 2k_{ne}\cos(\phi)\right]_{\phi_s}^{\phi_e}. \quad (4.7.14)$$

which can be rewritten as shown in Eq. 4.7.15.

$$\bar{F}_x = \left[\frac{N_t b f_t}{8\pi}(-k_t\cos(2\phi) + k_n(2\phi - \sin(2\phi))) + \frac{N_t b}{2\pi}(k_{te}\sin(\phi) - k_{ne}\cos(\phi))\right]_{\phi_s}^{\phi_e} \quad (4.7.15)$$

Similarly, the y and z direction mean cutting forces per revolution are:

$$\bar{F}_y = \left[\frac{N_t b f_t}{8\pi}(k_t(2\phi - \sin(2\phi)) + k_n\cos(2\phi)) - \frac{N_t b}{2\pi}(k_{te}\cos(\phi) + k_{ne}\sin(\phi))\right]_{\phi_s}^{\phi_e} \quad (4.7.16)$$

and

$$\bar{F}_z = \left[\frac{N_t b}{2\pi}(k_a f_t\cos(\phi) - k_{ae}\phi)\right]_{\phi_s}^{\phi_e}. \quad (4.7.17)$$

If we select 100% radial immersion (slotting) for the cutting tests, then $\phi_s = 0$ and $\phi_e = 180$ deg and Eqs. 4.7.15 through 4.7.17 simplify to:

$$\bar{F}_x = \frac{N_t b k_n}{4}f_t + \frac{N_t b k_{ne}}{\pi}, \quad (4.7.18)$$

$$\bar{F}_y = \frac{N_t b k_t}{4}f_t + \frac{N_t b k_{te}}{\pi}, \quad (4.7.19)$$

and

$$\bar{F}_z = -\frac{N_t b k_a}{\pi}f_t - \frac{N_t b k_{ae}}{2}. \quad (4.7.20)$$

4.7.2 Linear Regression

Given these expressions, we complete linear regressions (over chip thickness) to determine the six unknown cutting force coefficients: k_n, k_{ne}, k_t, k_{te}, k_a, and k_{ae} from measured (mean) force values. Note that the first term on the right hand side of the average force expressions in Eqs. 4.7.18 through 4.7.20 is a function of the feed per tooth, while the second term is not. These equations therefore match the (linear) slope-intercept form if f_t is the independent variable and the mean force is the dependent variable. In the x direction, for example, the slope is $\frac{N_t b k_n}{4}$ and the intercept is $\frac{N_t b k_{ne}}{\pi}$. See Fig. 4.7.1.

The form of the linear regression for the x direction is $\bar{F}_{x,i} = a_{0x} + a_{1x} f_{t,i} + E_i$, where $(f_{t,i}, \bar{F}_{x,i})$ are the data pairs, a_{0x} is the intercept, a_{1x} is the slope, and E_i is the error between the measured $\bar{F}_x$ values and the line $a_{0x} + a_{1x} f_t$. For $n > 2$ data pairs, the slope and intercept are determined by minimizing the sum of the errors squared [14].

$$\sum_{i=1}^{n} E_i^2 = \sum_{i=1}^{n} \left(\bar{F}_{x,i} - a_{0x} - a_{1x} f_{t,i}\right)^2$$

The slope and intercept expressions obtained from this minimization are provided in Eqs. 4.7.21 and 4.7.22.

$$a_{1x} = \frac{n \sum_{i=1}^{n} f_{t,i} \bar{F}_{x,i} - \sum_{i=1}^{n} f_{t,i} \sum_{i=1}^{n} \bar{F}_{x,i}}{n \sum_{i=1}^{n} f_{t,i}^2 - \left(\sum_{i=1}^{n} f_{t,i}\right)^2} \tag{4.7.21}$$

$$a_{0x} = \frac{1}{n} \sum_{i=1}^{n} \bar{F}_{x,i} - a_{1x} \frac{1}{n} \sum_{i=1}^{n} f_{t,i} \tag{4.7.22}$$

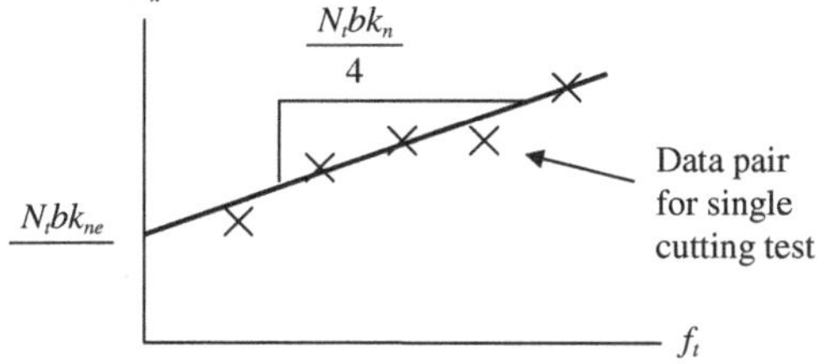

Fig. 4.7.1 Graph of Eq. 4.7.18, which relates the feed per tooth to the x direction mean force per revolution for slotting

To determine the quality of the linear fit to the data, we can calculate the coefficient of determination, r^2, where r is the correlation coefficient. The r^2 value describes how well the original uncertainty is explained by the linear model. For example, if $r^2 = 0.95$, then the line captures 95% of the data behavior. (There are exceptions where a high r^2 value does not guarantee a successful fit, but a visual analysis of the data and line is sufficient to identify these situations [14].) See Eq. 4.7.23, which is again specific to the x direction.

$$r_x^2 = \frac{\sum_{i=1}^{n}\left(\bar{F}_{x,i} - \frac{1}{n}\sum_{i=1}^{n}\bar{F}_{x,i}\right)^2 - \sum_{i=1}^{n}E_i^2}{\sum_{i=1}^{n}\left(\bar{F}_{x,i} - \frac{1}{n}\sum_{i=1}^{n}\bar{F}_{x,i}\right)^2} \tag{4.7.23}$$

Once the slope and intercept values are determined from the linear regressions for the x, y, and z direction mean force data (for slotting conditions), the cutting force coefficients are determined from Eqs. 4.7.24 through 4.7.26. In these expressions, the first a subscript denotes slope (1) or intercept (0), while the second subscript indicates the measurement direction (x, y, or z) as shown in Eqs. 4.7.21 and 4.7.22.

$$k_n = \frac{4a_{1x}}{N_t b} \quad k_{ne} = \frac{\pi \cdot a_{0x}}{N_t b} \tag{4.7.24}$$

$$k_t = \frac{4a_{1y}}{N_t b} \quad k_{te} = \frac{\pi \cdot a_{0y}}{N_t b} \tag{4.7.25}$$

$$k_a = -\frac{\pi \cdot a_{1z}}{N_t b} \quad k_{ae} = -\frac{2a_{0z}}{N_t b} \tag{4.7.26}$$

Example 4.7.1: Determination of cutting force coefficients To demonstrate the linear regression procedure for obtaining cutting force coefficient values from a particular tool-workpiece material pair, the MATLAB® program p_4_7_1_1.m is included on the companion CD. In this program, the cutting forces are first generated for the following conditions: $k_n = 190$ N/mm^2, $k_{ne} = 5$ N/mm, $k_t = 710$ N/mm^2, $k_{te} = 4$ N/mm, $k_a = 95$ N/mm^2, and $k_{ae} = 2$ N/mm; $b = 5$ mm, $f_t = \{0.05, 0.1, 0.15, 0.2, \text{and } 0.25\}$ mm/tooth, $\Omega = 5000$ rpm, $\phi_s = 0$, and $\phi_e = 180$ deg; and $N_t = 2$, $\beta = 30$ deg, and $d = 19$ mm for the square endmill. Figure 4.7.2 shows the forces for a single cutter revolution and feed per tooth of 0.25 mm, where (mean zero) Gaussian noise has been added. The mean forces for the x, y, and z directions are 136.1 N, 458.5 N, and −86.4 N, respectively. After the forces are generated and the mean values computed, the linear regression is completed to determine the slopes and intercepts for the x, y, and z directions; see Fig. 4.7.3. They are then used, together with the preselected N_t and b values, to determine the six cutting force coefficients as shown in

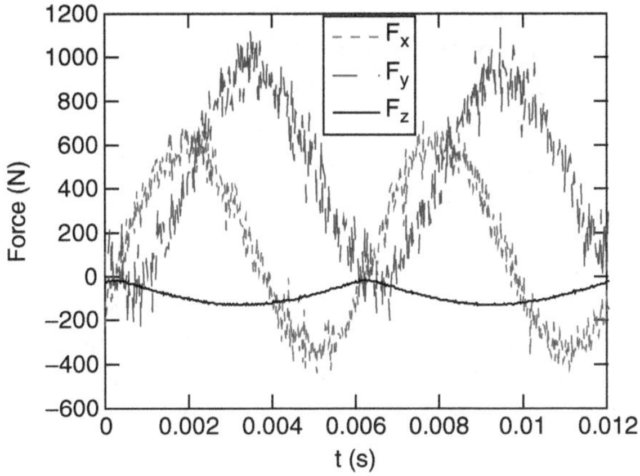

Fig. 4.7.2 Simulated cuttings forces for linear regression demonstration in EX. 4.7.1. The feed per tooth for the slotting cut is 0.25 mm/tooth

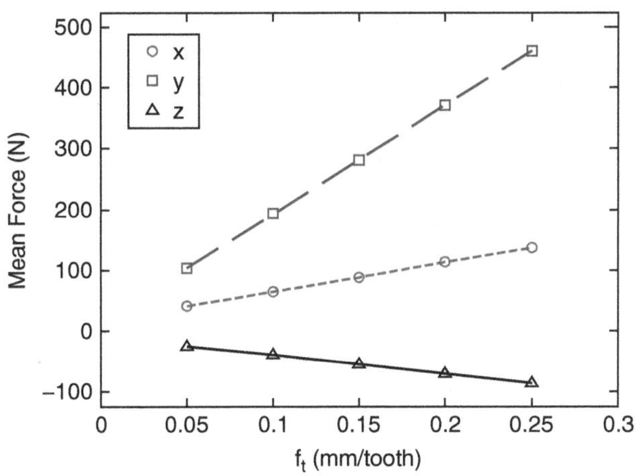

Fig. 4.7.3 Linear regression results for simulated forces in Ex. 4.7.1

Eqs. 4.7.24 through 4.7.26. Finally, the coefficients of determination are calculated.

4.7.3 Experimental Techniques

The required forces are typically measured using a table top cutting force dynamometer. The dynamometer is mounted to the machine table and aligned with the feed direction. Important considerations for force measurement include the force signal magnitudes and frequency content. As with any digital data collection, the range of the x, y, and z direction data acquisition channels should be selected so that adequate resolution is achieved. To set the maximum and minimum signal levels prior to machining, the simulation provided in p_4_7_1_1.m, for example, may be used with assumed coefficient values for the tool-workpiece material pair. Table 3.1.1 can be useful in estimating the cutting force coefficients.

Regarding frequency content, a dynamometer is also a dynamic system and therefore has its own associated frequency response. Naturally, if the cutting force has frequency content near a natural frequency of the dynamometer, then this content will be artificially amplified and the force data will be corrupted by the dynamometer dynamic response. The tooth passing frequency, defined in Eq. 4.1.14, provides a lower bound on the desired dynamometer bandwidth (i.e., the location of the first natural frequency for the dynamometer-workpiece combination). However, as shown in Fig. 4.1.15, multiple harmonics of the tooth passing frequency may be present, depending on the number of teeth and radial immersion. To determine the frequency response for a particular dynamometer-workpiece setup, impact testing can be applied. (It is necessary to include the mounted workpiece because it effectively mass loads the dynamometer and reduces it first natural frequency.) The impact hammer is used to excite the dynamometer in the x, y, and z directions and the corresponding force output is measured. The frequency-domain ratio of the output to input force can then be calculated and analyzed. To avoid data corruption by the dynamometer, the tooth passing frequency can be set such that there is no appreciable content at or beyond the first natural frequency of the dynamometer-workpiece combination. However, because decreases in cutting forces can be observed for higher cutting speeds [15], it may be preferable to test in the anticipated spindle speed range. In this case, one option is to filter the measured force using the inverse of the dynamometer force-to-input force frequency response. See, for example, the method proposed in [27]. An inverse filtering approach for a spindle-based torque dynamometer is also described in [28].

As a final note, it should of course be ensured that the cutting tests are stable. As described in the previous sections, the tool point frequency response function in the x and y directions should be measured and used as input, together with assumed cutting force coefficients, to one or both of the analytical stability lobe algorithms in order to estimate a stable axial depth of cut for the selected radial depth (slotting in Eqs. 4.7.24 through 4.7.26), number of cutter teeth, and spindle speed.

IN A NUTSHELL Of course, what constitutes "significant complexity" is in the eye of the beholder. Most of the information about stable and unstable cutting conditions can be correctly deduced using the approximation that the cutting force is proportional to the frontal area of the chip through a single cutting force coefficient. Increasing the accuracy of the prediction requires increasing the complexity of the model and the amount of required information. Knowing where to draw the line in order to obtain an answer that is good enough is the essence of engineering. Different readers will have varying objectives and will draw that line in different places.

Exercises

1. Compute the start and exit angles for the following milling cases.
   a) Up milling, 30% radial immersion
   b) Down milling, 40% radial immersion
2. Determine the tooth passing frequency for a cutter with three teeth rotating at 10000 rpm.
3. Calculate the maximum y direction force for a 40% radial immersion down milling cut carried out using a three tooth end mill; see Fig. e.4.3. The material-tool combination gives: $k_t = 720\,\text{N/mm}^2$ and $k_n = 200\ \text{N/mm}^2$ (aluminum alloy). Also, $b = 2.5$ mm and $f_t = 0.2$ mm/tooth. Assume a rigid cutting tool and workpiece.

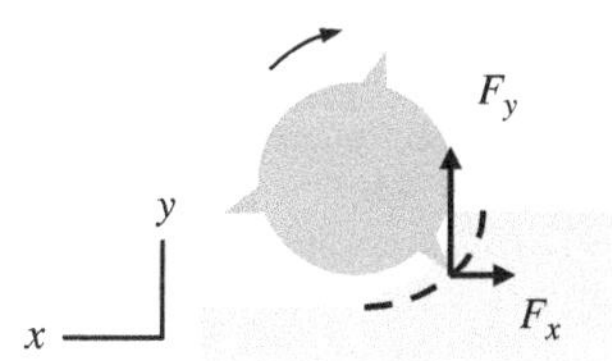

Fig. e.4.3 Down milling force geometry

4. For the average tooth angle milling stability analysis, complete parts a) through d). A 35% radial immersion up milling cut is to be performed using a square end mill with four teeth and the force angle, β, is 68 deg. See Fig. e.4.4.
   a) Determine the average angle of a tooth in the cut.
   b) Calculate the directional orientation factors.

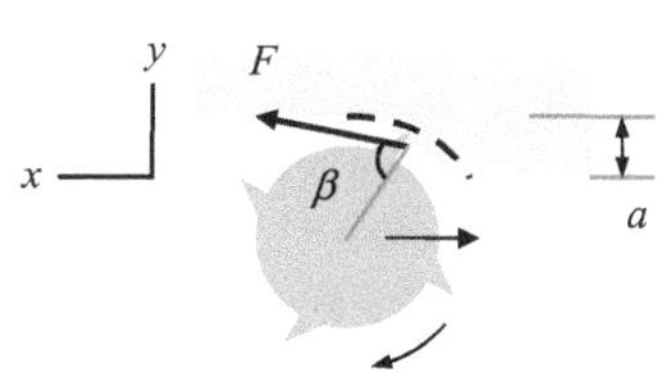

Fig. e.4.4 35% radial immersion up milling geometry

c) Compute the oriented frequency response function and identify the valid chatter frequency range(s) in Hz. The x direction dynamics are given by: $f_{nx} = 1000$ Hz, $k_x = 7 \times 10^6$ N/m, and $\zeta_x = 0.03$ and the y direction dynamics are: $f_{ny} = 1200$ Hz, $k_y = 6 \times 10^7$ N/m, and $\zeta_y = 0.04$. You may assume that these single degree of freedom parameters were obtained from a modal fit to frequency responses measured in the x and y directions with bandwidths of 0 to 2500 Hz.
d) Plot the first five stability lobes, $N = 0$ to 4. The workpiece material is 1020 carbon steel; see Table 3.1.1.

5. For the cut described in Exercise 4, use the Fourier approach to obtain the stability lobe diagram ($j = 0$ to 4).
6. Calculate the lowest axial depth for which a constant cutting force is obtained independent of the selected radial depth of cut. The helical square end mill has a diameter of 10 mm, six teeth, and a 42 deg helix angle.

References

1. Kumanchik, L. and Schmitz, T., 2007, Improved Analytical Chip Thickness Model for Milling, Precision Engineering, 31: 317–324.
2. Tlusty, J., W. Zaton, and F. Ismail, 1983, Stability Lobes in Milling, Annals of the CIRP, 32/1: 309–313.
3. Smith, S., and Tlusty, J., 1990, Update on High-Speed Milling Dynamics, Journal of Engineering for Industry, 112: 142–149.
4. Smith, S. and Tlusty, J., 1991, An Overview of Modeling and Simulation of the Milling Process, Journal of Engineering for Industry, 113: 169–175.
5. Altintas, Y. and Budak, E., 1995, Analytical Prediction of Stability Lobes in Milling, Annals of the CIRP, 44/1: 357–362.
6. Davies, M., Pratt, J., Dutterer, B., and Burns, T., 2000, The Stability of Low Radial Immersion Milling, Annals of the CIRP, 49/1: 37–40.
7. Davies, M., Pratt, J., Dutterer, B., and Burns, T., 2002, Stability Prediction for Low Radial Immersion Milling, Journal of Manufacturing Science and Engineering, 124/2: 217–225.
8. Bayly, P., Mann, B., Schmitz, T., Peters, D., Stépàn, G., Insperger, T., 2002, Effects of Radial Immersion and Cutting Direction on Chatter Instability in Endmilling, Proceedings of ASME International Mechanical Engineering Conference and Exposition, IMECE2002-34116, New Orleans, LA (on CD).
9. Mann, B., Bayly, P., Davies, M., and Halley, J., 2004, Limit Cycles, Bifurcations, and Accuracy of the Milling Process, Journal of Sound and Vibration, 227: 31–48.
10. Smith, S. and Tlusty, J., 1991, An Overview of Modeling and Simulation of the Milling Process, Journal of Engineering for Industry, 113: 169–175.
11. Tlusty, G., 2000, Manufacturing Equipment and Processes, Prentice-Hall, Upper Saddle River, NJ, Section 9.5.4.
12. Altintas, Y. and Lee, P., 1996, A General Mechanics and Dynamics Model for Helical End Mills, Annals of the CIRP, 45/1: 59–64.
13. Altintas, Y., 2000, Manufacturing Automation: Metal Cutting Mechanics, Machine Tool Vibrations, and CNC Design, Cambridge University Press, Cambridge, Section 2.8.1.
14. Chapra, S. and Canale, R., 1985, Numerical Methods for Engineers, McGraw-Hill Book Co., New York, NY, Section 10.1.

15. Duncan, G.S., 2006, Milling Dynamics Prediction and Uncertainty Analysis using Receptance Coupling Substructure Analysis, Ph.D. Dissertation, University of Florida, http://highspeedmachining.mae.ufl.edu/htmlsite/duncan_g.pdf.
16. Koenisgberger, F. and Sabberwal, A., 1961, An Investigation into the Cutting Force Pulsations during Milling Operations, International Journal of Machine Tool Design and Research, 1: 15–33.
17. Tlusty, J. and MacNeil, P., 1975, Dynamics of Cutting Forces in End Milling, Annals of the CIRP, 24/1: 21–25.
18. Kline, W., DeVor, R., and Lindberg, J., 1982, The Prediction of Cutting Forces in End Milling with Applications to Cornering Cuts, International Journal of Machine Tool Design and Research, 22: 7–22.
19. Fu, H., DeVor, R., and Kapoor, S., 1984, A Mechanistic Model for the Prediction of the Force System in Face Milling Operations, Journal of Engineering for Industry, 111: 27–36.
20. Yellowley, I., 1985, Observations of the Mean Values of Forces, Torque and Specific Power in the Peripheral Milling Process, International Journal of Machine Tool Design and Research, 25: 337–346.
21. Amarego, E. and Whitfield, R., 1985, Computer Based Modeling of Popular Machining Operations for Forces and Power Prediction, Annals of the CIRP, 34: 65–69.
22. Wang, J.-J., Liang, S., and Book, W., 1994, Convolution Analysis of Cutting Force Pulsation, Journal of Engineering for Industry, 116: 17–25.
23. Endres, W., DeVor, R., and Kapoor, S., 1995, A Dual-Mechanism Approach to the Prediction of Machining Forces, Part 1: Model Development, Journal of Engineering for Industry, 117: 526–533.
24. Budak, E., Altintas, Y., and Amarego, E., 1998, Prediction of Milling Force Coefficients from Orthogonal Cutting Data, Journal of Manufacturing Science and Engineering, 118: 216–224.
25. Jayaram, S., Kapoor, S., and DeVor, R., 2001, Estimation of the Specific Cutting Pressures for Mechanistic Cutting Force Models, International Journal of Machine Tools and Manufacture, 41: 265–281.
26. Wang, J.-J., and Zheng, C., 2002, An Analytical Force Model with Shearing and Ploughing Mechanisms for End Milling, International Journal of Machine Tools and Manufacture, 42: 761–771.
27. Jensen, S., Shin, Y., and Davies, P., 1996, Inverse Filtering of Unwanted System Dynamics in Cutting Force Measurement, Proceedings of the Dynamic Systems and Control Division, ASME International Mechanical Engineering Conference and Exposition, Atlanta, GA, pp. 167–174.
28. Smith, D., Smith, S., and Tlusty, J., 1998, High Performance Milling Torque Sensor, Journal of Manufacturing Science and Engineering, 120/3: 504–514.

Chapter 5
Surface Location Error in Milling

No amount of experimentation can ever prove me right; a single experiment can prove me wrong.

- Albert Einstein

In Chapter 4 we described analytical approaches to determining the stability behavior of milling. We showed stability lobe diagrams that identify stable and unstable combinations of spindle speed and axial depth of cut in a graphical format. We also developed time-domain simulations that predict forces and displacements for straight and helical teeth square endmills and helical teeth ball endmills. These simulations could also be employed to determine stability. In this chapter, we presume that stable cutting conditions have been selected and we investigate the influence of forced vibrations on part geometric accuracy. We refer to part errors that occur due to forced vibrations as surface location errors and again apply both analytical frequency-domain and time-domain approaches to their prediction.

5.1 Surface Location Error

As we've seen, the process dynamics can impose significant limitations on milling efficiency due to chatter, or self-excited vibrations that lead to large forces, displacements, and poor surface quality. However, productivity can also be limited by forced vibrations which cause surface location error, or workpiece geometric inaccuracies that result from dynamic displacements of the tool during stable milling [1-14]. Other limiting factors include, for example, machine tool quasi-static positioning errors, thermal errors, contouring errors, and tool wear [15], but we do not address these here.

A visual explanation of the surface location error phenomenon is provided in Fig. 5.1.1. Even under stable cutting conditions, the tool experiences forced/ synchronous vibrations which depend on the system frequency response

T.L. Schmitz, K.S. Smith, *Machining Dynamics*, DOI 10.1007/978-0-387-09645-2_5,

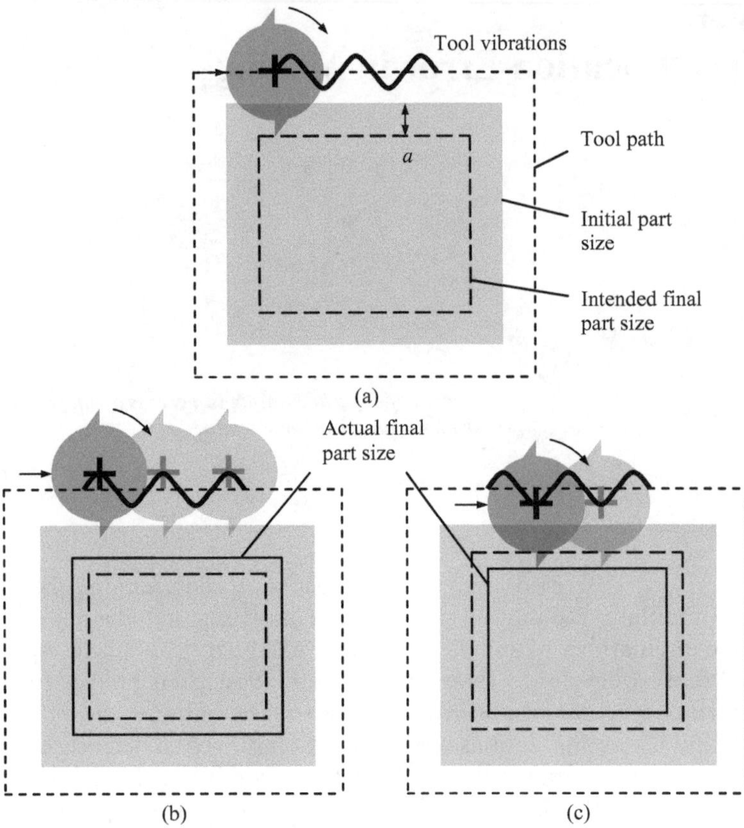

Fig. 5.1.1 Demonstration of surface location error. (a) Intended down milling cut geometry. (b) Undercut example. (c) Overcut example

function, FRF, and excitation frequency (or tooth passing frequency, see Eq. 4.1.14), as well as other process parameters, including the radial and axial depths of cut, feed per tooth, and force model coefficients (although the actual cutting force doesn't inherently obey our selected force model, we do require a model with corresponding coefficients for simulation purposes). In Fig. 5.1.1a, we conveniently assume that the tool vibration follows a sinusoidal profile in the feed direction while peripheral milling a square shape. The position of the tool in its periodic vibration cycle as is exits the down milling cut determines the actual location of the machined surface[1]. In Fig. 5.1.1b, due to the selected tooth passing frequency, the surface is undercut, i.e., less material is removed

[1] This also explains why a vibrating tool can leave a smooth surface. Because the tool vibration is synchronous with rotation, the tool is in the same position each time it creates the new surface.

than commanded. In this case, the tool vibration is away from the intended surface at each instant that a tooth is exiting the cut (only these cutter angles are shown and the lateral scale is greatly exaggerated). Figure 5.1.1c represents the overcut condition where more material is removed than commanded. Now the tool vibration is toward the intended surface when the tool is exiting the cut. Analogous representations could be provided for up milling, except we are interested in the tool location as it enters the cut.

The source of the surface location error behavior is the variation of the machine-spindle-holder-tool, and potentially the workpiece-fixture, FRF magnitude and phase with forcing frequency (as shown in Fig. 2.2.3, for example). Due to the change in phase with tooth passing frequency (spindle speed), the time lag between the force and vibration varies. Therefore, the location of the cutter in its vibration cycle when leaving the surface depends on the selected spindle speed. The dependence of surface location error on the phase lag between the forcing function and displacement causes significant variation near the natural frequency (considering a single degree of freedom system for simplicity) because for the lowly damped tool point FRFs typically observed in practice, the phase changes rapidly in this frequency range. To explore this behavior, we first describe a frequency-domain solution to surface location error and then demonstrate the spindle speed dependence with numerical examples.

IN A NUTSHELL The concept of surface location error may be surprising to some readers. In CNC programming packages, the tool is modeled as a cylinder (or perhaps a cylinder with a spherical end), the workpiece is modeled as a prismatic solid, and the workpiece is created by the relative motion of the cylinder with respect to the solid. However, the tool is not a cylinder. It is a collection of cutting edges that rotate together.

It is obvious to many that the tool and workpiece are not rigid. This is certainly one of the reasons for making a roughing pass followed by a finishing pass. This intuition reveals the static surface location error. However, actual surface generation is more complicated than a static deflection. The variable cutting force, even in stable machining, causes the tool to vibrate. The surface location is controlled by the position of the tool in its cycle of vibration at the time that a tooth is in a position to generate the final surface. The tool may exhibit large vibrations, yet still produce an accurately located surface. Alternately, these large vibrations may produce significant errors in the surface location.

By analogy, we might imagine that the vibrating tool is like a swing on a playground. The surface generation is like the moment of contact between the pusher and the swing. The pusher can stand far behind the swing and push as the swing reaches its peak displacement (poorly located surface) or the pusher may stand to the side and push the swing as it passes through its equilibrium position (perfectly located surface).

Of course, the cutting operation is more complicated. Naturally, the surface location error is spindle speed dependant. The spindle speed sets the frequency of the force exciting the vibration and, therefore, defines the resulting displacement according to the frequency response function. In addition, if the cutter teeth are helical, then different levels of the surface are generated at different instants in time. The surface location error along a line parallel to the tool's rotation axis, therefore, varies from the tip of the tool to the full axial depth of cut.

5.2 Frequency-Domain Solution

In chapter 4 we described the average tooth angle [16] and Fourier series [17] approaches to stability behavior prediction in milling. Both were frequency-domain methods that relied on knowledge of the FRF and force model coefficients. To complement these analytical tools, we now describe a frequency-domain solution to surface location error [18]. Together, these provide a comprehensive picture of the role of milling dynamics in process productivity.

In order to determine surface location error using a frequency-domain (or steady-state) approach, we make two basic assertions. First, although vibrations of the cutter occur in both the x and y directions, the y direction vibrations dominate the final surface location for an x direction feed. Second, regeneration can be neglected in stable machining. Based on these assumptions, the concept is to: 1) express the y direction cutting force in the frequency-domain, $F_y(\omega)$ using a Fourier series; 2) determine the frequency-domain y displacement, $Y(\omega)$, by multiplying $F_y(\omega)$ by the machine-spindle-holder-tool direct FRF (measured or modeled at the tool point) in the y direction, $\frac{Y(\omega)}{F_y(\omega)}$; and 3) inverse Fourier transform this result and sample at the cut entry (up milling) or exit (down milling) to find the surface location error. Note that, unlike time-domain simulation, the tool point FRF can be used directly without the requirement for a modal fit in this approach.

5.2.1 Fourier Force Model

If we apply the cutting force model provided in Eq. 5.2.1, which relates the tangential, F_t, and normal, F_n, cutting force components to the axial depth of cut, b, and chip thickness, h, $F_y(\phi)$ can be expressed as shown in Eq. 5.2.2. In this equation, the summations account for all possible teeth within the cut, a circular tool path is assumed, f_t is the feed per tooth, and $g(\phi_i)$ is the switching function previously defined in Eq. 4.3.10. Also, the angle of each tooth, i (N_t total), at any instant in time is $\phi_i = \omega t + \frac{2\pi}{N_t}(i-1)$ (rad), where ω is the spindle rotating frequency (in rad/s).

$$F_t(\phi) = k_t b h(\phi) + k_{te} b$$
$$F_n(\phi) = k_n b h(\phi) + k_{ne} b \tag{5.2.1}$$

$$F_y(\phi) = -b\begin{bmatrix} \frac{-k_t f_t}{2}\sum_{i=1}^{N_t} g(\phi_i)(1-\cos(2\phi_i)) + \frac{k_n f_t}{2}\sum_{i=1}^{N_t} g(\phi_i)\sin(2\phi_i) - \\ k_{te}\sum_{i=1}^{N_t} g(\phi_i)\sin(\phi_i) + k_{ne}\sum_{i=1}^{N_t} g(\phi_i)\cos(\phi_i) \end{bmatrix} \tag{5.2.2}$$

The equivalent Fourier series for the y direction cutting force can be written once the Fourier coefficients, a_n and b_n, are determined.

$$F_y(\phi) = \sum_{i=1}^{N_t}\left(a_0 + \sum_{n=1}^{\infty}(a_n\cos(n\phi_i) + b_n\sin(n\phi_i))\right)$$

The a_0 term, for example, can be found using Eq. 5.2.3, where the integral for a full revolution of the selected tooth may be divided into three parts. The three subsequent integrals are delineated by ϕ_1, which represents the cut entry angle in down milling or cut exit angle in up milling, and π rad, which defines the maximum angle that a tooth can be engaged in the cut (if ϕ is defined positive in a clockwise sense from the positive y axis). See Fig. 5.2.1. Considering a down milling cut, for example, only the middle of the three integrals in Eq. 5.2.3 is nonzero due to the switching function embedded in $F_y(\phi)$. Performing the relevant integration for down milling yields Eq. 5.2.4. For up milling, only the first integral in Eq. 5.2.3 is nonzero and the integration limits become zero to ϕ_1 in Eq. 5.2.4.

$$a_0 = \frac{1}{2\pi}\int_0^{2\pi} F_y(\phi)d\phi = \frac{1}{2\pi}\left[\int_0^{\phi_1} F_y(\phi)d\phi + \int_{\phi_1}^{\pi} F_y(\phi)d\phi + \int_{\pi}^{2\pi} F_y(\phi)d\phi\right] \tag{5.2.3}$$

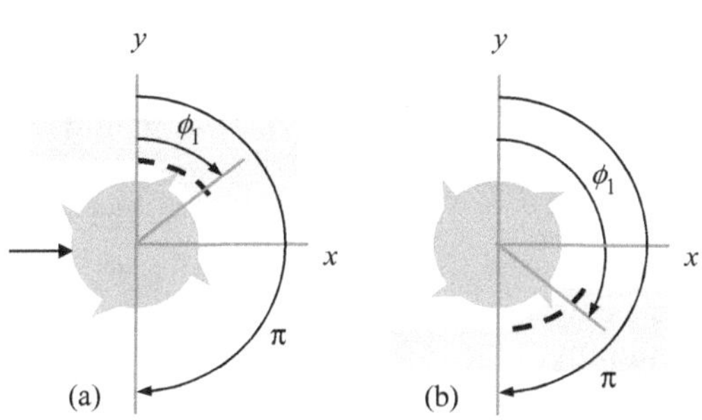

Fig. 5.2.1 Angles for Fourier series integrals. (a) Up milling. (b) Down milling

$$a_0 = -\frac{bN_t}{2\pi}\left[-\frac{k_t f_t \phi}{2} + \frac{k_t f_t}{4}\sin 2\phi - \frac{k_n f_t}{4}\cos 2\phi + k_{te}\cos\phi + k_{ne}\sin\phi\right]_{\phi_1}^{\pi} \quad (5.2.4)$$

The a_n coefficients are computed using Eq. 5.2.5 and the b_n coefficients using Eq. 5.2.6. Again, the integrals can be partitioned using ϕ_1 and π as shown in Eq. 5.2.3. Closed-form equations for the $n = 3, 4, 5, \ldots$ coefficients are determined by observing the recursive patterns after integration. See Appendix C.

$$a_n = \frac{1}{\pi}\int_0^{2\pi} F_y(\phi)\cos(n\phi)d\phi \quad (5.2.5)$$

$$b_n = \frac{1}{\pi}\int_0^{2\pi} F_y(\phi)\sin(n\phi)d\phi \quad (5.2.6)$$

To accurately represent milling forces, however, it is also necessary to account for the influence of the teeth helix angle, γ. This can be accomplished by sectioning the tool into A axial slices. Each slice is assumed to have a zero helix angle and the slices are rotated relative to one another by the angle $\chi = \frac{2db\tan(\gamma)}{d}$ (rad), where db is the slice height and d is the cutter diameter (see Fig. 4.5.3). The Fourier series is now written as $F_y(\phi) = \sum_{j=1}^{A}\sum_{i=1}^{N_t}\left(a_0 + \sum_{n=1}^{\infty}(a_n\cos(n\phi_i) + b_n\sin(n\phi_i))\right)$, where $\phi_i = \omega t + \frac{2\pi}{N_t}(i-1) - \chi(j-1)$. Naturally, a larger number of slices improves the force fidelity.

Example 5.2.1: Fourier series force dependence on number of coefficients As expected, the accuracy of the Fourier series force depends on the number, n, of coefficients included in the series. Figure 5.2.2 shows a comparison of the y direction force determined by time-domain simulation and its Fourier series for $n = 5$ and $n = 50$. The cutting conditions are: up milling, 25% radial immersion, $N_t = 4$, $d = 19$ mm, $\gamma = 30$ deg, $b = 1$ mm, $f_t = 0.1$ mm/tooth, $k_t = 700$ N/mm^2, $k_n = 210$ N/mm^2, $k_{te} = k_{ne} = 0$ N/mm, and $\omega = 300\pi$ rad/s (i.e., the spindle speed, Ω, is 9000 rev/min, or rpm). Figure 5.2.2 was generated using the MATLAB® program p_5_2_1_1.m included on the companion CD. The MATLAB® function `eval`, used to evaluate strings, was implemented to enable an arbitrary number of Fourier coefficients to be computed without requiring significant reprogramming.

Example 5.2.2: Frequency-domain surface location error calculations The four steps for frequency-domain surface location error prediction, namely:

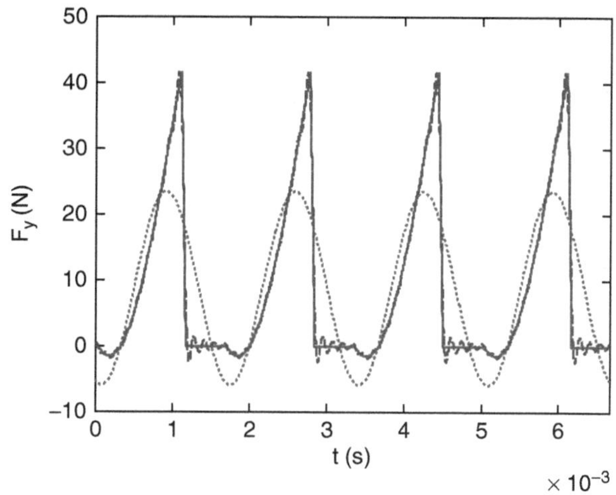

Fig. 5.2.2 Force reconstruction using time-domain (solid line) and Fourier series (dotted line: $n = 5$; dashed line: $n = 50$). Additional coefficients increase the force fidelity

- using a Fourier series to determine $F_y(t)$ and applying the discrete Fourier transform to obtain $F_y(\omega)$;
- calculating the frequency-domain y displacement using $Y(\omega) = \frac{Y(\omega)}{F_y(\omega)} F_y(\omega)$, where $\frac{Y(\omega)}{F_y(\omega)}$ is the y direction machine-spindle-holder-tool direct FRF (at the tool point);
- inverse Fourier transforming $Y(\omega)$ to obtain $y(t)$; and
- sampling it at the cut entry (for up milling) or exit (for down milling), are carried out in p_5_2_2_1.m. The function, p_5_2_2_2.m, called from within p_5_2_2_1.m, actually completes the individual surface location error computations.

To compare the surface location error (SLE) in up and down milling, simulations were completed for the following conditions: 50% radial immersion (up and down milling) at spindle speeds from 6900 rpm to 7800 rpm, $N_t = 4$, $\gamma = 30$ deg, $d = 12.7$ mm diameter, $f_t = 0.1$ mm/tooth, $b = 1$ mm, $k_t = 700$ N/mm^2, $k_n = 210$ N/mm^2, $k_{te} = k_{ne} = 0$ N/mm, symmetric structural dynamics with a stiffness of 1×10^7 N/m, 1% damping (i.e., a damping ratio, ζ, of 0.01), and 500 Hz natural frequency. The results are displayed in Fig. 5.2.3, where 50 terms were used in the Fourier series force model. It is seen that the surface location error yields an overcut surface in both instances. For up milling, the positive SLE means that cutter is radially deeper into the cut than commanded when creating the final surface. Similarly, for down milling, the negative SLE indicates that the cutter is farther into the cut than desired.

Because the frequency-domain surface location error simulation neglects regeneration, stable conditions are predicted in all instances (i.e., only forced vibrations are considered). Therefore, these calculations must be accompanied by the appropriate stability lobe diagram to select stable machining parameters. The Fourier series approach stability lobe diagram corresponding to the

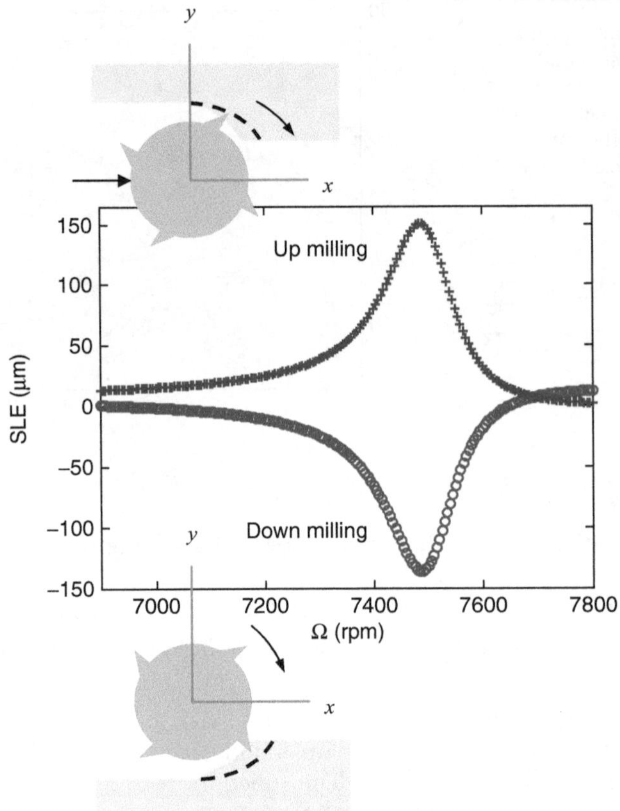

Fig. 5.2.3 Surface location error (SLE) for Ex. 5.2.2. Both up and down milling lead to an overcut condition

selected up milling scenario is provided in Fig. 5.2.4. The stability lobe diagram for 50% radial immersion down milling is nearly identical and is not shown. The test range for the surface location error simulations is indicated by the solid line. It is verified that the selected cutting conditions for Fig. 5.2.3 are in the stable zone. Figure 5.2.4 was generated using the MATLAB® program p_5_2_2_3.m.

To compare surface location error trends over a broader range, calculations were completed using the same system under 50% radial immersion up milling for spindle speeds from 6900 rpm to 7700 rpm and axial depths from 0.5 mm to 5 mm. See the dotted rectangle in Fig. 5.2.4. The surface location error contours (lines of constant error) are provided in Fig. 5.2.5, where 50 terms were again used for the Fourier series force. An interesting aspect of this figure is the high sensitivity of surface location error to spindle speeds near 7500 rpm. Using Eq. 4.3.7, we see that the best speeds for increasing axial depth of cut without chatter are:

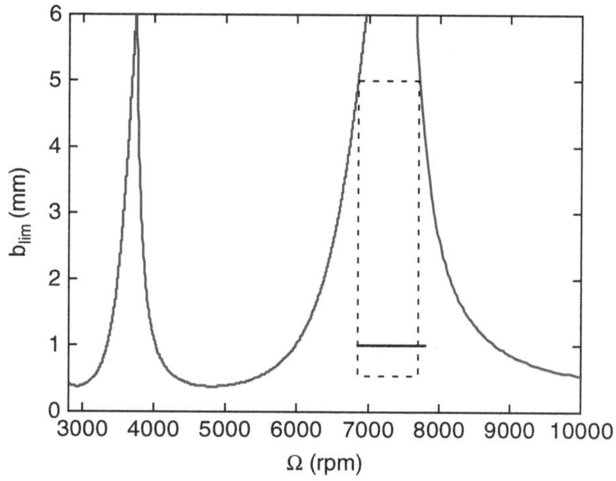

Fig. 5.2.4 Stability lobe diagram for Ex. 5.2.2 (50% radial immersion up milling). Two test ranges are also identified: (solid line) 6900 rpm to 7800 rpm at 1 mm axial depth; and (dotted line rectangle) 6900 rpm to 7700 rpm with axial depths from 0.5 mm to 5 mm

$$\Omega_{best} = \frac{f_n \cdot 60}{(N+1) \cdot N_t} = \frac{500 \cdot 60}{(N+1) \cdot 4} = \frac{7500}{(N+1)}.$$

For the right-most ($N = 0$) lobe highlighted by the rectangular simulation range in Fig. 5.2.4, the best speed is therefore in the high slope surface location error range. This high slope indicates that small errors in our knowledge of the system dynamics or spindle speed could lead to significant changes in the predicted error and affect our ability to compensate by tool path adjustments, for example. We also see that the error tends to increase with axial depth. This result is expected given that the force magnitude, and consequently the

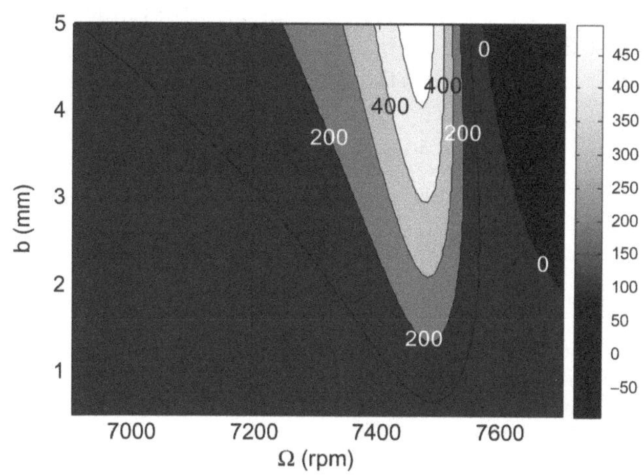

Fig. 5.2.5 Surface location error contours for rectangular stable zone identified in Fig. 5.2.4. High sensitivity of the error to spindle speed is observed near 7500 rpm, the traditional best speed for increased chatter-free axial depth of cut

vibration level, scales with axial depth. While these results do not preclude the use of the best speeds equation to select a preferred spindle speed, it does demonstrate that combining consideration of surface location error with stability may lead to a slightly different spindle speed choice to balance the two requirements. This is explored further at the end of this section. Figure 5.2.5 was generated using the MATLAB® program p_5_2_2_4.m.

A natural question to ask is if this same variation in surface location error also occurs in higher N value stability lobes. To answer this question, let's complete simulations for the spindle speed range from 2000 rpm to 10000 rpm. This span encompasses the first three best speeds (N = 0, 1, and 2) at 7500 rpm, 3750 rpm, and 2500 rpm. Using a 25% radial immersion to raise the critical stability limit (the depth at which the cut is stable for all spindle speeds) and enable an axial depth of 0.75 mm, but maintaining consistency in all other conditions with respect to Fig. 5.2.5, produces the error variation displayed in Fig. 5.2.6 (p_5_2_2_5.m). We see that the surface location error shows sensitivity to spindle speed at each of the best speeds.

IN A NUTSHELL It can be seen that over broad segments of the spindle speed range, the surface location error is quite small. However, there are also narrow bands in which the surface location error is large or changes rapidly from large negative to large positive values, for example. Unfortunately, these sections of extreme speed sensitivity are located within the stable zones of the stability lobe diagram. It is for this reason that the combined ability to choose stable cutting parameters as well as predict the surface location error is so important.

As noted in Section 5.1, it is the change in phase with frequency that varies the time lag between the force and vibration and causes the surface location error dependence on spindle speed. Further, the variation in phase with tooth passing

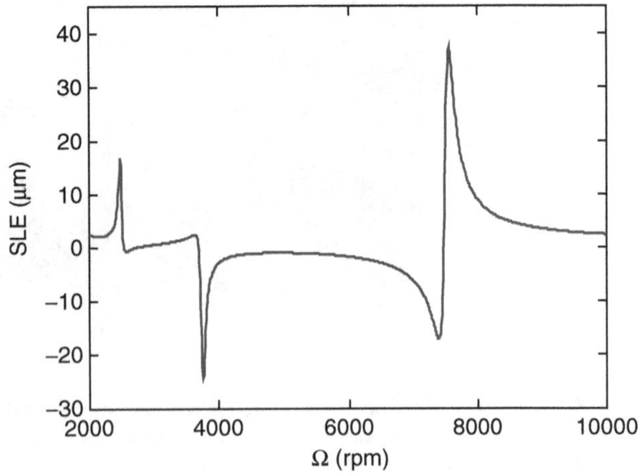

Fig. 5.2.6 Variation in surface location error from 2000 rpm to 10000 rpm for the system described in Ex. 5.2.2. The up milling radial immersion is 25% and the axial depth is 0.75 mm. The behavior is periodic with error sensitivity near the best speeds identified in Eq. 4.3.7

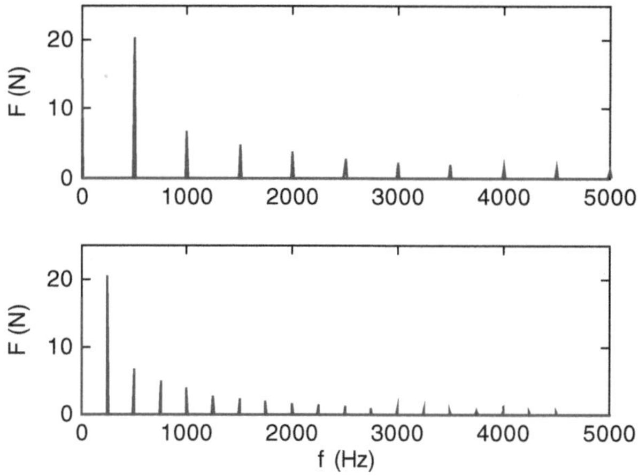

Fig. 5.2.7 Force spectrum for 7500 rpm (top) and 3750 rpm (bottom)

frequency (zero to -180 deg for a single degree of freedom system) is strongest near the natural frequency. The repetitive behavior observed in Fig. 5.2.6 occurs as increasing harmonics of the fundamental tooth passing frequency excite the system resonance. Figure 5.2.7 shows the spectra of the y direction cutting force for 7500 rpm and 3750 rpm (p_5_2_2_6.m). At 7500 rpm (top), the first harmonic, or fundamental peak, is coincident with the 500 Hz natural frequency. At 3750 rpm, (bottom) the first harmonic occurs at 250 Hz, but the second harmonic now matches the natural frequency. Similarly, the third harmonic excites resonance at 2500 rpm (not shown). The surface location error magnitude decreases with each increasing N value because the energy in the subsequent harmonics reduces. Figure 5.2.8 is included to demonstrate this phenomenon pictorially.

As a final activity before considering the effect of the helix angle on surface location error, let's investigate the influence of radial depth of cut. We'll use the same dynamic system with a spindle speed range of 6900 rpm to 7800 rpm and vary the radial engagement for down milling from 50% to 20% in decrements of 10%. The corresponding starting angles are {90, 101.5, 113.6, and 126.9} deg, respectively, as calculated using Eq. 4.1.5 and the exit angle is always 180 deg for down milling based on the circular tool path approximation. Let's select an axial depth below the 50% radial immersion critical stability limit. We can see from Fig. 5.2.4 that $b = 0.35$ mm ensures stable conditions for any spindle speed so we'll apply that value (although Fig. 5.2.4 is for up milling, the stability limit is similar for down milling at the same radial immersion in this case). The results, which were obtained using p_5_2_2_7.m, are provided in Fig. 5.2.9. We see that the transition from overcutting (negative error for down milling) to undercutting (positive error for down milling) near the $N = 0$ best spindle speed shifts to the left as the radial

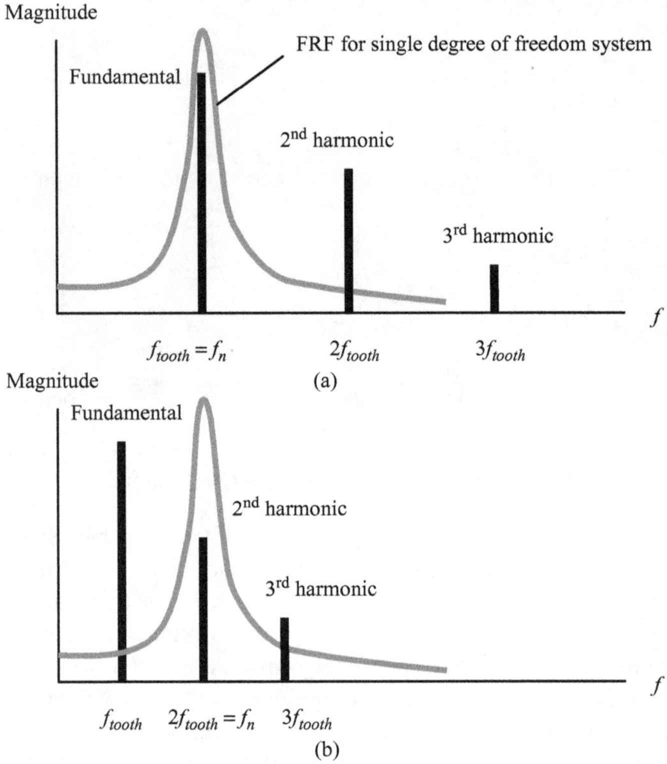

Fig. 5.2.8 Description of periodic surface location error variation. (a) The tooth passing frequency is coincident with the natural frequency (resonant force vibrations) for the $N = 0$ best spindle speed. (b) The second harmonic is located at resonance for the $N = 1$ best spindle speed

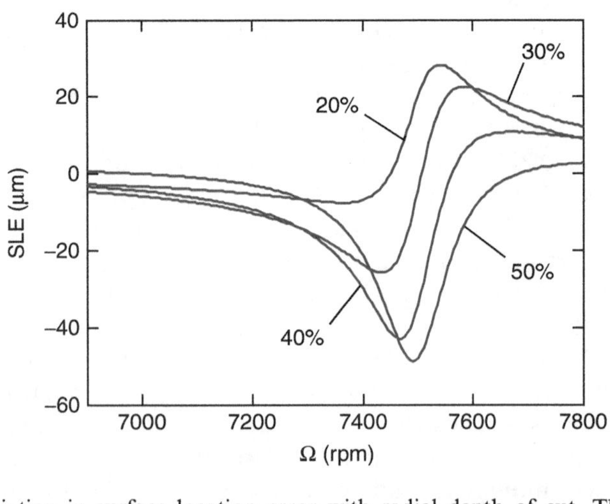

Fig. 5.2.9 Variation in surface location error with radial depth of cut. The four curves represent percent radial immersion cases of 50% to 20% down milling with a constant axial depth of 0.35 mm. The system dynamics are the same as were defined in Ex. 5.2.2

immersion is reduced. Also, the shape of the error profile changes and the peak-to-peak magnitude decreases. The latter occurs because the force level reduces with smaller radial depth of cut.

5.2.2 Variation in Surface Location Error with Axial Location

As discussed in Section 4.5, the effect of the helical cutting edge geometry is that the full length of the cutting edge does not enter (or exit) the cut at the same instant. The edge nearest the free end of the tool enters first and there is an increasing delay of the cut entry (and exit) for points on the edge that are farther from the free end (toward the spindle). The helical square endmill geometry considered here is shown in Fig. 5.2.10.

Due to the entry delay for up milling and exit delay for down milling, the surface location error varies with axial location (z direction) along the helical cutting edge length. Effectively, this occurs because the surface location is determined by the time dependent y vibration, but all points along the surface (in the z direction) are not generated simultaneously. Relative to the free end of the cutter, the angular delay along the tool axis is $\chi = \frac{2|z|\tan(\gamma)}{d}$ (rad), where $|z|$ is the absolute value of the distance from the end. The corresponding time delay is $\frac{\chi}{\frac{\Omega \cdot 2\pi}{60}}$ (s), where Ω is the spindle speed in rpm. The surface along the cutter axis (from the free end toward the spindle) is therefore produced at progressively later points in time while the cutter vibration state varies continuously. This leads to a periodic variation in the surface location error with the z value, as shown in [14].

We can use the MATLAB® program p_5_2_2_8.m to investigate this behavior. Similar to the time-domain simulations in Sections 4.5 and 4.6, the tool is discretized into axial slices. The angular delay with z location is handled by changing the times at which the $y(t)$ vector is sampled to determine surface location error. For axial slices not at the tool point, the time vector is sampled later than the cut entry (up milling) or exit (down milling) by `dX*(cnt-1)/(omega/60*2*pi), where cnt = 1, 2, 3`... is the index of the current axial slice (equal to 1 at the tool

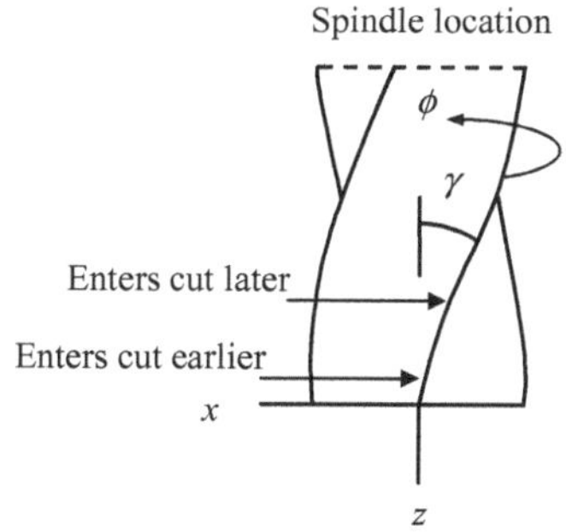

Fig. 5.2.10 The helical endmill geometry causes the cutting edge to enter and exit the cut at later instants in time when moving from its free end toward the spindle (the feed direction is to the left for this figure)

point), the delay angle per slice is `dX = 2*db*tan(gamma*pi/180)/d` (rad), `db` is the slice width in the z direction, `gamma` is the helix angle (deg), and `d`is the tool diameter. Figure 5.2.11 shows the results for 50% radial immersion up milling with the same specifications provided previously. The spindle speed is 7500 rpm and the axial depth of cut is 12 mm; the stability of this cut was verified using p_5_2_2_3.m. Clearly, the surface location error is strongly dependent on the axial location for this example, where the $|z| = 0$ location corresponds to the tool's free end and the orientation is the same as shown in Fig. 5.2.10. We should state explicitly here that all previous figures reported results for the error only at the free end of the cutting tool. For comparison purposes, p_5_2_2_8.m was used to calculate the surface location error for $b = 6$ mm at 3750 rpm, the $N = 1$ best speed from Fig. 5.2.4. This result is provided in Fig. 5.2.12. We see that the general profile is maintained relative to Fig. 5.2.11, although the error magnitude is decreased (the smaller axial depth gives smaller force and deflection) and the spatial period (in z) is reduced.

Let's explore what happens to surface location error and its axial variation if we select the axial depth for constant cutting force, as described in Section 4.5. For the selected tool, the constant cutting force axial depth is:

$$b = \frac{d \cdot \phi_p}{2\tan(\gamma)} = \frac{12.7 \cdot 90\frac{\pi}{180}}{2\tan(30)} = 17.3\ \text{mm}.$$

Unfortunately, this axial depth is unstable for the given FRF and force model, even at the $N = 0$ best speed of 7500 rpm. If the helix angle is increased to 45 deg, however, the constant force depth becomes 10.0 mm, which is stable at 7500 rpm. In this case, the constant force eliminates the surface location error variation with axial depth due to the time invariance of the vibration, i.e., constant force yields constant deflection (after the initial transients have

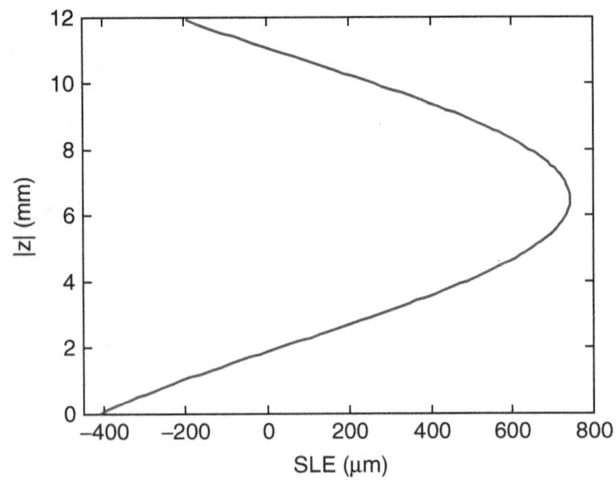

Fig. 5.2.11 Surface location error variation with axial location for $b = 12$ mm, $\Omega = 7500$ rpm, and $\gamma = 30$ deg (helix angle). The $|z| = 0$ position corresponds to the tool's free end (see Fig. 5.2.10)

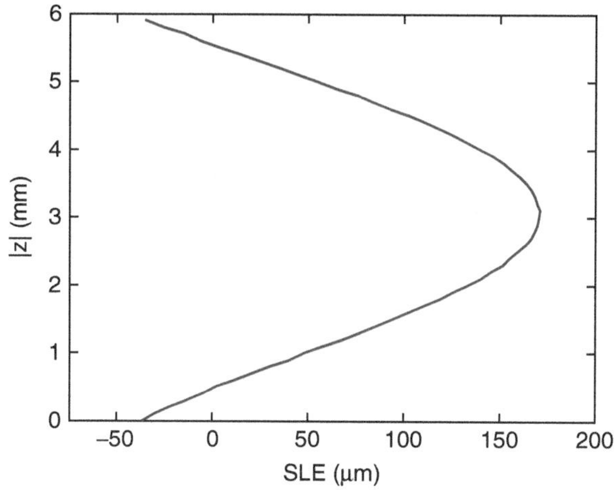

Fig. 5.2.12 Surface location error variation with axial location for $b = 6$ mm, $\Omega = 3750$ rpm, and $\gamma = 30$ deg

decayed) so the time at which the cutter enters or exits the cut does not change the error. We must realize that the resulting surface location error still is not zero, even for the constant force condition. The error is simply determined from the ratio of the force to the stiffness, where $k = 1 \times 10^7$ N/m for this example. The MATLAB® program p_5_2_2_9.m was used to generate Fig. 5.2.13, which displays the y direction force, $F_y = 284$ N at steady state, and corresponding deflection, $y = \frac{284}{1\times10^7} = 2.8 \times 10^{-5}$ m $= 28$ μm. For the selected up milling 50% radial immersion cut, this gives an overcut surface. However, the error is small

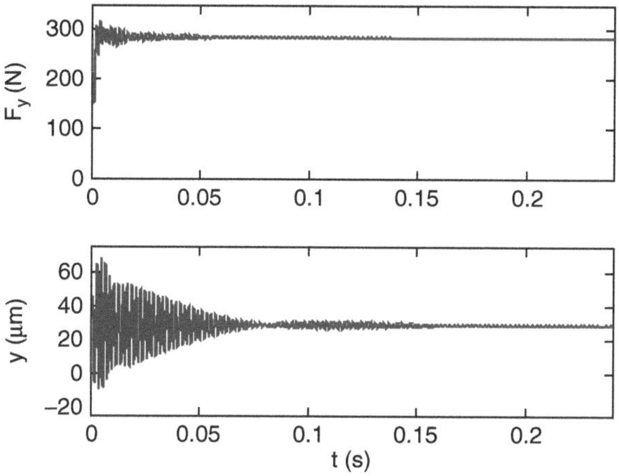

Fig. 5.2.13 Constant y direction cutting force and displacement when $b = 10.0$ mm for a helix angle of 45 deg on the four tooth, 12.7 mm diameter endmill

relative to the 12 mm axial depth, non-constant force results shown in Fig. 5.2.11.

IN A NUTSHELL The helix of the tool allows the surface location error to vary along the axis of an endmill. For this reason, a perfectly ground, exactly centered, and correctly balanced endmill that is rotating in a spindle with no error motions may still produce a non-straight sidewall during peripheral milling. The error along the wall that is parallel to the axis of the tool is a record of the tool's dynamic displacement as it rotates because different levels of the wall (bottom to top) are created at successive instants in time during the tool's rotation.

5.2.3 *Combining Stability and Surface Location Error in a Single Diagram*

As a final activity in this section, let's combine the stability and surface location error data in a single diagram. We'll consider the following conditions: 50% radial immersion up milling at spindle speeds from 2800 rpm to 10000 rpm, $N_t = 4$, $\gamma = 30$ deg, $d = 12.7$ mm diameter, $f_t = 0.1$ mm/

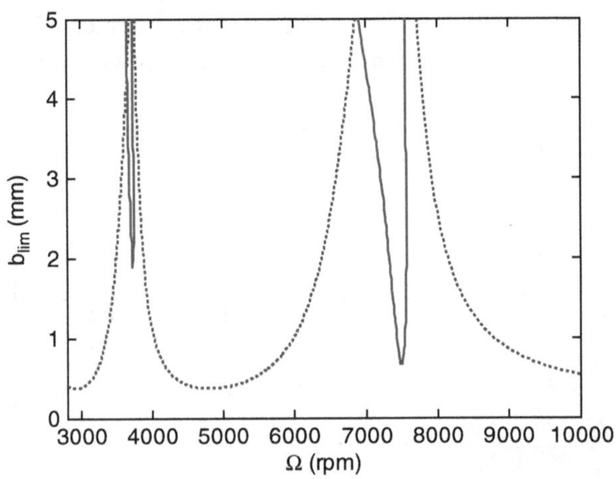

Fig. 5.2.14 Stability (dotted line) and surface location error (solid line) information combined in a single "super" diagram. The solid contour lines identify the {spindle speed, axial depth} combinations where the error level is 100 μm. The simulation parameters are: 50% radial immersion up milling, $N_t = 4$, $\gamma = 30$ deg, $d = 12.7$ mm diameter, $f_t = 0.1$ mm/tooth, $k_t = 700$ N/mm^2, $k_n = 210$ N/mm^2, $k_{te} = k_{ne} = 0$ N/mm, and symmetric structural dynamics with a stiffness of 1x10^7 N/m, 1% damping, and 500 Hz natural frequency

tooth, b = 0.4 mm to 5 mm (for the surface location error calculations), k_t = 700 N/mm^2, k_n = 210 N/mm^2, k_{te} = k_{ne} = 0 N/mm, and symmetric structural dynamics with a stiffness of 1×10^7 N/m, 1% damping, and 500 Hz natural frequency. The results are displayed in Fig. 5.2.14, where 15 terms were used in the Fourier series force model[2]. The MATLAB® program p_5_2_2_10.m was executed to generate the figure, which displays the stability boundary (dotted line) in addition to surface location error contours (solid lines) at a constant error level of 100 μm; note that the surface location error was calculated at the free end of the tool for this example. The error level is larger within the contours similar to Fig. 5.2.5. This "super" diagram shows that a portion of the stable zone (inside the error contours) is inaccessible if the user desires the surface location error to be less than 100 μm. The information provided in Fig. 5.2.14 could be used at the process planning stage, for example, to select machining conditions that satisfy both stability and accuracy requirements.

5.3 Cycloidal Tool Path Time-Domain Simulation

Similar to Sections 4.4 and 4.5., we now detail a time-domain milling simulation based on the 'Regenerative Force, Dynamic Deflection Model' described by Smith and Tlusty [6]. The simulation includes the contribution of the tool vibrations to the instantaneous chip thickness and provides predictions for both force and deflection in the x (feed) and y directions. Vibrations along the tool axis, or z direction, are not considered. We also model the cycloidal motion of the cutter teeth, rather than assuming a circular tool path. Other instances of cycloidal tool path simulations from the literature are provided in [19–21], for example. Our approach [22] is similar in nature to that described in [21].

We begin the milling simulation by first defining the cutting parameters, including spindle speed, Ω, feed per tooth, f_t, radial and axial depths of cut, a and b, respectively, and the number of teeth, N_t. We then describe the system dynamics using the modal mass, m, damping, c, and stiffness, k, values for any number of modes in the x (feed) and y directions. As detailed in Section 2.5, these values are typically obtained from impact tests followed by a modal fitting procedure, such as the peak picking method.

We determine the forces and deflections by numerical integration over small steps in time, $dt = \frac{60}{SR \cdot \Omega}$ (sec), where SR is the number of steps per cutter revolution and Ω is expressed in rpm. In each time step, we rotate the cutter

[2] In general, it is not necessary to use a large number of terms to represent the force. It is usually only necessary that the first few harmonics be characterized since higher order harmonics often have little impact on the system behavior.

by an angle, $d\phi = \frac{360}{SR}$ (deg). We then calculate the current nominal coordinates of each tooth on the cutter (Cx_j, Cy_j) according to Eq. 5.3.1, where r is the cutter radius, ϕ_j is the tooth angle, j is the tooth number which varies from 1 to N_t, $df = \frac{N_t f_t}{SR}$ is the incremental feed during the time step dt, and x_{tool} and y_{tool} are the tool center coordinates determined in the previous time step (set equal to zero initially).

$$\begin{aligned} Cx_j &= r \sin \phi_j + df + x_{tool} \\ Cy_j &= r \cos \phi_j + y_{tool} \end{aligned} \tag{5.3.1}$$

In order to calculate the instantaneous chip thickness at each time step, we compare the (Cx_j, Cy_j) coordinates of the current tooth (i.e., point **C** in Fig. 5.3.1) to the surface coordinates recorded during the prior tooth passage at the same angular orientation. However, because we are not applying the circular tool path assumption, it is not required that a data point exist at this angle from the prior pass. Therefore, we must complete a search to determine the two points from the previous tooth passage which bound this angle; we refer to these points as **A** and **B** in Fig. 5.3.1. We then carry out linear interpolation between points **A** and **B** to determine point **D**, which lies on the line between point **C** and the cutter origin [23]. The coordinates of point D, (Dx_j, Dy_j), are given in Eq. 5.3.2:

$$\begin{aligned} Dx_j &= \frac{\tan(\phi_j) \cdot Ax_j C^* - \tan(\phi_j) \cdot Ay_j + \tan(\phi_j) \cdot Cy_j - Cx_j}{\tan(\phi_j) \cdot C^* - 1}, \\ Dy_j &= Ay_j - Ax_j C^* + Dx_j C^* \end{aligned} \tag{5.3.2}$$

where $C^* = \frac{Ay_j - By_j}{Ax_j - Bx_j}$. We include the nonlinearity that is exhibited when the vibration amplitude is large enough that a tooth leaves the cut by setting the chip thickness, h_j, equal to zero if:

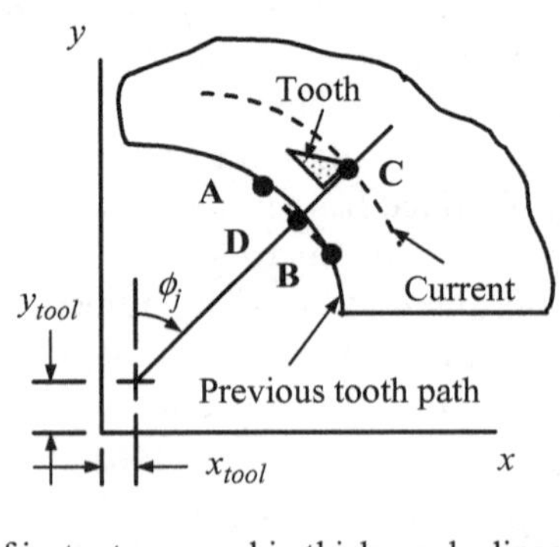

Fig. 5.3.1 Determination of instantaneous chip thickness by linear interpolation for cycloidal tool path

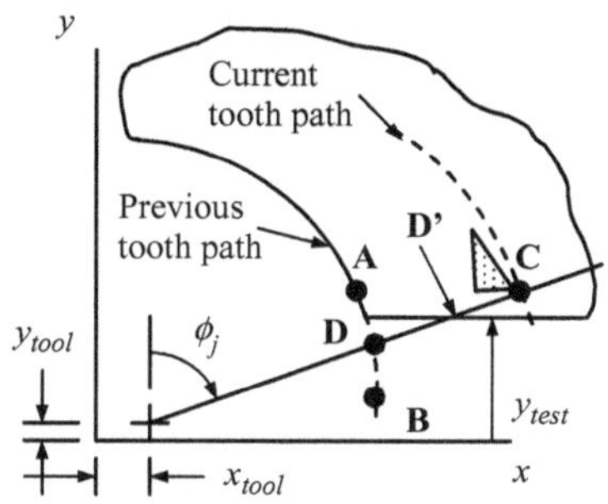

Fig. 5.3.2 Reduced instantaneous chip thickness at cut exit in up milling

$$\sqrt{(Cx_j - x_{tool})^2 + (Cy_j - y_{tool})^2} < \sqrt{(Dx_j - x_{tool})^2 + (Dy_j - y_{tool})^2}. \quad (5.3.3)$$

We must also query two other conditions for the chip thickness calculation. First, we must determine if the current tooth is bounded by the specified radial immersion. Second, we must verify that the chip thickness has not been reduced during cut entry for down milling or cut exit for up milling. The chip thickness reduction that occurs at the cut exit for up milling, for example, is exhibited in Fig. 5.3.2.

To determine if the current tooth is bounded by the selected radial depth of cut (i.e., engaged in the cut), we use the value y_{test}, which gives the y direction coordinate of the desired surface as shown in Fig. 5.3.2. For up milling with less than or equal to 50% radial immersion, cutting occurs if Cy_j is greater than y_{test}. This situation is depicted in Fig. 5.3.2. If the up milling radial immersion is greater than 50%, then Dy_j must be greater than y_{test} if cutting is to occur (note that y_{test} is negative in this case). For down milling, Cy_j must be less than y_{test} if the radial immersion is less than or equal to 50% and cutting is to take place (y_{test} is again negative). If the radial immersion is greater than 50%, it is required that Dy_j be less than y_{test} if cutting is to occur. In each case, provided the chip thickness is not reduced at the cut exit (up milling) or entry (down milling), as shown in Fig. 5.3.2, and the tooth has not vibrated out of the cut (Eq. 5.3.3), then h_j is calculated according to Eq. 5.3.4:

$$h_j = \sqrt{(Cx_j - Dx_j)^2 + (Cy_j - Dy_j)^2}. \quad (5.3.4)$$

To check if the chip thickness reduction condition is met, we again compare the tooth coordinates to y_{test}. The thickness reduction occurs if the following circumstances are satisfied: 1) up milling, less than or equal to 50% radial immersion: Dy_j is less than y_{test}; 2) up milling, greater than 50% radial immersion: Cy_j is less than y_{test}; 3) down milling, less than or equal to 50% radial immersion: Dy_j is greater than y_{test}; and 4) down milling, greater than 50% radial immersion: Cy_j is greater than y_{test}. In these cases, we can no longer use

Eq. 5.3.4 to compute the instantaneous chip thickness. Rather, we must consider point **D′** identified in Fig. 5.3.2. The coordinates of this point, (Dx'_j, Dy'_j), are provided in Eq. 5.3.5.

$$\begin{aligned} Dx'_j &= (y_{test} - y_{tool}) \tan\phi_j + x_{test} \\ Dy'_j &= y_{test} \end{aligned} . \quad (5.3.5)$$

Under these conditions, we calculate the chip thickness using Eq. 5.3.6 for up or down milling with less than or equal to 50% radial immersion or Eq. 5.3.7 for greater than 50% radial immersion.

$$h_j = \sqrt{\left(Cx_j - Dx'_j\right)^2 + \left(Cy_j - Dy'_j\right)^2} \quad (5.3.6)$$

$$h_j = \sqrt{\left(Dx'_j - Dx_j\right)^2 + \left(Dy'_j - Dy_j\right)^2} \quad (5.3.7)$$

In any case that the computed chip thickness is greater than zero, we calculate the tangential and normal force components, $F_{t,j}$ and $F_{n,j}$, respectively, for tooth j according to Eq. 5.3.8, where we have included the edge effects described in Section 4.7:

$$\begin{aligned} F_{t,j} &= k_t b h_j + k_{te} b \\ F_{n,j} &= k_n b h_j + k_{ne} b \end{aligned} , \quad (5.3.8)$$

where k_t and k_n are the force model cutting (shearing) coefficients and k_{te} and k_{ne} are the edge (rubbing/plowing) coefficients. Next, we project the forces $F_{t,j}$ and $F_{n,j}$ onto the x and y directions using Eq. 5.3.9. We then sum the x and y direction forces over all teeth engaged in the cut at the given instant in time, $F_x = \sum_{j=1}^{N_t} F_{x,j}$ and $F_y = \sum_{j=1}^{N_t} F_{y,j}$. We use these force values to determine the instantaneous displacements x_{tool} and y_{tool} for the next time step by numerical integration of the modal equations of motion with the appropriate modal parameters. If multiple vibration modes are included, we sum the displacement contributions from each mode to determine the total displacement. Provided the modal parameters were determined from a direct FRF measurement (or model), we use the same forces for each vibration mode as shown in Section 3.5.

$$\begin{aligned} F_{x,j} &= -F_{t,j}\cos(\phi_j) - F_{n,j}\sin(\phi_j) \\ F_{y,j} &= F_{t,j}\sin(\phi_j) - F_{n,j}\cos(\phi_j) \end{aligned} \quad (5.3.9)$$

In the case of a helical cutting edge, we segment the tool into multiple slices along its axis, each of which is treated as having a zero helix angle. We sum the forces for all slices to determine the total normal and tangential cutting force components for that particular simulation time step (and cutter angular orientation). We then apply Eq. 5.3.9 to project the forces onto the x and y directions, sum the forces over all the teeth engaged in the cut, and complete the numerical integration. The difference, $\Delta\phi$ (deg), between the tooth angle, ϕ_j, for tooth j on slice k and the angle for the same tooth j on slice $k+1$ (located farther away from the tool tip by a distance b/SA) is provided in Eq. 5.3.10, where γ is the helix angle, SA is the number of axial slices, and d is the cutter diameter.

$$\Delta\phi = \frac{2b\tan(\gamma)}{SA \cdot d} \cdot \frac{180}{\pi} \tag{5.3.10}$$

Example 5.3.1: Comparison of time-domain simulation results to Ex. 5.2.2 In order to demonstrate the capabilities of the cycloidal tool path time-domain simulation (p_5_3_1_1.m) described in the previous paragraphs, let's compare results with those obtained from the frequency-domain analysis in Ex. 5.2.2. We'll use the same specifications: 50% radial immersion up milling, $N_t = 4$, $\gamma = 30$ deg, $d = 12.7$ mm diameter, $f_t = 0.1$ mm/tooth, $k_t = 700$ N/mm^2, $k_n = 210$ N/mm^2, $k_{te} = k_{ne} = 0$ N/mm, and symmetric structural dynamics with a stiffness of 1×10^7 N/m, 1% damping, and 500 Hz natural frequency. The stability lobe diagram for this situation is provided in Fig. 5.2.4. As a first step, we will verify the stability behavior at $\Omega = 6000$ rpm and 7500 rpm for an axial depth of 4 mm. Figure 5.3.3 displays the y direction force and

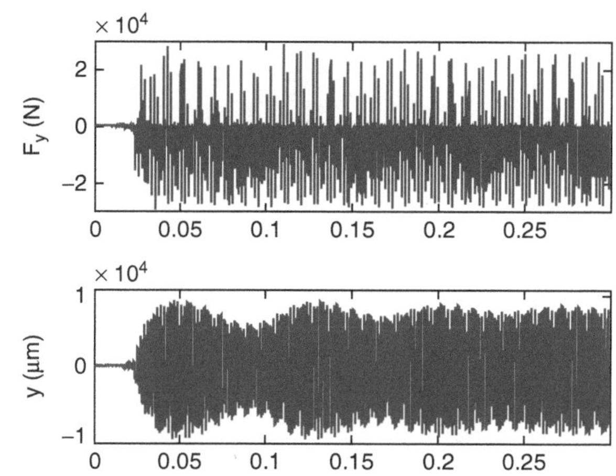

Fig. 5.3.3 Example 5.3.1 y direction force (top) and displacement (bottom) versus time results. Unstable behavior for $\Omega = 6000$ rpm and $b = 4$ mm is observed

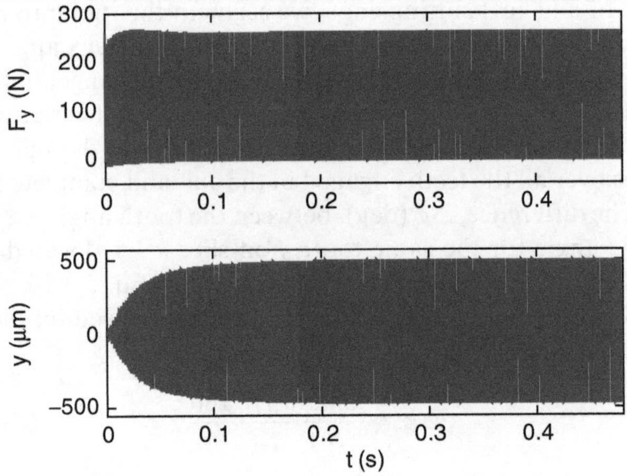

Fig. 5.3.4 Example 5.3.1 y direction force (top) and displacement (bottom) versus time results. Stable behavior for $\Omega = 7500$ rpm and $b = 4$ mm is seen

displacement for 6000 rpm. As expected, the cut is strongly unstable. Figure 5.3.4 shows the stable result for 7500 rpm.

In addition to the time plots, we can also use the simulation to display the x versus y tool path. Figure 5.3.5 presents the results for the $\Omega = 7500$ rpm 50% radial immersion up milling cut with $b = 4$ mm at the tool point, i.e., the axial slice nearest the free end of the tool. Only the portion of the tool path where the teeth enter the material (i.e., the "top" of the tool for the up milling case) is

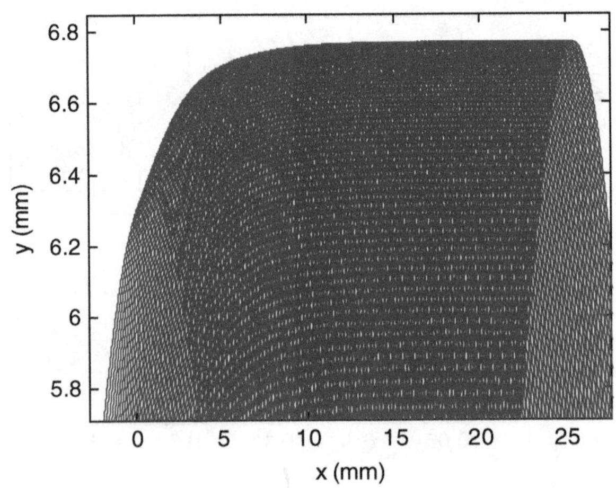

Fig. 5.3.5 Tool path for $\Omega = 7500$ rpm 50% radial immersion up milling cut with $b = 4$ mm. The x versus y teeth coordinates are shown for the axial slice nearest the free end of the cutter

included. The tool center is nominally located at $y = 0$, while the x position varies with the time dependent feed. We can see that the top of the path is initially at the tool radius of 6.35 mm (near $x = 0$). After the entry transients, the y displacement approaches 6.77 mm at the path apex where the machined surface is located. This indicates an overcut condition because more material is removed than commanded for the up milling cut. Note that the material to be cut away is located above the tool in Fig. 5.3.5. Other axial slices can also be selected using the `plot_depth` variable in p_5_3_1_1.m, where a value of one designates the slice at the tool point and larger integer values (up to `steps_axial`) specify slices nearer the spindle face.

The benefit of this figure is that it can be used to isolate the machined surface. Using a trimming algorithm to identify only the extreme points on the tool path, which define the machined surface geometry, the surface location error and roughness average, *Ra* [24], are determined to be 420 μm and 0.2 μm, respectively, from Fig. 5.3.5. The discrete roughness average equation is provided in Eq. 5.3.11, where n is the number of points that define the machined surface. Figure 5.3.6 displays the individual points and surface (solid line that connects the points) for the selected axial slice (`plot_depth = 1`). The small slope in the line indicates that steady state has not quite been reached. However, the (overcut) surface location error value of 420 μm agrees with the tool point frequency-domain solution results previously reported in Fig. 5.2.5 (read the contour value at the coordinates $\Omega = 7500$ rpm and $b = 4$ mm to verify this statement). The surface location error is determined by comparing the y coordinate of the mean of this line to the tool radius (the commanded surface location). For the up milling case shown, if the line is positioned above the tool radius, more material is removed than commanded and an overcut surface is obtained. The trimming algorithm used to identify the machined surface at the selected axial slice proceeds by:

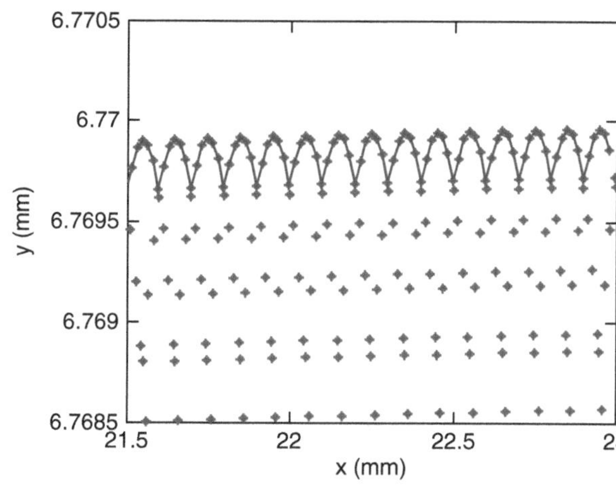

Fig. 5.3.6 The machined surface geometry (line) is defined by isolating the extreme points from the tool path. The surface location error is determined by comparing the mean of this line to the tool radius (the commanded surface location). For the up milling case shown, if the line is positioned above the tool radius, more material is removed than commanded and an overcut surface is obtained

- arranging all points on the tool path in ascending x (feed direction) values;
- selecting a point and comparing its y value to the y values of the next two points;
- keeping the higher point from the next two points for up milling and lower point for down milling;
- incrementing to the next point and repeating the comparison process; and
- repeating the entire exercise multiple times to "bubble up" (up milling) or "trickle down" (down milling) to the final surface.

Although this is not a particularly elegant solution, it is effective. We do not have a defined stopping condition for the procedure, but experience has shown that 50 to 100 iterations are generally adequate.

$$Ra = \frac{\sum_{i=1}^{n} |y_i - \bar{y}|}{n} \tag{5.3.11}$$

Once the surface location error has been determined for all axial slices, the change in error with axial depth can be interrogated as shown in the Section 5.2; see Figs. 5.2.11 and 5.2.12. The time-domain simulation results (circles) are superimposed on the frequency-domain solution (line) in Fig. 5.3.7 for the conditions described in Ex. 5.3.1. For the time-domain solution, five axial slices were used so that $db = 4/5 = 0.8$ mm. The points are placed at the midpoint of each slice in the figure, i.e., {0.4, 1.2, 2, 2.8, and 3.6} mm. The MATLAB® program p_5_3_1_2.m was used to generate the frequency-domain results in Fig. 5.3.7. The time-domain results were obtained from p_5_3_1_1.m by sequentially plotting and trimming the tool path for `plot_depth = 1, 2, ..., 5`.

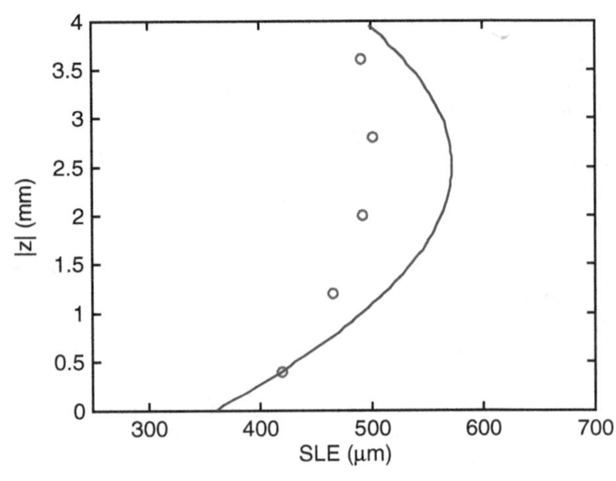

Fig. 5.3.7 Variation in surface location error with axial location. The circles represent the time-domain result for each axial slice; they are located at the midpoint of each of the five slices. The line displays the frequency-domain solution

IN A NUTSHELL As with stability, the use of time-domain simulation to predict surface location error eliminates many of the simplifying assumptions. The tool path may be modeled as a cycloid instead of a circle. Unequal teeth spacing may be applied (see Section 6.3). The surface produced by an unstable cut may be predicted. For the analytical solution, these features are difficult to incorporate. For time-domain simulation, on the other hand, they are relatively easy to include.

Exercises

1. Determine the value of the mean y direction cutting force for the following cuts using Eq. 5.2.3. The aluminum alloy-four tooth cutter combination gives: $k_t = 790$ N/mm^2 and $k_n = 190$ N/mm^2, $k_{te} = 8$ N/mm, and $k_{ne} = 4$ N/mm. Also, $b = 5$ mm and $f_t = 0.15$ mm/tooth. Assume a rigid cutting tool and workpiece.

   a) Up milling, 30% radial immersion
   b) Down milling, 40% radial immersion

2. Plot the y direction force over one cutter revolution for: down milling, 50% radial immersion, $N_t = 2$, $d = 19$ mm, $\gamma = 30$ deg, $b = 2$ mm, $f_t = 0.2$ mm/tooth, $k_t = 730$ N/mm^2, $k_n = 205$ N/mm^2, $k_{te} = k_{ne} = 0$ N/mm, and $\Omega = 10000$ rpm. Use the Fourier series approach and show results for both five and 25 terms.
3. Calculate the surface location error for the following conditions: 25% radial immersion down milling, spindle speeds from 11000 rpm to 13000 rpm, $N_t = 4$, $\gamma = 30$ deg, $d = 12.7$ mm diameter, $f_t = 0.15$ mm/tooth, $b = 2$ mm, $k_t = 700$ N/mm^2, $k_n = 210$ N/mm^2, $k_{te} = k_{ne} = 2$ N/mm, and symmetric structural dynamics with a stiffness of 8×10^6 N/m, $\zeta = 0.02$, and 800 Hz natural frequency. For the Fourier computations, use 15 terms, five axial steps, and a spindle speed resolution of 20 rpm. At a spindle speed of 12140 rpm, is the surface overcut or undercut?
4. Using time-domain simulation, determine the surface location error at the free end of the cutter for the same conditions described in Exercise 3. Use a spindle speed equal to the best speed calculated from Eq. 4.3.7 for the $N = 0$ (rightmost) lobe. Carry out your simulation for 40 revolutions with 500 steps per tooth.

References

1. Kline, W., DeVor, R., and Shareef, I., 1982, The Prediction of Surface Accuracy in End Milling, Journal of Engineering for Industry, 104: 272–278.
2. Kline, W., DeVor, R., and Lindberg, J., 1982, The Prediction of Cutting Forces in End Milling with Application to Cornering Cuts, International Journal of Machine Tool Design Research, 22: 7–22.

3. Tlusty, J., 1985, Effect of End Milling Deflections on Accuracy, in: R.I. King (Ed.), Handbook of High Speed Machining Technology, Chapman and Hall, New York, pp. 140–153.
4. Sutherland, J. and DeVor, R., 1986, An Improved Method for Cutting Force and Surface Error Prediction in Flexible End Milling Systems, Journal of Engineering for Industry, 108: 269–279.
5. Montgomery, D. and Altintas, Y., 1991, Mechanism of Cutting Force and Surface Generation in Dynamic Milling, Journal of Engineering for Industry, 113/2: 160–168.
6. Smith, S. and Tlusty, J., 1991, An Overview of Modeling and Simulation of the Milling Process, Journal of Engineering for Industry, 113/2: 169–175.
7. Altintas, Y., Montgomery, D., and Budak, E., 1992, Dynamic Peripheral Milling of Flexible Structures, Journal of Engineering for Industry, 114/2: 137–145.
8. Tarng, Y., Liao, C., and Li, H., 1994, A Mechanistic Model for Prediction of the Dynamics of Cutting Forces in Helical End Milling, International Journal of Modeling and Simulation, 14/2: 92–97.
9. Schmitz, T. and Ziegert, J., 1999, Examination of Surface Location Error Due to Phasing of Cutter Vibrations, Precision Engineering, 23/1: 51–62.
10. Altintas, Y., 2000, Manufacturing Automation, Cambridge University Press, Cambridge, UK.
11. Mann, B., Bayly, P, Davies, M. and Halley, J., 2004, Limit Cycles, Bifurcations, and Accuracy of the Milling Process, Journal of Sound and Vibration, 277: 31–48.
12. Schmitz, T., Couey, J., Marsh, E., Mauntler, N., and Hughes, D., 2007, Runout Effects in Milling: Surface finish, Surface Location Error, and Stability, International Journal of Machine Tools and Manufacture, 47: 841–851.
13. Yun, W.-S., Ko, J., Cho, D.-W., and Ehmann, K., 2002, Development of a Virtual Machining System, Part 2: Prediction and Analysis of a Machined Surface Error, International Journal of Machine Tools and Manufacture, 42: 1607–1615.
14. Tlusty, G., 2000, Manufacturing Equipment and Processes, Prentice-Hall, Upper Saddle River, NJ, Section 9.5.5, Fig. 9.50.
15. Schmitz, T., Ziegert, J., Canning, J.S., and Zapata, R., 2007, Case Study: A Comparison of Error Sources in High-Speed Milling, Precision Engineering, 32: 126–133.
16. Tlusty, J., Zaton, W., and Ismail, F., 1983, Stability Lobes in Milling, Annals of the CIRP, 32/1: 309–313.
17. Altintas, Y. and Budak, E., 1995, Analytical Prediction of Stability Lobes in Milling, Annals of the CIRP, 44/1: 357–362.
18. Schmitz, T. and Mann, B., 2006, Closed Form Solutions for Surface Location Error in Milling, International Journal of Machine Tools and Manufacture, 46: 1369–1377.
19. Ikua, B.W., Tanaka, H., Obata, F., and Sakamoto, S., 2001, Prediction of Cutting Forces and Machining Error in Ball End Milling of Curved Surfaces – I. Theoretical Analysis, Precision Engineering, 25/4: 266–73.
20. Ranganath, S., Narayanan, K., and Sutherland, J., 1999, The Role of Flank Face Interference in Improving the Accuracy of Dynamic Force Predictions in Peripheral Milling, Journal of Manufacturing Science and Engineering, 121: 593–599.
21. Campomanes, M.L. and Altintas, Y., 2003, An Improved Time Domain Simulation for Dynamic Milling at Small Radial Immersions, Journal of Manufacturing Science and Engineering, 125: 416–422.
22. Schmitz, T., Couey, J., Marsh, E., Mauntler, N., and Hughes, D., 2007, Runout Effects in Milling: Surface finish, Surface Location Error, and Stability, International Journal of Machine Tools and Manufacture, 47: 841–851.
23. Chau, W.-M., 1992, Accuracy of Milling Operations Based on Dynamic Models and Simulation, M.S. Thesis, University of Florida.
24. American Society of Mechanical Engineers, 2003, ASME B46.1-2002, Surface Texture (Surface Roughness, Waviness, and Lay), New York, NY.

Chapter 6
Special Topics in Milling

If the facts don't fit the theory, change the facts.

- Albert Einstein

In Chapters 4 and 5 we analyzed the milling process for both stability and surface location error through frequency and time-domain approaches. Using frequency-domain analyses, we: 1) generated stability lobe diagrams that identify stable and unstable combinations of spindle speed and axial depth of cut; and 2) completed surface location error predictions over the same parameter space. We developed time-domain simulations for the circular tool path approximation that predict forces and displacements for square and ball endmills, including the effects of helical teeth geometries. We next extended the square endmill time-domain simulation to incorporate the actual cycloidal tool path and investigated both stability and surface location error for various cutting conditions. Finally, we compared stability and surface location error solutions between the frequency and time-domain analyses. In this chapter, we continue our investigation of milling by exploring the frequency content of stable and unstable milling signals and enhancing our time-domain simulations to include runout of the cutter teeth and variable teeth spacing. We then discuss low radial immersion milling and the corresponding stability behavior. We conclude the chapter with some comments regarding stability boundary uncertainty evaluation for the frequency-domain analyses.

6.1 Frequency Content of Milling Signals

Let's return to the average tooth angle analysis for milling stability from Section 4.3 [1]. The relationships used to determine the spindle speed dependent axial depth of cut limit, b_{lim}, are repeated in Eqs. 6.1.1 through 6.1.3.

$$b_{\text{lim}} = \frac{-1}{2K_s \text{Re}[FRF_{orient}]N_t^*} \qquad (6.1.1)$$

T.L. Schmitz, K.S. Smith, *Machining Dynamics*, DOI 10.1007/978-0-387-09645-2_6,

$$\frac{f_c}{\Omega N_t} = N + \frac{\varepsilon}{2\pi} \tag{6.1.2}$$

$$\varepsilon = 2\pi - 2\tan^{-1}\left(\frac{\mathrm{Re}[FRF_{orient}]}{\mathrm{Im}[FRF_{orient}]}\right) \tag{6.1.3}$$

As a reminder, we restate the variable definitions here:

- K_s, specific force – process dependent coefficient that relates the resultant cutting force to the uncut chip area;
- FRF_{orient}, oriented frequency response function – x and y direction frequency response functions, or FRFs, "weighted" by the direction orientation factors, which are determined from the two step projection of the cutting force (due to a tooth located at the average between the start and exit angles) first onto the x and y directions and then onto the average surface normal;
- N_t^*, average number of teeth in the cut – it is possible for multiple teeth to be engaged simultaneously and for the number of teeth in the cut to vary during a single revolution. The average value $N_t^* = \frac{\phi_e - \phi_s}{\frac{360}{N_t}}$ captures this behavior, where ϕ_e and ϕ_s are the exit and start angles (deg), respectively, and N_t is the number of teeth on the cutter;
- f_c, chatter frequency (Hz) – frequency at which self-excited vibrations will occur if the stability limit is exceeded;
- Ω, spindle speed (rev/s) – the rotating frequency of the spindle for milling;
- N, lobe number – $N = 0, 1, 2, \ldots$ indexed from right to left (higher to lower spindle speeds) in a stability lobe diagram; and
- ε, phase between the current and previous tooth vibrations (rad) – varies between π rad (180 deg) and 2π rad (360 deg), where $\varepsilon = 3\pi/2$ rad (270 deg) is the least favorable value.

We use Eqs. 6.1.1 through 6.1.3 to generate milling stability lobe diagrams for the average tooth angle approach by: 1) determining the oriented FRF and identifying the valid chatter frequency range(s), i.e., where the real part of FRF_{orient} is negative; 2) solving for ε over the valid frequency range(s); 3) finding the average number of teeth in the cut for the selected radial immersion; 4) calculating $b_{\lim}$ over the valid frequency range(s); 5) selecting an $N = 0, 1, 2, \ldots$ value (representing N waves of vibration between teeth) and calculating the associated spindle speeds over the valid frequency range(s); and 6) plotting Ω vs. $b_{\lim}$ for each N value. As we discussed previously, any (Ω, b) pair leads to stable or unstable behavior depending on whether it appears above or below the stability boundary, respectively.

In our discussions of both stability and surface location error, we showed spectra of the cutting force signals which included content not only at the tooth passing frequency, but also at multiple harmonics due to the "impulse train" nature of typical milling force profiles. The notable constant force exceptions are: 1) slotting with an even number of teeth, where $N_t > 2$; and 2) particular axial depths of cut for helical square endmills which are determined from $b = \frac{d \cdot \phi_p}{2\tan(\gamma)}$, where d is the cutter diameter, $\phi_p = \frac{2\pi}{N_t}$ (rad) is the pitch angle, and γ is the helix angle. We expressed the tooth passing frequency as a function of the spindle speed (rpm) and number of teeth, $f_{tooth} = \frac{\Omega \cdot N_t}{60}$ (Hz). We can expect, then, that during stable cutting conditions we should observe content at f_{tooth} and integer multiples of this value.

"What about unstable cuts?", we may ask. We've already shown in Section 4.3, through our description of the average tooth angle approach, that each stability lobe can be described as a mapping of the oriented FRF onto the (Ω, b_{lim}) parameter space. See Fig. 4.3.4, for example, where the chatter frequency is scanned through its potential values to obtain the stability limit for each N value. We know, therefore, that the chatter frequency (should chatter occur) is not a fixed value. It depends on the spindle speed and occurs within the valid chatter frequency range from FRF_{orient}. We also found that increased axial depths of cut were available at the "best speeds" of $\Omega_{best} = \frac{f_n \cdot 60}{(N+1) \cdot N_t}$ (rpm), where f_n is expressed in Hz, although multiple vibration modes can lead to competing lobes and more complicated stability lobe diagrams. The purpose of this best speeds equation is to match the tooth passing frequency to the system natural frequency so that the tooth-to-tooth surface undulations are in phase, i.e., $\varepsilon = 360$ deg, and the force variation due to chip thickness modulation is minimized. Given this circumstantial information, we can expect that unstable cuts will exhibit content not only at f_{tooth} and its harmonics, but also some chatter frequency.

FOR INSTANCE Stable cuts sound different from those that chatter. Stable cuts generate sound at the tooth passing frequency, runout frequency (see Section 6.2), and multiples of these. For that reason, stable cuts sound "clean"; they produce "pure" tones. Unstable cuts, however, also emit sound at the chatter frequency, which is generally not a multiple of the tooth passing or runout frequencies. For this reason, unstable cuts sound "harsh" or "raspy". They exhibit a mixture of frequencies that are not harmonically related.

To explore the frequency content of milling signals in more detail, as well as how we might capitalize on this information, let's update the average tooth angle MATLAB® code we developed previously. In p_6_1_1_1.m, we'll add new figures to show the chatter and tooth passing frequencies, as well as the phase between surface undulations from one tooth to the next, as a function of spindle speed. We'll then see how we can use the chatter frequency for an unstable (Ω, b_{lim}) combination to select a new spindle speed and converge on the preferred $\varepsilon = 360$ deg phase relationship [2-3] in Ex. 6.1.1.

Example 6.1.1: Selecting new spindle speeds using the chatter frequency For simplicity, we'll consider the single degree of freedom model shown in Fig. 6.1.1. A single mode in the x (feed) direction is identified with $f_n = 800$ Hz, $k = 5 \times 10^6$ N/m, and $\zeta = 0.01$. We'll assume the y direction is rigid. The cutter has four teeth and the tool-material pair yields a specific force of $K_s = 2000$ N/mm^2 and force angle of $\beta = 72$ deg for the 50% radial immersion up milling cut.

Our first task is to determine the directional orientation factor, μ_x. As detailed in Section 4.3, two steps are required to calculate this value. First, the force is projected onto the x direction. Second, this result is projected onto the average surface normal. As shown in Fig. 6.1.2, projection of the force onto the x direction gives $F_x = F\cos(\beta - 45)$. The projection of this result onto the average surface normal is $F_n = F_x\cos(45) = F\cos(\beta - 45)\cos(45)$. Using Eq. 4.3.5, we obtain $FRF_{orient} = \mu_x FRF_x = \cos(\beta - 45)\cos(45)FRF_x$. This result is displayed in Fig. 6.1.3 for the selected x direction dynamics. Figure 6.1.4 shows the valid chatter frequency range ($\mathrm{Re}[FRF_{orient}] < 0$) for the oriented FRF (top) and corresponding $N = 0$ stability limit versus spindle speed in rev/s (bottom). Note the $\frac{1}{N_t}$ relationship between frequency and spindle speed from the top to bottom subplots. This mapping supports both the best speeds equation $\Omega_{best} = \frac{f_n}{(N+1)\cdot N_t} = \frac{800}{(0+1)\cdot 4} = 200$ rev/s and the worst speeds equation $\Omega_{worst} = \frac{f_c}{N_t} \cdot \frac{1}{N + \frac{\varepsilon}{2\pi}} = \frac{f_n(1+\zeta)}{4} \cdot \frac{1}{0 + \frac{3\pi/2}{2\pi}} = \frac{800(1+0.01)}{4}\frac{1}{0+\frac{3}{4}} = 269.3$ rev/s since increased

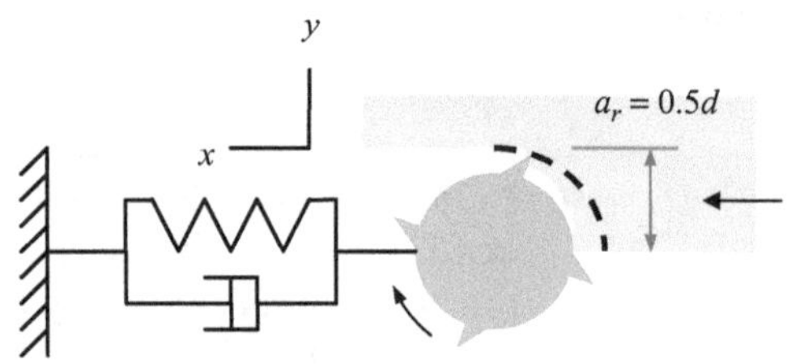

Fig. 6.1.1 50% radial immersion up milling model for Ex. 6.1.1

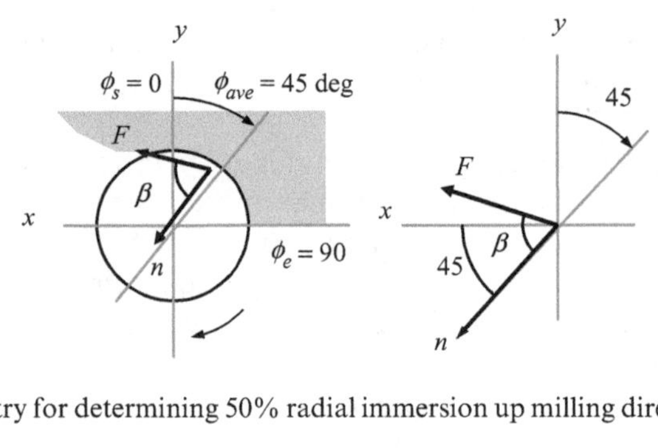

Fig. 6.1.2 Geometry for determining 50% radial immersion up milling directional orientation factor, μ_x

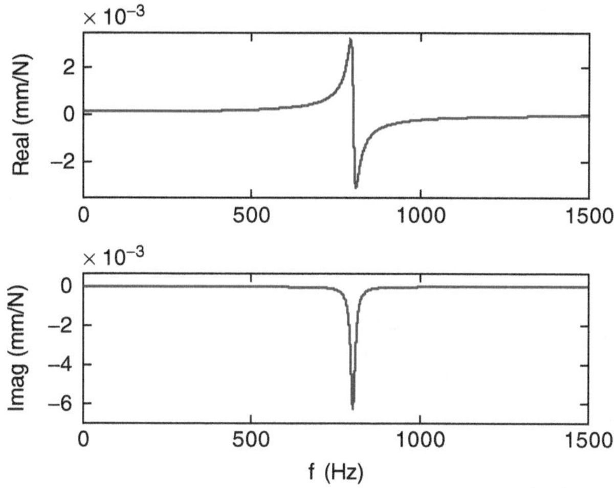

Fig. 6.1.3 Oriented frequency response function for Ex. 6.1.1

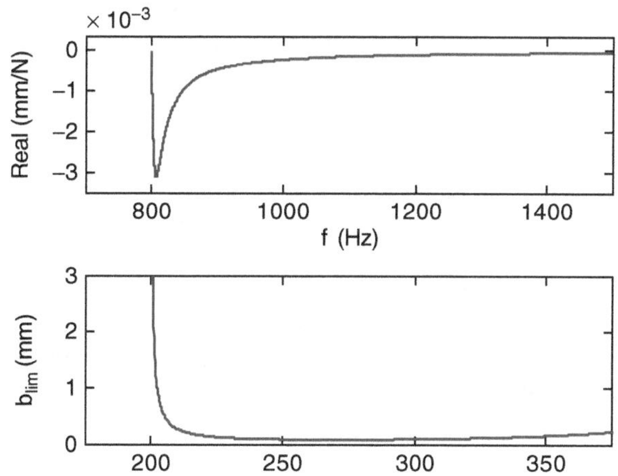

Fig. 6.1.4 Valid chatter frequency range for oriented FRF (top). Corresponding (Ω, $b_{\lim}$) parameter space for $N = 0$ mapping (bottom)

axial depths are obtained near 200 rev/s and the minimum allowable depth, equal to the critical stability limit, occurs at 269.3 rev/s.

In Fig. 6.1.5, we have included the $N = 0$ to 4 stability lobes and converted the spindle speed to rpm in the top subplot. The bottom subplot shows the potential chatter frequencies (solid lines) and harmonics of the tooth passing frequency (dotted) as a function of spindle speed. The latter naturally grow linearly with spindle speed. The rightmost dotted line represents f_{tooth}, while

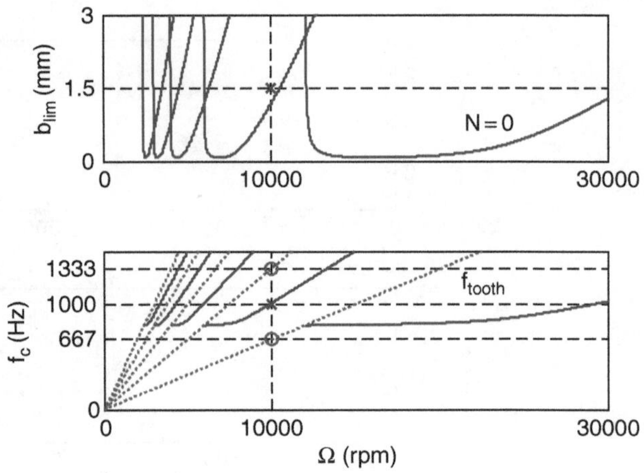

Fig. 6.1.5 (Top) Stability lobes for $N = 0$ to 4; the selected cutting condition of (10000 rpm, 1.5 mm) is indicated by the 'x'. (Bottom) The tooth passing frequencies (dotted lines) and chatter frequencies (solid) are shown as a function of spindle speed. The tooth passing frequency content at 667 Hz and 1333 Hz is identified by the circles, while the chatter frequency at 1000 Hz is located by the 'x'

subsequent dotted lines moving from right to left indicate increasing harmonics of $2f_{tooth}$ to $5f_{tooth}$. For this single degree of freedom case, the valid chatter frequencies begin at $f_n = 800$ Hz and then increase with spindle speed. This reiterates the behavior observed in Fig. 6.1.4, which shows the mapping from chatter frequency to spindle speed between the top and bottom subplots. Figure 6.1.5 also identifies selected points at 10000 rpm. We'll discuss these next.

Let's assume a cut was attempted at $\Omega = 10000$ rpm, $b = 1.5$ mm (suppose that we did not know if the cut would be stable or unstable). This cut is identified by the 'x' at $\Omega = 10000$ rpm and $b = 1.5$ mm in the top subplot. We would expect frequency content at $f_{tooth} = \frac{\Omega \cdot N_t}{60} = \frac{10000 \cdot 4}{60} = 667$ Hz, $2f_{tooth} = 2\frac{10000 \cdot 4}{60} = 1333$ Hz, and so on, if the cut was stable since only forced vibrations would be present. Circles in the bottom subplot indicate these frequencies within the plot's vertical limits. However, because the operating point (10000 rpm, 1.5 mm) is located in the unstable zone of the stability lobe diagram (top of Fig. 6.1.5), content will also be observed at 1000 Hz due to the self-excited vibration. This point is marked with an 'x' on the chatter frequency line in the bottom subplot. Note that this relationship between chatter frequency and spindle speed is defined in Eq. 6.1.2.

Returning to the best speeds equation, its purpose is to match the tooth passing frequency to the chatter frequency in order to drive the operating condition towards a tooth-to-tooth undulation phase relationship of $\varepsilon =$ 360 deg. Given the chatter frequency of 1000 Hz identified in the bottom subplot of Fig. 6.1.5, we should select a new spindle speed of:

$$\Omega = \frac{f_c \cdot 60}{(N+1) \cdot N_t} = \frac{1000 \cdot 60}{(0+1) \cdot 4} = 15000 \text{ rpm}$$

for the $N = 0$ lobe and:

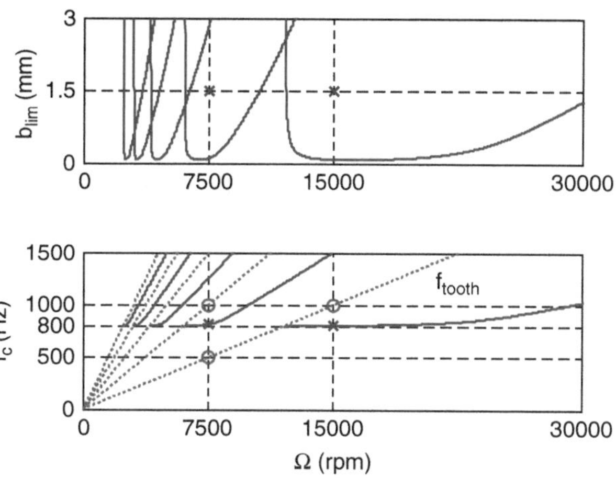

Fig. 6.1.6 (Top) A first spindle speed regulation scenario gives new speeds at 7500 rpm and 15000 rpm (N = 1 and 0 lobes, respectively) based on the 1000 Hz chatter frequency from Fig. 6.1.5. (Bottom) The new cutting conditions are also unstable and yield chatter frequencies of 818 Hz (N = 1) and 806 Hz (N = 0)

$$\Omega = \frac{f_c \cdot 60}{(N+1) \cdot N_t} = \frac{1000 \cdot 60}{(1+1) \cdot 4} = 7500\,\text{rpm}$$

for the $N = 1$ lobe. The tooth passing frequency for 15000 rpm is $f_{tooth} = \frac{15000 \cdot 4}{60} = 1000\,\text{Hz}$, as expected. Similarly, the tooth passing frequency for 7500 rpm is $f_{tooth} = \frac{7500 \cdot 4}{60} = 500\,\text{Hz}$ and the second harmonic is 1000 Hz. The points are shown as the circles in the bottom subplot of Fig. 6.1.6. However, because the cuts at (15000 rpm, 1.5 mm) for the $N = 0$ lobe and (7500 rpm, 1.5 mm) for the $N = 1$ lobe are again unstable – see the top subplot of Fig. 6.1.6 – we will observe content at the corresponding chatter frequencies. These are 818 Hz and 806 Hz for the $N = 0$ and 1 lobe spindle speed adjustments, respectively. Each of these points is marked by an 'x' in the bottom subplot of Fig. 6.1.6. If we make a second spindle speed adjustment for the $N = 0$ lobe, for example, the new value is $\Omega = \frac{806 \cdot 60}{(0+1) \cdot 4} = 12090\,\text{rpm}$. Figure 6.1.7 shows this final

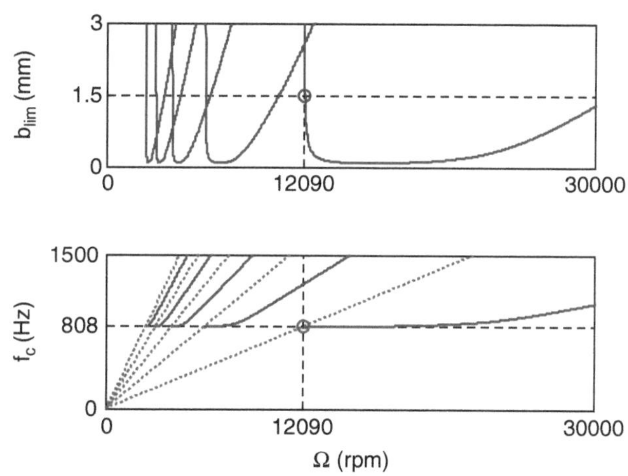

Fig. 6.1.7 (Top) Based on the 806 Hz chatter frequency at (15000 rpm, 1.5 mm) from Fig. 6.1.6, the new spindle speed is 12090 rpm. The new cut is identified by a circle and is stable. (Bottom) Because the cut is stable, the milling signal only includes content at f_{tooth} = 806 Hz (indicated by the circle) and integer harmonics, which occur outside the axis limits

regulation into the stable zone. On the other hand, the $N = 1$ lobe would yield a new spindle speed of $\Omega = \frac{818 \cdot 60}{(1+1) \cdot 4} = 6135\,\text{rpm}$. Although this selection corresponds to the gap between the $N = 1$ and 2 lobes, the axial depth (gain) is still too high and the cut would again be unstable.

As we noted previously, we use the best speeds equation to converge on the $\varepsilon = 360$ deg phase relationship. Let's repeat the previous analysis but show the spindle speed versus ε relationship for the two regulations. Figure 6.1.8 displays the (Ω, b_{lim}) parameter space (top), as well as the (Ω, ε) relationship (bottom), again for the $N = 0$ to 4 lobes. We see that the tooth-to-tooth undulation phase relationship is initially 185 deg, or nearly directly out of phase, for the (10000 rpm, 1.5 mm) cutting condition.

As shown in Fig. 6.1.6, the first adjustment based on the 1000 Hz chatter frequency leads to new spindle speeds of 15000 rpm ($N = 0$ lobe) or 7500 rpm ($N = 1$). The corresponding (Ω, ε) diagram is provided in Fig. 6.1.9 (bottom). We see that the phase relationship between subsequent tooth passages is $\varepsilon = 229$ deg for $\Omega = 7500$ rpm and 290 for 15000 rpm. Both speeds are unstable (top). For the second spindle speed adjustment ($N = 0$ lobe), however, the $\Omega = \frac{806 \cdot 60}{(0+1) \cdot 4} = 12090\,\text{rpm}$ speed gives stable cutting conditions. The corresponding ε value is 357 deg, very near the "best" tooth-to-tooth undulation phase relationship of 360 deg. This case is shown in Fig. 6.1.10.

To conclude this example, let's describe potential sources of the required chatter frequency information, which we can use to converge on preferred spindle speeds for improvements in stable axial depths of cut. Essentially, any signal that is derived from the milling process is acceptable. However, the most common choices include force, displacement, and the audio signal emitted by the cutting process [3] due to acceptable signal-to-noise ratios in most instances. The cutting force is a natural option since it directly communicates not only the tooth passing frequency and harmonics, but also the chatter frequency for

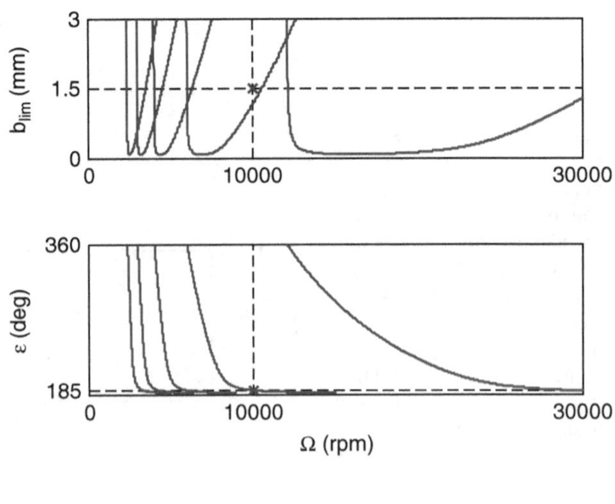

Fig. 6.1.8 (Top) Stability lobes for $N = 0$ to 4; the selected cutting condition of (10000 rpm, 1.5 mm) is indicated by the 'x'. (Bottom) The tooth-to-tooth undulation phase is shown as a function of spindle speed. The 185 deg phase for the unstable cut is identified by the 'x'

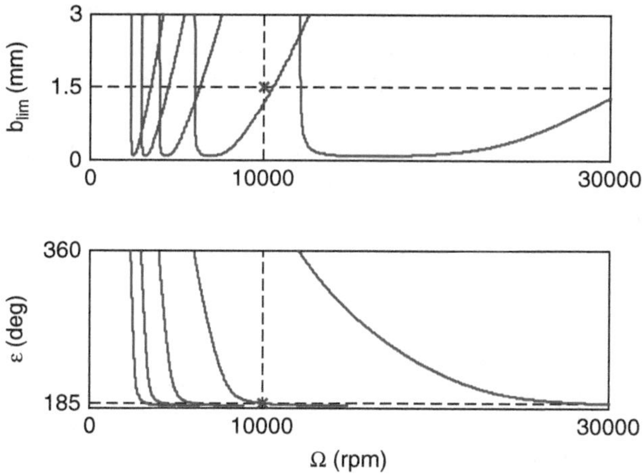

Fig. 6.1.9 (Top) The first spindle speed regulation gives new values at 7500 rpm and 15000 rpm (N = 1 and 0 lobes, respectively) based on the 1000 Hz chatter frequency from Fig. 6.1.5. (Bottom) The new cutting conditions are also unstable and yield ε values of 229 deg (N = 1 lobe) and 290 deg (N = 0)

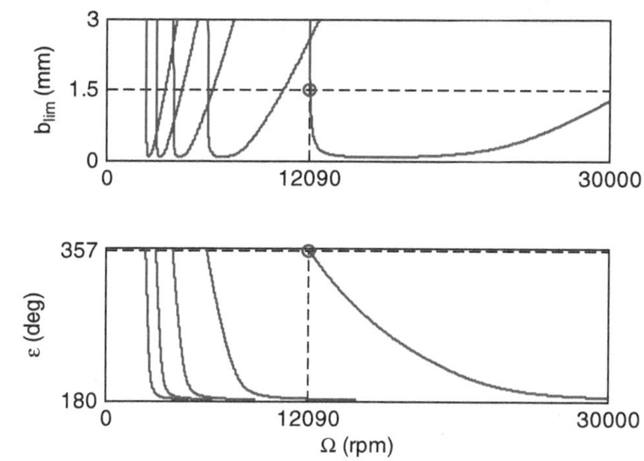

Fig. 6.1.10 (Top) For the 806 Hz chatter frequency observed in Fig. 6.1.6 for (15000 rpm, 1.5 mm), the new spindle speed is 12090 rpm. The stable cut is identified by the circle. (Bottom) The ε value is 357 deg for the stable cut

unstable conditions. The challenge, however, is obtaining this signal. Mounting a tabletop dynamometer between the workpiece and pallet/tombstone on the milling center is possible, but inherent drawbacks are: 1) the dynamometer influences the system dynamic response; 2) the dynamometer has a limited bandwidth so that content at higher frequencies may be corrupted by the dynamometer response (see Section 4.7); 3) the cost is significant for commercially available cutting force dynamometers/amplifiers; and 4) this approach is not well suited to industrial applications. The information could also be obtained from the torque signal emitted by a spindle-based torque dynamometer [4]. However, the same difficulties apply. The tool displacement signal could also be used, but the tool-workpiece interface is notoriously difficult to instrument. Displacement can be recorded on the tool shank, but non-contact sensors will generally be affected by coolant, for example, and this approach is again not particularly well suited to shop floor conditions. The use of a microphone to record the cutting process audio signal offers a good compromise and has been implemented in a control system to sense and correct unstable cutting conditions using the techniques described in the previous paragraphs [5].

A natural challenge, regardless of the transduction scheme, is setting a limit on what magnitude of spurious frequency content (not at the tooth passing frequency or harmonics) constitutes chatter. This limit selection remains largely experience-based, although some efforts have been made toward automating the process of chatter identification. See [6-9], for example, which depend on the statistical distribution in the once-per-revolution (or once-per-tooth) sampled signal. The fundamental concept is that stable milling signals will repeat with the once-per-revolution/tooth sampling because the vibration is synchronous with the force. For self-excited vibrations, on the other hand, vibration occurs at the system natural frequency (corresponding to the most flexible mode) which is incommensurate[1] with the forcing frequency in general. When the unstable milling signal is sampled at once-per-revolution/tooth, therefore, it will not repeat [10]. The statistical variation in the unstable (and asynchronous) sampled signal will be larger than for the stable sampled signal.

IN A NUTSHELL It is surprising that, given the complexity of milling (including the tooth passing frequency content and system dynamics), there exists such a simple strategy for directing unstable cuts into stable zones. The basic steps are: 1) record the frequency content of the cutting signal; 2) ignore any frequency components caused by the teeth passing or runout; 3) if there is a significant frequency content remaining, define it as chatter (this is the "chatter detector"); and 4) choose a new spindle speed so that the tooth passing frequency is equal to the detected chatter frequency. This approach identifies a stable speed if one is available at the selected axial and radial depths of cut.

---

[1] In other words, the ratio of the natural frequency to forcing frequency cannot be expressed as a ratio of whole numbers [60].

Example 6.1.2: Selecting new spindle speeds in the presence of competing lobes To continue with our study of milling frequency content in the presence of chatter, let's consider a slightly more complicated dynamic system with two modes modeled in both the x and y directions. The symmetric modal dynamics are described by: $f_{n1} = 800$ Hz, $k_1 = 5 \times 10^6$ N/m, and $\zeta_1 = 0.01$ (first mode); and $f_{n2} = 900$ Hz, $k_2 = 9 \times 10^6$ N/m, and $\zeta_2 = 0.02$ (second mode). We'll consider the 25% radial immersion up milling cut depicted in Fig. 6.1.11, which is completed using a four tooth cutter in a tool-material pair that exhibits a specific force of $K_s = 700$ N/mm^2 and force angle of $\beta = 66$ deg.

The exit angle is calculated using Eq. 4.1.4:

$$\phi_e = \cos^{-1}\left(\frac{r - \frac{r}{2}}{r}\right) = \cos^{-1}(0.5) = 60\,\text{deg}$$

and the start angle for the up milling cut is zero. To find μ_x, we first project F onto the x direction to obtain $F_x = F\cos(\beta - 60)$, where β is given in deg. The projection of this result onto the average surface normal, which occurs at the angle $\phi_{ave} = \frac{0+60}{2} = 30\,\text{deg}$, is $F_n = F_x\cos(60) = F\cos(\beta - 60)\cos(60)$. For μ_y, projection onto the y direction gives $F_y = F\cos(150 - \beta)$. The projection of this result onto the average surface normal is $F_n = F_y\cos(150) = F\cos(150 - \beta)\cos(150)$. The directional orientation factors are therefore $\mu_x = \cos(\beta - 60)\cos(60)$ and $\mu_y = \cos(150 - \beta)\cos(150)$.

This oriented FRF is shown in Fig. 6.1.13 for the symmetric two mode dynamics. The top subplot in Fig. 6.1.14 displays the negative real part of FRF_{orient} so that the two valid chatter frequency ranges may be identified, while the bottom subplot provides the $N = 0$ stability limit versus spindle speed in rev/s (bottom). The two sections correspond to the two chatter frequency ranges and the overall stability limit is the lowest point from the pair at

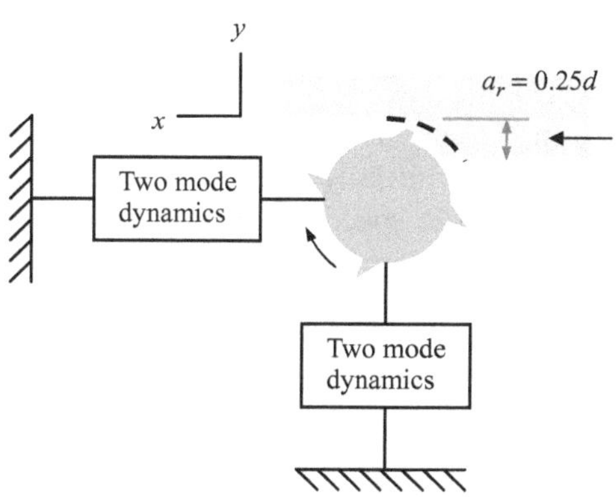

Fig. 6.1.11 25% radial immersion up milling model for Ex. 6.1.2

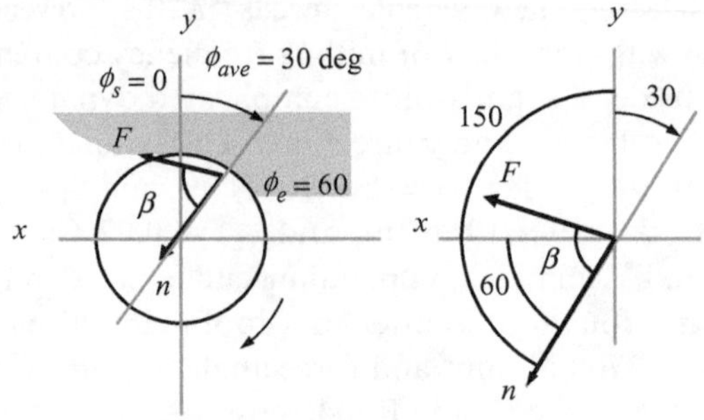

Fig. 6.1.12 Geometry for determining 25% radial immersion up milling directional orientation factors

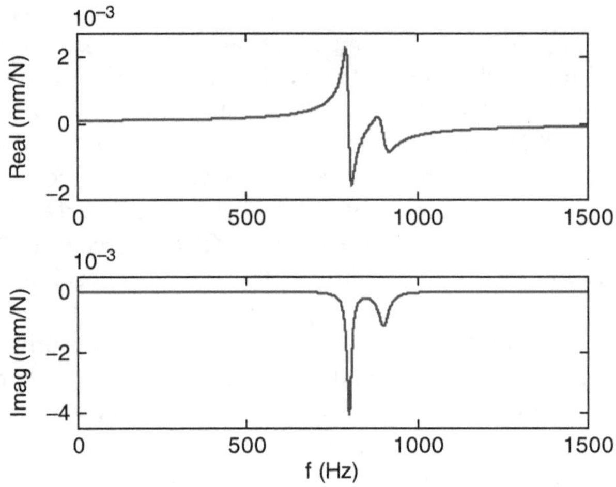

Fig. 6.1.13 Oriented frequency response function for Ex. 6.1.2

each spindle speed where they overlap. You may recall that this situation is referred to as "competing" lobes, as described in Section 4.3.

If the cutting conditions are initially selected to be 20000 rpm (let's say this is the maximum available spindle speed) with an axial depth of 2 mm, chatter would be observed. Similar to the previous example, however, we know that if we can obtain the chatter frequency from an appropriate transducer then we can use this information to select a new spindle speed. In Fig. 6.1.15, the unstable cutting conditions are identified by an 'x' in the top subplot. In the bottom subplot, we see that the chatter frequency for 20000 rpm is 919 Hz; we also know that the tooth passing frequency is $f_{tooth} = \frac{20000 \cdot 4}{60} = 1333$ Hz. These frequencies are marked with an 'x' and circle, respectively. Similar to the competing lobes, the chatter frequency curve for each N value has two distinct

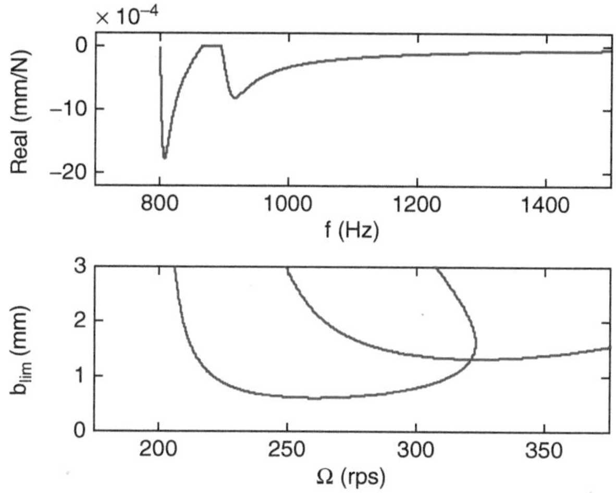

Fig. 6.1.14 Valid chatter frequency range for oriented FRF (top). Corresponding (Ω, b_{lim}) parameter space for $N = 0$ mapping (bottom). Competing lobes are seen due to the two valid chatter frequency ranges

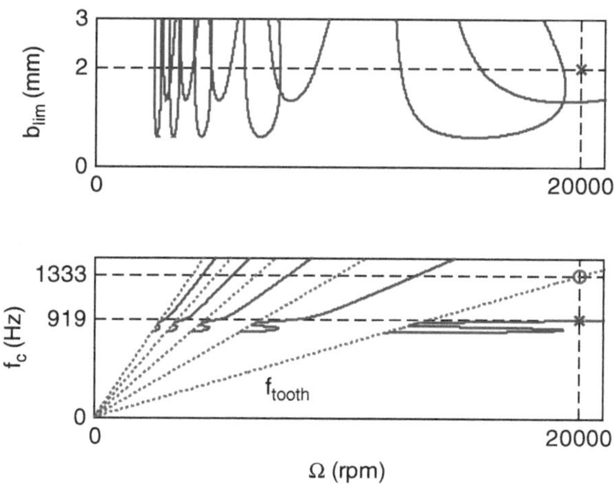

Fig. 6.1.15 (Top) Stability lobes for $N = 0$ to 4; the selected cutting condition of (20000 rpm, 2 mm) is indicated by the 'x'. (Bottom) The tooth passing frequency content at 1333 Hz is identified by the circle, while the chatter frequency at 919 Hz is marked by the 'x'

sections. Because the stability limit for the second section of the $N = 0$ lobe is exceeded in this instance, the chatter frequency also occurs on the second section of the chatter frequency curve for $N = 0$.

Based on our previous discussions, we know that the Fourier transform of the time-domain milling signal for the unstable cut will contain content at f_c, as well as f_{tooth} and its harmonics. The corresponding phase relationship for

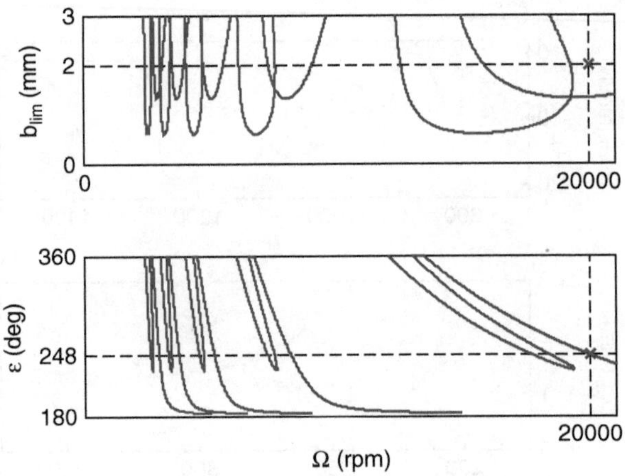

Fig. 6.1.16 (Top) Stability lobes for $N = 0$ to 4; the unstable cutting condition of (20000 rpm, 2 mm) is indicated by the 'x'. (Bottom) The tooth-to-tooth undulation phase of 248 deg for the unstable cut is identified by the 'x'

surface undulations between subsequent teeth is provided in Fig. 6.1.16. From the bottom subplot, we see that $\varepsilon = 248$ deg for the initial, unstable milling conditions. Further, this value is obtained from the second section of the undulation phase curve for $N = 0$.

Based on the 919 Hz chatter frequency from Fig. 6.1.15, the new spindle speed for $N = 0$ is $\Omega = \frac{919 \cdot 60}{(0+1) \cdot 4} = 13785$ rpm. The updated operating condition is shown in the top subplot of Fig. 6.1.17, where we see that the cut is again unstable. We also observe that this initial regulation moved the spindle speed to the left of the second competing lobe because chatter occurred in this section from the first cutting condition. The corresponding chatter frequency

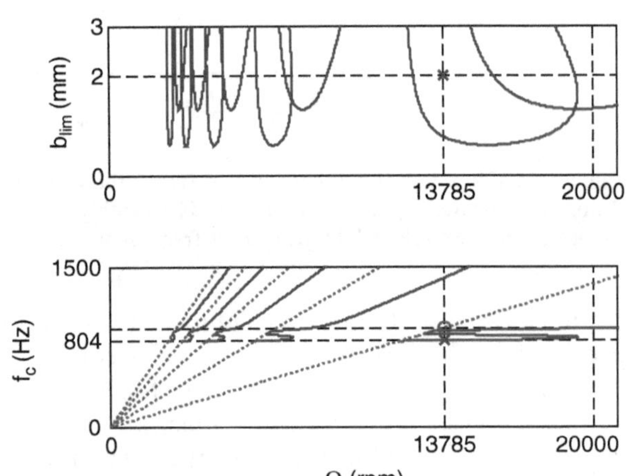

Fig. 6.1.17 (Top) The first spindle speed regulation gives a new speed at 13785 rpm ($N = 0$) based on the 919 Hz chatter frequency from Fig. 6.1.15. The new cutting condition is also unstable. (Bottom) The new cutting condition is also unstable and gives a chatter frequency of 804 Hz ($N = 1$); this value is located by the 'x'. The tooth passing frequency at 919 Hz is identified with a circle

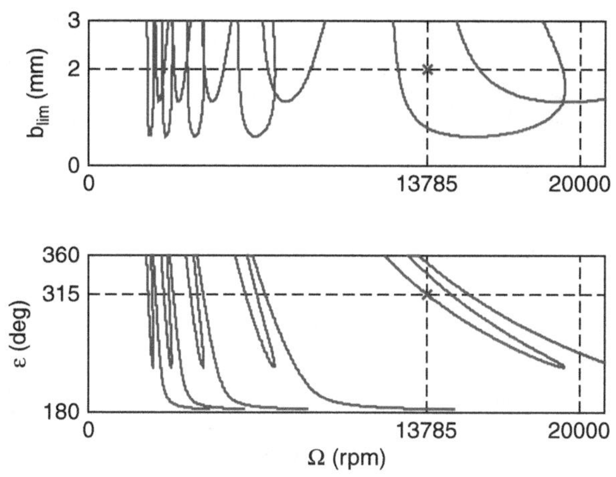

Fig. 6.1.18 (Top) The first spindle speed regulation gives a new speed at 13785 rpm ($N = 0$). This new condition is again unstable. (Bottom) The undulation phase is 315 deg for the new unstable cutting condition, (13785 rpm, 2 mm)

is 804 Hz. As seen in the bottom subplot, this chatter frequency is obtained from the lower portion of the first section of the curve because the stability limit imposed by the first competing lobe is exceeded to the left of the reversal in speed direction at 19384 rpm. The undulation phase is determined from Fig. 6.1.18. We see that $\varepsilon = 315$ deg and the value is again obtained from the first section of the undulation phase curve following the same logic as for the chatter frequency.

A second regulation to $\Omega = \frac{804 \cdot 60}{(0+1) \cdot 4} = 12060$ rpm provides a stable cutting condition, however. The new spindle speed is located to the left of the first competing lobe for $N = 0$, where increases in the stable axial depth may be obtained. The new operating point and corresponding tooth passing frequency (804 Hz) are shown in Fig. 6.1.19. The undulation phase is displayed in Fig. 6.1.20 (358 deg). Figures 6.1.13 through 6.1.20 were obtained using p_6_1_2_1.m, which is included on the companion CD.

6.2 Runout

One complication that we did not address in the previous section is that content other than the tooth passing frequency, its harmonics, and the chatter frequency may be present in the milling signal spectrum (i.e., the magnitude of the Fourier transformed time-domain signal). Even in the absence of external noise sources, such as pumps on the machine tool and other equipment on the shop floor, for example, runout (or eccentricity) leads to content in the milling

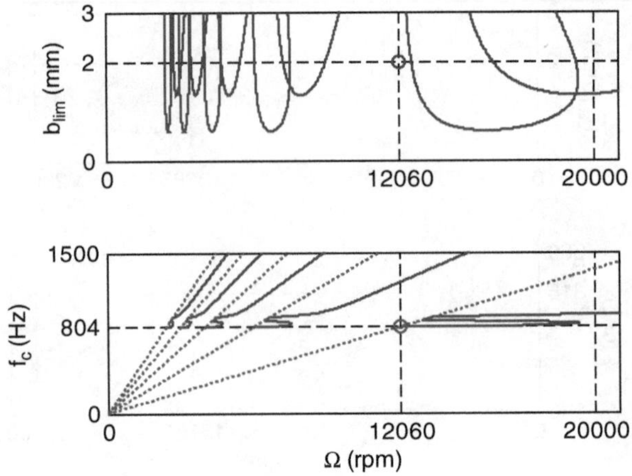

Fig. 6.1.19 (Top) The second spindle speed regulation gives a new stable speed of 12060 rpm ($N = 0$) based on the 804 Hz chatter frequency from Fig. 6.1.17. (Bottom) The tooth passing frequency for the stable cut is 804 Hz (circle)

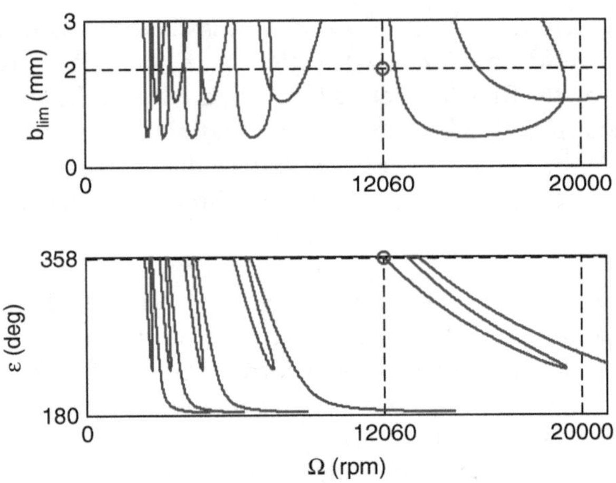

Fig. 6.1.20 (Top) The second spindle speed regulation provides a new stable speed of 12060 rpm ($N = 0$). (Bottom) The undulation phase is 358 deg for the stable cutting condition, (12060 rpm, 2 mm)

signal which is synchronous with spindle speed. Here we include the following possibilities under the generic heading of runout:

- axis of rotation errors of the spindle, including radial excursions of the spindle centerline during rotation [11-13];
- an offset between the holder centerline and spindle axis of rotation;
- an offset between the tool centerline and holder centerline; and

- radii variation between cutter teeth due to imperfect grinding for a solid tool or errors in the cutting edge placement for inserted cutters.

For demonstration purposes, let's focus on variation in the teeth radii as shown in Fig. 6.2.1. Naturally, the associated variation in chip thickness due to the "big tooth" is synchronous with spindle rotation. This generates frequency content at the runout frequency, $f_{ro} = \frac{\Omega}{60} = \frac{f_{tooth}}{N_t}$ (Hz), where Ω is expressed in rpm and f_{tooth} in Hz. We also observe content at the higher harmonics $2f_{ro}$, $3f_{ro}$, ... in general.

Example 6.2.1: Frequency content in the presence of runout If we return to Ex. 6.1.1 and plot the chatter frequency, tooth passing frequency and harmonics, and runout frequency with its first few harmonics (using p_6_2_1_1.m), we see that the third runout harmonic, $3f_{ro}$, intersects the chatter frequency curve (solid line). We also observe that $4f_{ro}$ (dot-dashed line) exactly overlaps f_{tooth} (dotted line). The latter is expected because $N_t = 4$. According to Fig. 6.2.2 (bottom subplot), we cannot discern between a potential chatter frequency and the third runout harmonic at 808 Hz, as indicated by the diamond. This is an important issue because one strategy to aid in isolating the chatter frequency for unstable cuts is applying a comb filter, i.e., a frequency-domain filter that passes all content except at selected equally spaced frequencies, to reject the tooth passing frequency and its harmonics [3-4]. The same technique cannot be applied to remove runout content if it coincides with possible chatter frequencies [14].

Let's use Eq. 6.1.2 to determine when this intersection between the chatter frequency and runout harmonics may occur. Restating this equation, we have that $\frac{f_c}{\Omega N_t} = N + \frac{\varepsilon}{2\pi}$. We also have the relationship $nf_{ro} = n\Omega$ where $n = 1, 2, 3, \ldots$ and Ω is expressed in rev/s in both instances. We are looking for instances when the chatter frequency is equal to a runout harmonic, or $f_c = nf_{ro} = n\Omega$. Substituting for f_c from Eq. 6.1.2 and solving for n, we obtain:

$$n = N_t\left(N + \frac{\varepsilon}{2\pi}\right). \tag{6.2.1}$$

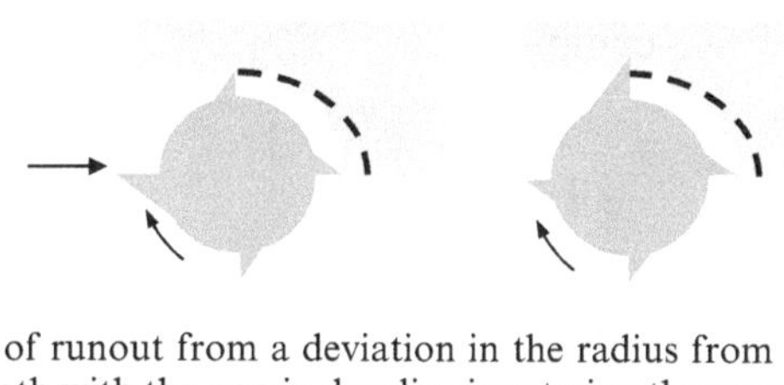

Fig. 6.2.1 Example of runout from a deviation in the radius from tooth to tooth in a solid endmill. (Left) A tooth with the nominal radius is entering the up milling cut where the chip thickness is zero according to the circular tool path approximation. (Right) The "big tooth" has a non-zero chip thickness for the same cutter angle

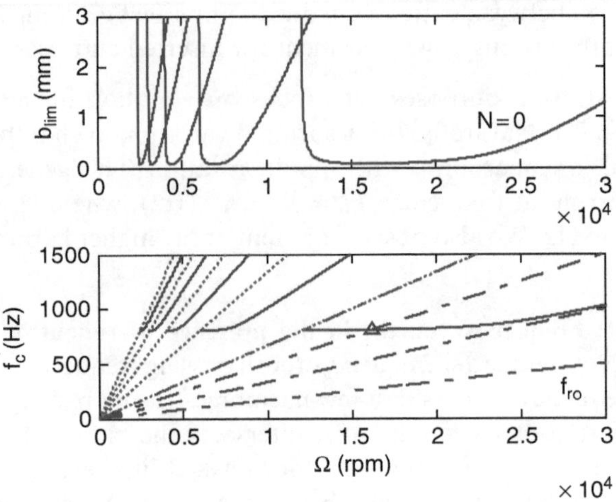

Fig. 6.2.2 An intersection between the third runout harmonic and the chatter frequency curve is seen at 808 Hz (marked by the diamond). This impedes our ability to identify potential chatter frequencies in the milling signal spectrum

Because $\pi < \varepsilon < 2\pi$ rad (see Fig. 3.3.2, for example), only particular integer n values yield acceptable ε values for a selected number of teeth and lobe number. We'll explore this in Ex. 6.2.2.

Example 6.2.2: Identifying intersections between runout harmonics and chatter frequencies Consider a four tooth cutter that exhibits runout. For the $N = 0$ lobe, we have that $n = 4\left(\frac{\varepsilon}{2\pi}\right)$ from Eq. 6.2.1. If $n = 1$, we obtain $\varepsilon = \frac{\pi}{2}$; therefore, no intersection is possible because this is not an acceptable ε value. For $n = 2$, we find that $\varepsilon = \pi$. However, because $\varepsilon \to \pi$ only as $f_c \to \infty$, this asymptotic approach is not practically important. For $n = 3$, on the other hand, $\varepsilon = \frac{3\pi}{2}$ and an intersection is possible as shown in Fig. 6.2.2. We also know from Section 4.3 that this is the least favorable tooth-to-tooth undulation phase relationship (for a single degree of freedom system). Note that if we know ε and f_c, then we can calculate the corresponding spindle speed for a selected lobe number by rearranging Eq. 6.1.2. Specifically, we compute:

$$\Omega = \frac{f_c}{\left(N + \frac{\varepsilon}{2\pi}\right)N_t} = \frac{808}{\left(0 + \frac{\frac{3\pi}{2}}{2\pi}\right)4} = 269.3\,\text{rev/s} = 16160\,\text{rpm}.$$

When $n = 4$, we have an overlap with f_{tooth} and intersection is only possible for $\varepsilon = 2\pi$, which represents the most favorable undulation phase.

Runout in milling affects the instantaneous chip thickness as demonstrated in Fig. 6.2.1. Therefore, the force is also influenced and, subsequently, the vibration. Well-known outcomes include premature cutting edge failure and increased machined surface roughness. We can investigate these effects using updated versions of our time-domain simulations.

6.5.1 Simulation Modification

Incorporating the effects of runout in the cutter teeth, as depicted in Fig. 6.2.1, is straightforward. We'll first consider the circular tool path time-domain simulation for helical square endmills introduced in Section 4.5. Including runout requires that we modify the chip thickness, `h`, calculation to be:

```
h = ft*sin(phia*pi/180) + surf(cnt4, phi_counter) - n +
    RO(cnt3);
```

where `ft` is the feed per tooth, `phia` is the tooth angle for the current tooth and axial slice, `surf` is the array the contains the surface position for the previous tooth at each axial slice, `n` is the current vibration along the instantaneous surface normal direction, and `RO` is the vector that contains the tooth-to-tooth runout values. To establish a convention for the simulation and match general measurement procedures, we'll normalize the `RO` vector entries to the largest tooth radii (i.e., we'll set the dial indicator to zero at the largest tooth and measure the deviation in radii, if any, for the other teeth). This means that one `RO` entry will be zero and all other values will be zero (for no runout) or less than zero if runout is present. Naturally, the number of entries in this vector is N_t.

In addition to changing the chip thickness calculation, the `surf` array updating must include the runout effect. Specifically, we use the following line when the computed chip thickness is greater then zero (i.e., the current tooth is cutting).

```
surf(cnt4, phi_counter) = n - RO(cnt3);
```

This approach neglects variation in runout along the tool axis, which is clearly possible for actual cutters. To include axial dependence, the `RO` vector could be redefined as an array with a column for each tooth and a row for each axial slice, for example. We leave this activity as an exercise. As an example of the effect of runout, let's show the force profile for the helical endmill described in Ex. 4.5.1 together with the force variation in the presence of tooth-to-tooth runout.

Example 6.2.3: Comparison of forces with and without runout In this example, we compare the cutting forces produced by helical endmills with and without runout. We'll consider a 30% radial immersion up milling cut (zero start angle and 66.4 deg exit angle). There are two identical modes in both the x and y directions. The modal parameters are: $f_{n1} = 800$ Hz, $k_{q1} = 2 \times 10^7$ N/m, and $\zeta_{q1} = 0.05$; and $f_{n2} = 1000$ Hz, $k_{q2} = 1.5 \times 10^7$ N/m, and $\zeta_{q2} = 0.03$. The workpiece material is an aluminum alloy and it is machined with a four tooth, 19 mm diameter, 45 deg helix square endmill using a feed per tooth of 0.15 mm/tooth. The cutting force coefficients are $k_t = 520$ N/mm^2 and $k_n = 300$ N/mm^2. The corresponding stability diagram is shown in Fig. 4.5.5. The resultant

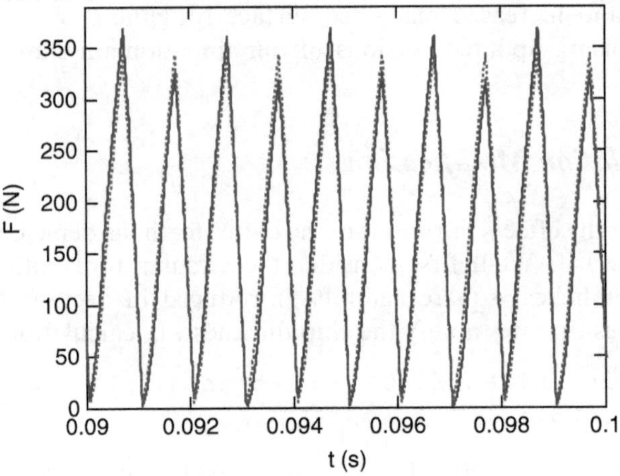

Fig. 6.2.3 Comparison of the resultant cutting force without runout (dotted line) and with runout (solid line) for the circular tool path simulation

cutting force for an axial depth of 5 mm at a spindle speed of 15000 rpm with no runout is shown as the dotted line in Fig. 6.2.3. The force with runout, `RO = [ 0 -10 0 -15]` μm, is shown as the solid line. At $t = 0.0977$ s, the runout force is smaller by 25.5 N (7.4% decrease from the nominal force). At $t = 0.0947$ s, the runout force is larger by 25.1 N (7.3% increase). This behavior matches the "big-little" teeth profile described by the selected runout values. Figure 6.2.3 was produced using the MATLAB® program p_6_2_3_1.m.

Let's now consider the cycloidal tool path simulation. We define the tooth dependent runout values in vector form, again using the same convention (normalized to a maximum value of zero). The only other changes to the program described in Section 5.3 are updating the radius with the runout values, `r = d/2 + RO`, and indexing `r` by the appropriate `Nt` counter each time it appears [15].

Example 6.2.4: Comparison of cycloidal and circular tool path results In this example, we compare the cutting forces for the cycloidal tool path simulation (p_6_2_4_1.m) to those observed in Ex. 6.2.3. The conditions are identical. The results with and without runout, `RO = [ 0 -10 0 15]` μm, are shown in Fig. 6.2.4. We see that the force levels are similar to those seen in Fig. 6.2.3. Perhaps more interesting, however, is a comparison of the machined surface profiles predicted by the cycloidal simulation. Figure 6.2.5 shows the surface without runout; the corresponding roughness average is 0.08 μm. When including runout, the roughness average increases to 0.66 μm (greater than eight times larger) as seen in Fig. 6.2.6. Note that the surface location error at the tool's free end is essentially unaffected: 87.4 μm overcut without runout and 86.1 μm when including runout effects. Additional information regarding the influence of runout on machining behavior can be found in [15-34]. In [15], for example,

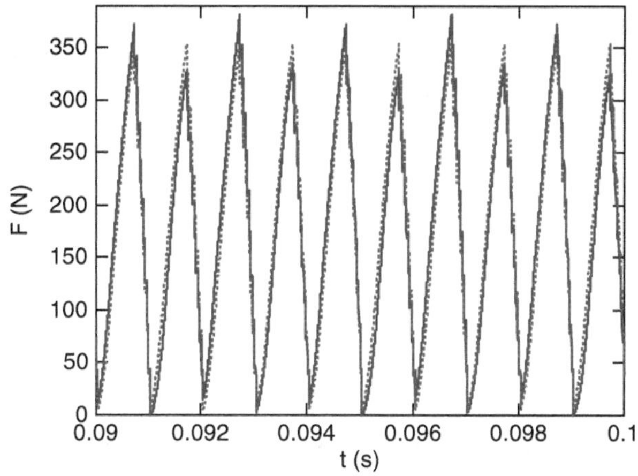

Fig. 6.2.4 Comparison of the resultant cutting force without runout (dotted line) and with runout (solid line) for the cycloidal simulation

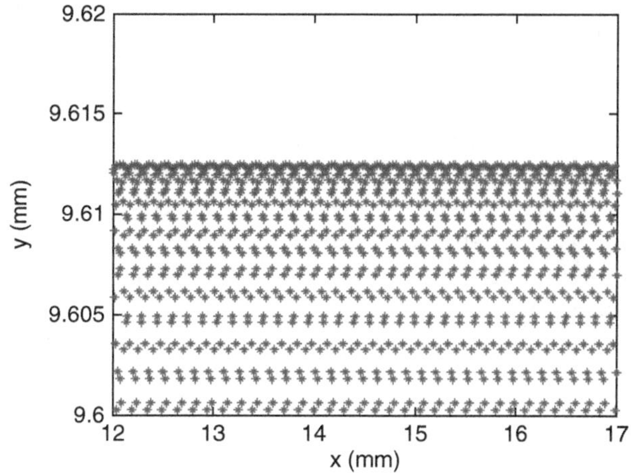

Fig. 6.2.5 Machined surface profile for Ex. 6.2.4 with no runout. The '*' symbols indicate the simulated cutting edge locations. The surface is shown by the solid line; it is defined using the "bubble up" algorithm described in Section 5.3. Note that the remaining material is located above the line for the up milling cut

the influences of runout, feed per tooth, and teeth spacing on surface roughness, surface location error, and stability are explored.

6.3 Variable Teeth Spacing

In all our previous stability analyses, we have assumed that the teeth are equally spaced around the cutter periphery (i.e., constant teeth pitch). It is of course also possible to place the teeth with unequal, or non-proportional, spacing.

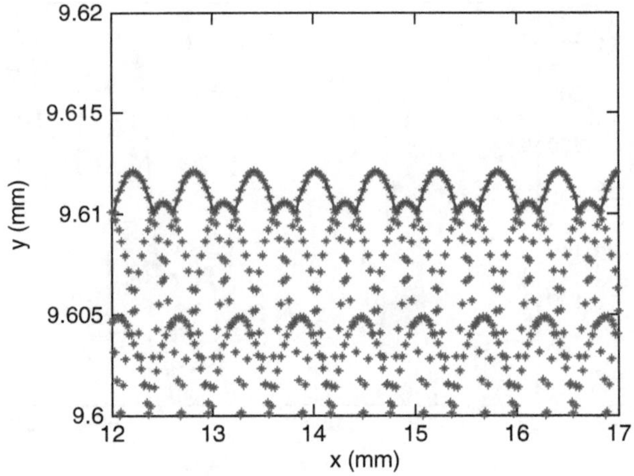

Fig. 6.2.6 Machined surface profile for Ex. 6.2.5 with runout. The roughness average is approximately eight times higher than the surface in Fig. 6.2.5

Several researchers have implemented variable teeth spacing to interrupt regeneration of surface waviness (caused by tool vibrations) and, therefore, modify stability behavior [35–43]. Although we could argue that determining the stability lobe diagram and using it to select cutting parameters is a valid approach, the use of variable teeth spacing can be applied in situations where it is inconvenient to adjust the spindle speed, such as a transfer line, or the larger stable lobes available at higher spindle speeds are inaccessible due to significant tool wear at the corresponding spindle (and cutting) speeds. In these cases, changing the teeth spacing can yield stable zones where they would not otherwise exist. Varying the helix angle from tooth to tooth has also been implemented to reduce chatter [44], but we will focus on variable teeth spacing here.

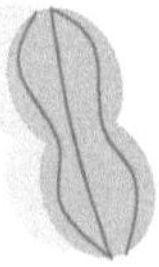

IN A NUTSHELL Because non-proportional tooth spacing disturbs regeneration of waviness, it can improve milling stability. Not all spacing selections increase stability, however, and whether or not there is improvement depends on the system dynamics, tooth spacing, and spindle speed. In addition, stability improvement by this technique generally requires an accompanying feed reduction. Constant spindle speed with non-proportional tooth spacing produces a non-constant feed per tooth. Because the maximum permissible chip load is a function of the cutting edge strength, the maximum feed for a tool with non-proportionally spaced teeth is controlled by the tooth with the largest preceding angle. All other teeth are essentially under used and the required feed reduction must be recovered through an increased axial depth of cut to simply break even in terms of the material removal rate.

6.7.1 Simulation Updating

As described in Section 4.3, there is a periodic time delay between teeth engagements due to the (typically) uniform teeth spacing. Changing the teeth spacing, therefore, varies this time delay and interrupts the surface regeneration periodicity. We can explore this through minor modifications to our time-domain simulations. First, let's consider the circular tool path code. The first modification is to directly define the angles of the individual teeth starting from tooth 1 located at an initial angle of zero. The teeth are ordered by increasing clockwise angles as described in Section 4.4 and shown in Fig. 6.3.1. Teeth angles of {0 95 180 275} deg are depicted, where teeth 2 and 4 have been advanced by five degrees relative to their uniform spacing angles of 90 degrees. To describe the teeth spacing, a new vector is defined in p_6_3_1_1.m.

```
tooth_angle = [0 95 180 275];
```

Due to the variable teeth angles, the feed per tooth also changes from one tooth to the next. This is handled using the following code, where `ft_mean` is the mean feed per tooth value (m/tooth) and `theta` is the angle between teeth (deg); it is defined using the MATLAB® `diff` function.

```
ft_mean = 0.15e-3;
theta = diff([tooth_angle 360]);
for cnt = 1:Nt
    ft(cnt) = (ft_mean*theta(cnt)*Nt)/360;
end
```

Based on these modifications, two additional changes are made. First, the `teeth` vector, which is used to index the appropriate `phi` entry, is redefined using the following `for` loop. The `round` function is necessary because the teeth entries must be integer values.

```
for cnt = 1:Nt
    teeth(cnt) = round(tooth_angle(cnt)/dphi) + 1;
end
```

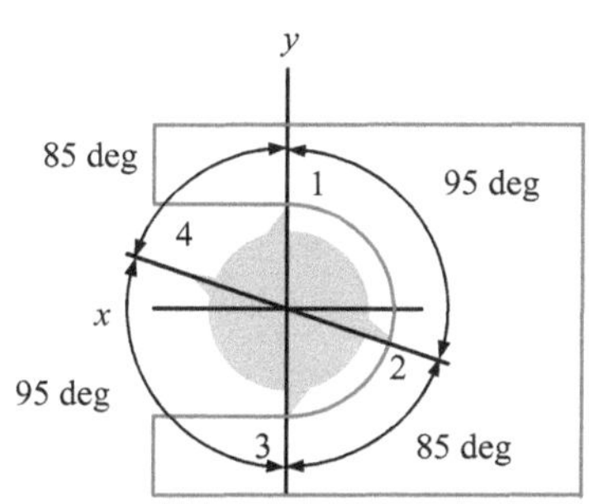

Fig. 6.3.1 Variable teeth spacing example with four teeth at angles {0 95 180 275} deg

Finally, each time the feed per tooth appears, it must now be indexed using the tooth (`Nt`) counter: `ft(cnt3)`. Note that the runout vector entries, `RO`, must now correspond to the appropriate entries in `tooth_angle`.

Example 6.3.1: Comparison of uniform and variable teeth spacing forces Let's compare the cutting forces produced by endmills with uniform and variable teeth spacing. We'll consider the same 30% radial immersion up milling cut (zero start angle and 66.4 deg exit angle) as in the previous section. Again, there are two identical modes in the x and y directions with modal parameters: $f_{n1} = 800$ Hz, $k_{q1} = 2 \times 10^7$ N/m, and $\zeta_{q1} = 0.05$; and $f_{n2} = 1000$ Hz, $k_{q2} = 1.5 \times 10^7$ N/m, and $\zeta_{q2} = 0.03$. The workpiece material is an aluminum alloy machined with a four tooth, 19 mm diameter square endmill using a feed per tooth of 0.15 mm/tooth. The cutting force coefficients are $k_t = 520$ N/mm^2 and $k_n = 300$ N/mm^2. We'll consider tools with zero helix angles (straight teeth) and tooth angles of {0 90 180 270} deg and {0 95 180 275} deg.

Figure 6.3.2 shows the resultant forces for an axial depth of 3 mm at 15000 rpm. The figure is produced using p_6_3_1_1.m, where the teeth numbers are marked according to the convention shown in Fig. 6.3.1. Both cuts are clearly stable; however, we see that the time between one tooth's exit and the next tooth's entry (where the force is zero) varies periodically with the variable teeth spacing. As we'd expect, the time is smaller for the decreased spacing between teeth 1 and 4 and, similarly, teeth 3 and 2.

Next let's explore the global stability behavior of the variable teeth spacing tool relative to the uniform teeth spacing tool. We obtain the stability lobe diagram for the latter using p_6_3_1_2.m; see Fig. 6.3.3. To establish a baseline for the comparison of the time-domain and frequency-domain solutions, let's

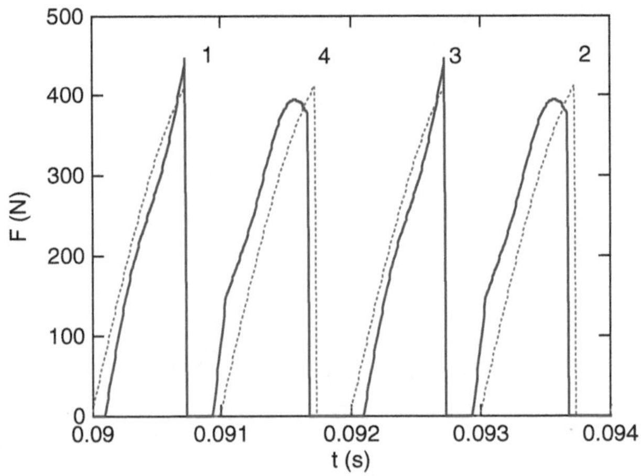

Fig. 6.3.2 Resultant cutting force for uniform (dotted line) and variable (solid line) teeth spacing with angles {0 95 180 275} deg. The helix angle is zero, the axial depth of cut is 3 mm, and the spindle speed is 15000 rpm for the 30% radial immersion up milling cut. The teeth numbers are also identified (as depicted in Fig. 6.3.1)

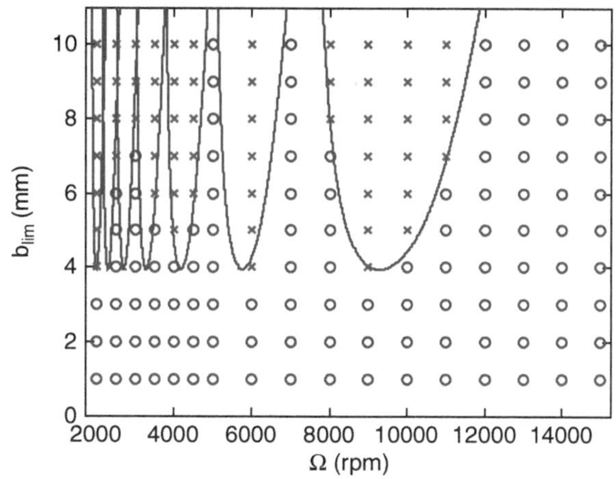

Fig. 6.3.3 Stability limits for uniform teeth spacing cutter. The frequency-domain solution for the same tool geometry is identified by the solid line. Using the circular tool path time-domain simulation, stable results are identified by circles and unstable results by 'x' symbols

select a grid of points and complete time-domain simulations using p_6_3_1_1.m. For each {Ω, *b*} combination, we determine stability from the corresponding force and displacement profiles. The stable cuts are marked with a circle in Fig. 6.3.3 and the unstable cuts with an 'x'. Good agreement is observed.

Implementing the variable teeth spacing {0 95 180 275} deg yields the results provided in Fig. 6.3.4. Again, the variable teeth spacing stability is compared to the frequency-domain uniform teeth spacing solution. A significant increase in the allowable axial depth of cut is observed in the region near 4000 rpm.

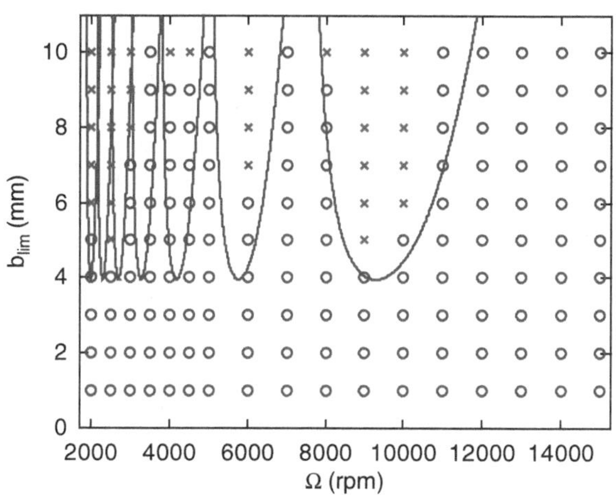

Fig. 6.3.4 Stability limits for variable teeth spacing, {0 95 180 275} deg, cutter. The frequency-domain solution for the uniform teeth spacing cutter is again identified by the solid line. Using the circular tool path time-domain simulation for the variable teeth spacing cutter, stable results are identified by circles and unstable results by 'x' symbols. Improved stability is seen near 4000 rpm

Although we would not expect the same behavior for different system dynamics, this example does demonstrate the potential gains made available by selection of appropriate tooth angles.

In a similar manner, we can model variable teeth spacing in the cycloidal tool path time-domain milling simulation by specifying the appropriate teeth angles in the vector `teeth`. This vector serves the same function as `tooth_angle` in the circular tool path simulation. See the MATLAB® program p_6_2_4_1.m included on the companion CD

6.4 Low Radial Immersion Milling

In 1998, Davies *et al.* reported that "undesirable vibrations observed in partial immersion cuts seem inconsistent with existing theory" [10]. Using a Poincaré sectioning technique (once-per-revolution sampling) combined with capacitive measurements of the tool shank displacements in the x (feed) and y directions during cutting, they found that some unstable low radial immersion cuts gave discrete clusters of once-per-revolution sampled points when plotted in the x-y plane, while others presented elliptical distributions. They subsequently showed that this behavior was the manifestation of two different types of instability [45–46]. Traditional quasiperiodic chatter, also referred to as Hopf bifurcation[2], leads to the elliptical distribution of periodically sampled points. The second instability type, encountered during low (less than 25%) radial immersions, is a period doubling, or flip, bifurcation. It reveals itself as two tightly grouped clusters of sampled points as opposed to a single group of points for the synchronous vibrations that occur during stable cutting with forced vibrations only. Subsequent modeling efforts are described in [47-53] and include temporal finite element analysis, time-domain simulation, a multi-frequency analytical solution, and the semi-discretization approach. These techniques give improved accuracy for the predicted stability limit over the average tooth angle and frequency-domain approaches (Chapter 4) in very low radial immersion (less than 10%) cases. However, they do not offer the convenient closed-form expressions for the stability boundary.

IN A NUTSHELL The cutting force in low radial immersion milling resembles a series of impacts. In between impacts, the tool experiences free vibration. The new cutting force depends on where the tool is in its decaying free vibration cycle when the tooth next encounters the previous surface. The result is the formation of additional stable areas near what was previously the worst case for stability.

---

[2] In the analysis of dynamic systems, a bifurcation represents the sudden appearance of a qualitatively different solution for a nonlinear system as some parameter is varied [61].

Example 6.4.1: Low radial immersion stability We can use time-domain simulation to explore the Hopf and flip bifurcations. By modifying the cycloidal tool path code to include once-per-revolution sampling, we can observe the two instabilities in x (feed direction) versus y displacement plots [10]. We'll use single degree of freedom, symmetric dynamics ($f_n = 720$ Hz, $k = 4.1 \times 10^5$ N/m, and $\zeta = 0.009$) for a 5% radial immersion up milling cut (zero start angle and 25.8 deg exit angle). The workpiece material is an aluminum alloy machined with a single tooth, 45 degree helix, 8 mm diameter square endmill using a feed per tooth of 0.1 mm/tooth. The cutting force coefficients are $k_t = 644$ N/mm^2 and $k_n = 238$ N/mm^2. These conditions mimic those reported in [53] to enable convenient comparison. The code added in p_6_4_1_1.m to enable once-per-revolution sampling follows, where `xpos` and `ypos` are the tool displacements in the x and y directions, respectively, `steps_rev` is the number of simulation steps per revolution of the cutter, and `xsample` and `ysample` are the sampling results.

```
xsample = xpos(1:steps_rev:length(xpos));
ysample = ypos(1:steps_rev:length(ypos));
```

Figure 6.4.1 shows the stability limit obtained using the frequency-domain solution (Section 4.3) as a solid line. The region shown corresponds to the left side of the $j = 0$ (rightmost) lobe. The results of time-domain simulations are identified by: circles (stable); 'x' symbols (Hopf bifurcation); and triangles (flip bifurcation). A narrow band of increased stability is seen between 27000 rpm and 28000 rpm. This is accompanied by the spindle speed range from 29000 rpm and 32000 rpm which exhibits flip bifurcation behavior. Three points are selected for further study. The {28000 rpm, 2 mm} point demonstrates the traditional Hopf instability; see Fig. 6.4.2 for the time history and Fig. 6.4.3

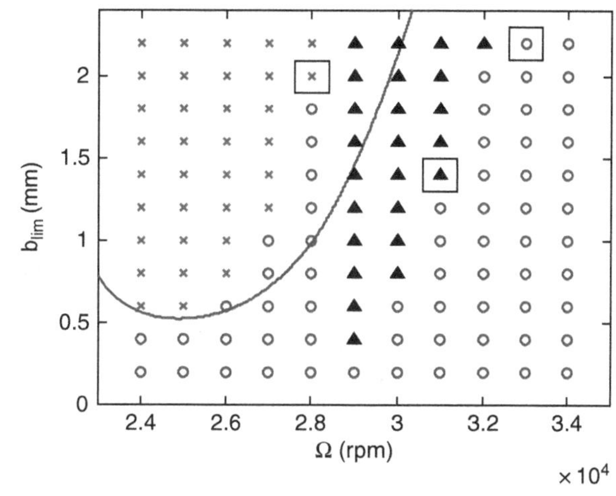

Fig. 6.4.1 The frequency-domain solution stability limit (solid line) is compared to time-domain simulation results for the $j = 0$ lobe. For the time-domain simulations, stable cuts are represented by circles and unstable cuts by 'x' symbols (Hopf) and triangles (flip)

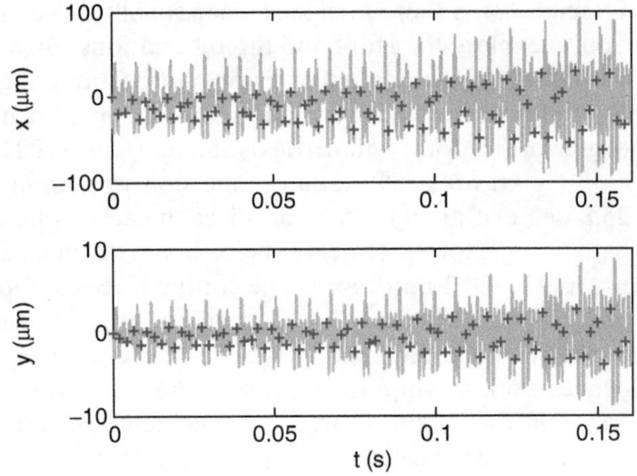

Fig. 6.4.2 Time history for the x (feed) and y direction displacements obtained from the {28000 rpm, 2 mm} cut in Fig. 6.4.1. The once-per-revolution sampled data ('+' symbols) demonstrate quasiperiodic, Hopf instability

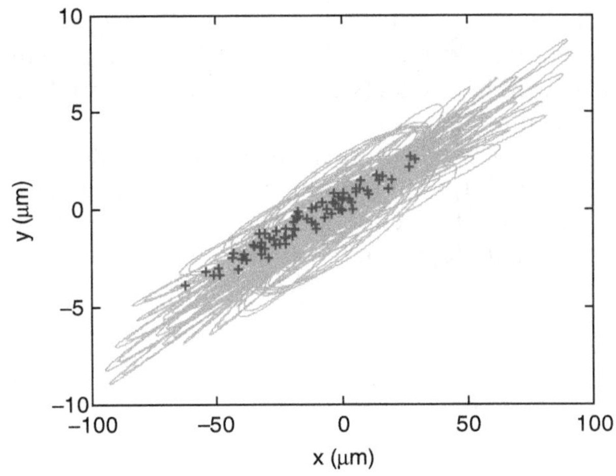

Fig. 6.4.3 Plot of x versus y direction displacements obtained from the {28000 rpm, 2 mm} cut. The elliptical distribution of the once-per-revolution sampled data ('+' symbols) indicates Hopf instability

for the x vs. y plot. As discussed previously, the once-per-revolution sampled data appears as an elliptical distribution for Hopf instability. Conversely, Figs. 6.4.4 and 6.4.5 show the flip bifurcation for the {31000 rpm, 1.4 mm} operating parameters. The synchronously sampled data now occur in two clusters after the initial transients attenuate in the x vs. y plot (Fig. 6.4.5). A stable cut is represented by the {33000 rpm, 2.2 mm} spindle speed, axial depth of cut pair. As expected, Figs. 6.4.6 and 6.4.7 display repetitive behavior from one revolution to the next (forced vibration only).

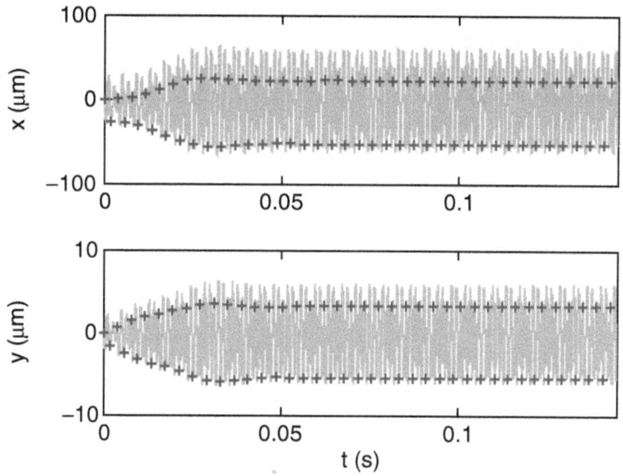

Fig. 6.4.4 Time history for the x (feed) and y direction displacements obtained from the {31000 rpm, 1.4 mm} cut. The once-per-revolution sampled data ('+' symbols) demonstrate a flip bifurcation

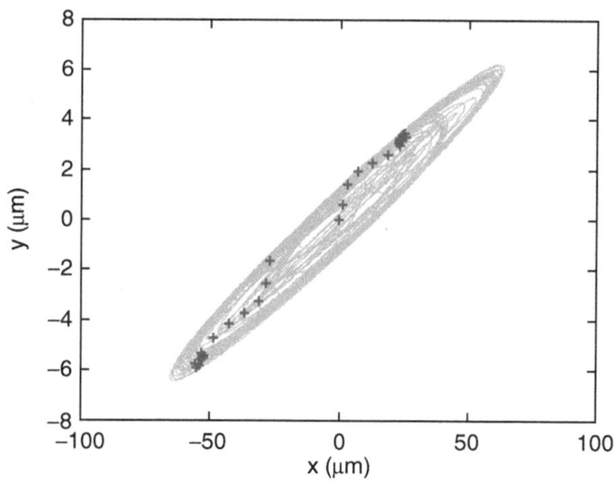

Fig. 6.4.5 Plot of x versus y direction displacements obtained from the {31000 rpm, 1.4 mm} cut. The two clusters of once-per-revolution sampled data ('+' symbols) identify a flip instability

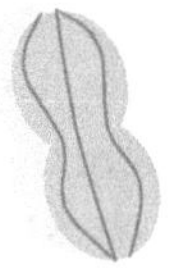

IN A NUTSHELL For those concerned with detailed process modeling, the exact nature of the milling instability (Hopf or flip bifurcation) is extremely interesting. For practical machining applications, we just need to consider the radial depth of cut. As long as the radial depth is high, then the stability lobe algorithms described in Section 4.3 are valid. When the radial depth of cut is low, additional stable zones appear that "split" the higher radial depth stability lobes.

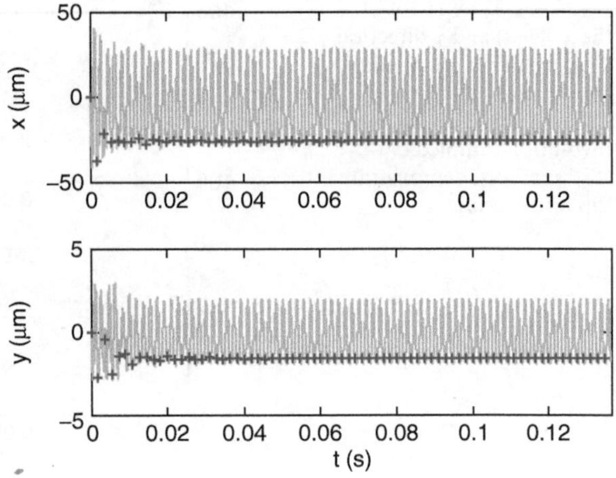

Fig. 6.4.6 Time history for the x (feed) and y direction displacements obtained from the {33000 rpm, 2.2 mm} cut. The once-per-revolution sampled data ('+' symbols) demonstrate stable behavior

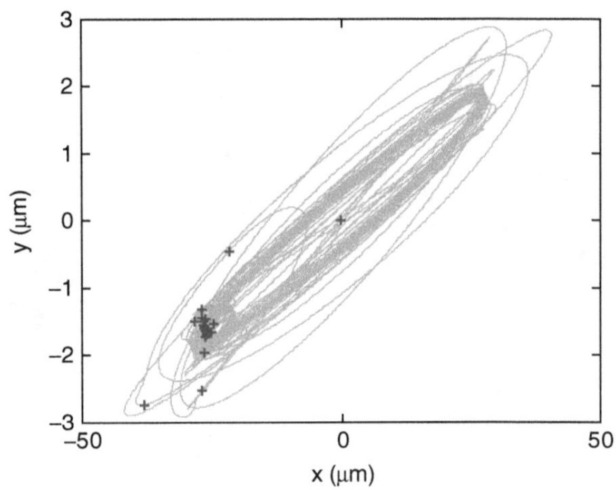

Fig. 6.4.7 Plot of x versus y direction displacements obtained from the {33000 rpm, 2.2 mm} cut. The single cluster of once-per-revolution sampled data ('+' symbols) indicates stable operation

6.5 Uncertainty Propagation

In the stability lobe diagrams we've displayed so far, the stability boundary has been represented by a single line. This indicates step-like behavior, where the cut is stable below the line and unstable above. If the inputs were perfectly known and the theory exactly captured the physical behavior, then this step behavior could be true (although in practice, even experienced machinists could disagree over whether a particular cut was stable or unstable near the stability limit). However, no measured quantity, such as the cutting force coefficients or tool point frequency response function, is perfectly known. Instead, there are uncertainties associated with these inputs. Additionally, the stability algorithms incorporate approximations that limit their accuracy.

As stated in the National Institute of Standards and Technology Technical Note 1297 [54], “the result of a measurement is only an approximation or estimate of the value of the specific quantify in question, that is, the measurand, and thus the result is complete only when accompanied by a quantitative statement of its uncertainty”. The inclusion of a defensible uncertainty statement enables the user to determine his/her confidence in the measurement and its usefulness in decision making. This concept can be extended to simulation results based on measured input quantities. Again, the user requires some indication of the reliability of the analysis output to gage its usefulness.

Guidelines for evaluating the uncertainty in measurement results are described in [54-57], for example. Often the measurand is not observed directly, but is expressed as a mathematical function of multiple input quantities. In this case, the fundamental steps in uncertainty estimation are to define the measurand, identify the input uncertainty contributors and their distributions, and propagate the uncertainties through the measurand using either analytical (Taylor series expansion) or sampling (such as Monte Carlo or Latin hypercube) approaches.

Identification of the uncertainty in the stability limit for both the average tooth angle [1] and frequency-domain [58] solutions is described in [59]. In this work, Monte Carlo simulation was applied to propagate uncertainties in the measured tool point FRF, cutting force coefficients (determined using the method outlined in Section 4.7), and radial depth of cut through the two approaches. In Monte Carlo simulation, random samples from the input variable distributions are selected and the output is computed over many iterations. The mean and standard deviation in the output are then reported. For the stability analyses, this requires that a new diagram is computed in each of the iterations. It is then necessary to identify the stability limit distribution at each spindle speed within the range of interest. The uncertainty is therefore spindle speed dependent and forms an envelope around the mean stability boundary as depicted in Fig. 6.5.1. To interpret the uncertainty region shown in the figure, we can state that it represents the axial depths, at the corresponding spindle speeds, where the cuts can either be stable or unstable. Above the upper bound,

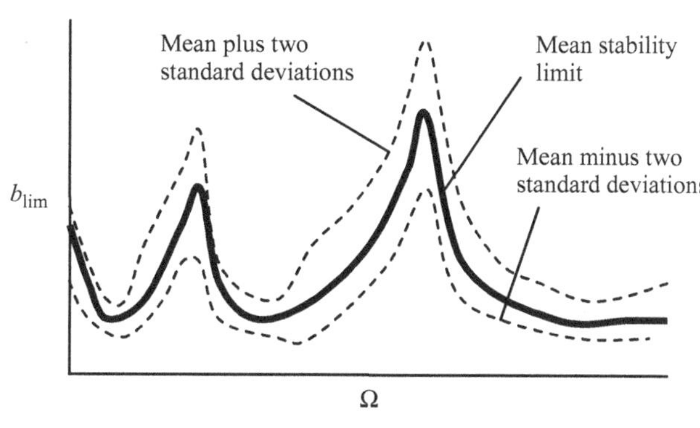

Fig. 6.5.1 Representation of two standard deviation uncertainty limits superimposed on the mean stability boundary determined from Monte Carlo simulation [59]

cuts are expected to be unstable, while cuts should be stable below the lower bound (with 95% confidence due to the selection of plus/minus two standard deviations about the mean). Naturally, we could apply the same approach to determining the uncertainty in surface location error predictions using the frequency-domain technique described in Section 5.2.

IN A NUTSHELL Uncertainties in the input data are part of the reason that simplified expressions for cutting forces and stability lobes are widely applied. It does not make much sense to attempt to model the location of the stability boundary with great accuracy when the force model coefficients, cutting geometry, system dynamics, etc. are only approximately known. Certainly there is an incentive (as in all of engineering) to strive for improved accuracy in models, but it is also sensible to avoid implementing models that are more complicated than the uncertainties in the required inputs warrant.

Exercises

1. For parts a) through d), indicate the action of an automatic spindle speed regulation system for chatter avoidance in milling. The chatter avoidance system operates by: 1) sampling the sound signal produced by the cutting process using a microphone; 2) computing the Fourier transform of the microphone signal; and 3) analyzing the spectrum content. Based on the spectrum content, a new spindle speed is recommended if chatter is sensed.

   a) The cutter has eight teeth, the spindle speed is 3000 rpm, and the frequency spectrum shows a large peak at 400 Hz. The maximum available spindle speed is 7500 rpm.
   b) The cutter has six teeth, the spindle speed is 4200 rpm, and the frequency spectrum shows a large peak at 380 Hz. The maximum available spindle speed is 5000 rpm.
   c) The cutter has four teeth, the spindle speed is 10000 rpm, and the frequency spectrum shows a large peak at 820 Hz. The maximum available spindle speed is 10000 rpm.
   d) The cutter has four teeth, the spindle speed is 30000 rpm, and the frequency spectrum shows a large peak at 1520 Hz. The maximum available spindle speed is 30000 rpm. The spindle speed is regulated once and the spectrum of the second cut shows a large peak at 2280 Hz.

2. Calculate the ε value(s) in radians which correspond to coincidence(s) between the runout harmonics and chatter frequencies for the $N = 0$ lobe when using a cutter with three teeth.
3. Chatter was observed for milling at 7000 rpm with a 4 mm axial depth of cut. The helical square end mill had four teeth. Using the diagrams provided in

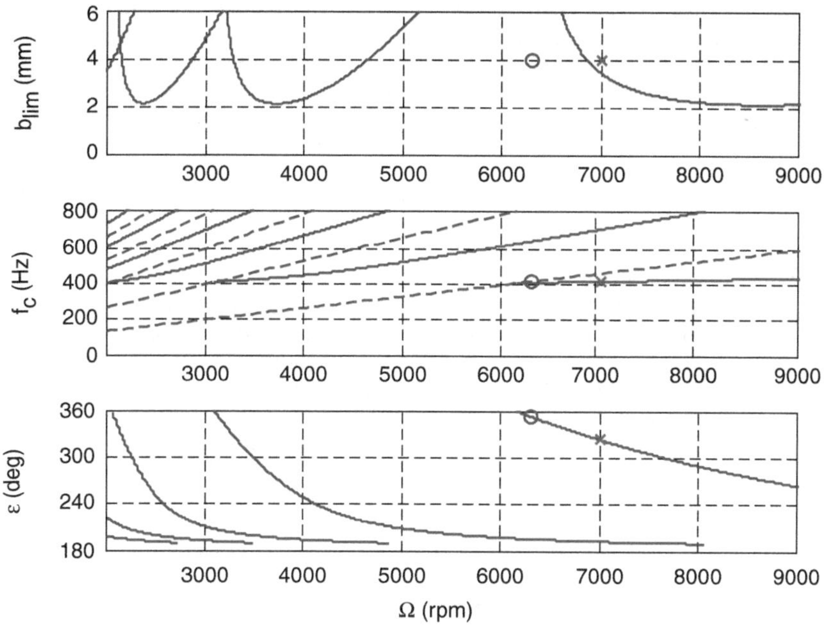

Fig. e.6.3 Stability, chatter frequency, and undulation phase diagrams

Fig. e.6.3, describe the automatic spindle speed regulation step(s) required to arrive at a stable cut. List the chatter frequency(s) encountered, the spindle speed(s) selected, and the corresponding ε value(s).

4. Consider a 30% radial immersion up milling cut. The tool dynamics are described by two identical modes in both the x and y directions (assume the workpiece is rigid). The modal parameters are: $f_{n1} = 800$ Hz, $k_{q1} = 2 \times 10^7$ N/m, and $\zeta_{q1} = 0.05$; and $f_{n2} = 1000$ Hz, $k_{q2} = 1.5 \times 10^7$ N/m, and $\zeta_{q2} = 0.03$. The workpiece material is an aluminum alloy and it is machined with a four tooth, 12.7 mm diameter, 30 deg helix square endmill using a feed per tooth of 0.2 mm/tooth. The variable teeth spacing for this cutter is `teeth= [0 95 180 275]` deg. The cutting force coefficients are $k_t = 520$ N/mm^2, $k_n = 300$ N/mm^2, and $k_{te} = k_{ne} = 0$. The axial depth of cut is 5 mm and the spindle speed is 15000 rpm. Determine the roughness average (in μm) for runout values of `RO = [0 0 -20 0]` μm using the cycloidal tool path time-domain simulation. Compare this value to the roughness average with zero runout.

References

1. Tlusty, J., W. Zaton, and F. Ismail, 1983, Stability Lobes in Milling, Annals of the CIRP, 32/1: 309–313.
2. Smith, S. and Tlusty, J., 1992, Stabilizing Chatter by Automatic Spindle Speed Regulation, Annals of the CIRP, 41/1: 433–436.

3. Delio, T., Tlusty, J., and Smith, S., 1992, Use of Audio Signals for Chatter Detection and Control, Journal of Engineering for Industry, 114: 146–157.
4. Smith, D., Smith, S., and Tlusty, J., 1998, High Performance Milling Torque Sensor, Journal of Manufacturing Science and Engineering, 120/3: 504–514.
5. Smith, S. and Delio, T., 1992, Sensor-based Chatter Detection and Avoidance by Spindle Speed Selection, Journal of Dynamic Systems, Measurement, and Control, 114/3: 486–492.
6. Schmitz, T., Davies, M., Medicus, K., Snyder, J., 2001, Improving High-Speed Machining Material Removal Rates by Rapid Dynamic Analysis, Annals of the CIRP, 50/1: 263–268.
7. Schmitz, T., Medicus, K., and Dutterer, B., 2002, Exploring Once-per-revolution Audio Signal Variance as a Chatter Indicator, Machining Science and Technology, 6/2: 215–233.
8. Schmitz, T., 2003, Chatter Recognition by a Statistical Evaluation of the Synchronously Sampled Audio Signal, Journal of Sound and Vibration, 262/3: 721–730.
9. Cheng, C.-H., Duncan, G.S., and Schmitz, T., 2007, Rotating Tool Point Frequency Response Prediction using RCSA, Machining Science and Technology, 11/3: 433–446.
10. Davies, M., Dutterer, B., Pratt, J., and Schaut, A., 1998, On the Dynamics of High-Speed Milling with Long, Slender Endmills, Annals of the CIRP 47/1: 55–60.
11. Tlusty, J., 1959, Systems and Methods of Machine Tool Testing, Microtechnic, 13: 162–178.
12. Bryan, J., Clouser, R., and Holland, B., 1967, Spindle Accuracy, American Machinist, 149–64.
13. American National Standards Institute, American Society of Mechanical Engineers, 1985, ANSI/ASME B89.3.4 M, Axes of Rotation: Methods for Specifying and Testing, New York.
14. Smith, S. and Winfough, W.R., 1994, The Effect of Runout Filtering on the Identification of Chatter in the Audio spectrum of Milling, Transactions of the NAMRI/SME, 22: 173–178.
15. Schmitz, T., Couey, J., Marsh, E., Mauntler, N., and Hughes, D., 2007, Runout Effects in Milling: Surface finish, Surface Location Error, and Stability, International Journal of Machine Tools and Manufacture, 47: 841–851.
16. Kline, W. and DeVor, R., 1983, The Effect of Runout on Cutting Geometry and Forces in End Milling, International Journal of Machine Tool Design and Research, 23/2-3: 123–140.
17. Armarego, E. and Deshpande, N., 1991, Computerized End Milling Force Predictions with Cutting Models Allowing Eccentricity and Cutter Deflections, Annals of the CIRP, 40/1: 25–29.
18. Altintas, Y. and Chan, P., 1992, In-Process Detection and Suppression of Chatter in Milling, International Journal of Machine Tools and Manufacture, 32/3: 329–47.
19. Liang, S. and Wang, J., 1994, Milling Force Convolution Modeling for Identification of Cutter Axis Offset, International Journal of Machine Tools and Manufacture, 34/8: 1177–1190.
20. Feng, H.-Y. and Menq, C.-H., 1994, The Prediction of Cutting Forces in the Ball-End Milling Process – I. Model Formulation and Model Building Procedure, International Journal of Machine Tools and Manufacture, 34/5: 697–710.
21. Feng, H.-Y. and Menq, C.-H., 1994, The Prediction of Cutting Forces in the Ball-End Milling Process – II. Cut Geometry Analysis and Model Verification, International Journal of Machine Tools and Manufacture, 34/5: 711–719.
22. Yan, D., El-Wardany, T., and Elbestawi, M., 1995, A Multi-Sensor Strategy for Tool Failure Detection in Milling, International Journal of Machine Tools and Manufacture, 35/3: 383–398.
23. Stevens, A. and Liang, S., 1995, Runout Rejection in End Milling through Two-Dimensional Repetitive Force Control, Mechatronics, 5/1: 1–13.
24. Hekman, K. and Liang, S., 1997, In-Process Monitoring of End Milling Cutter Runout, Mechatronics, 7/1: 1–10.

25. Baek, D., Ko, T., and Kim, H., 1997, A Dynamic Surface Roughness Model for Face Milling, Precision Engineering, 20/3: 171–178.
26. Zheng, H., Li, X., Wong, Y., and Nee, A., 1999, Theoretical Modeling and Simulation of Cutting Forces in Face Milling with Cutter Runout, International Journal of Machine Tools and Manufacture, 39/12: 2003–2018.
27. Yun, W.-S. and Cho, D.-W., 2001, Accurate 3-D Cutting Force Prediction using Cutting Condition Independent Coefficients in End Milling, International Journal of Machine Tools and Manufacture, 41/4: 463–478.
28. Baek, D., Ko, T., and Kim, H., 2001, Optimization of Feedrate in a Face Milling Operation using a Surface Roughness Model, International Journal of Machine Tools and Manufacture, 41/3: 451–462.
29. Mezentsev, O., Zhu, R., DeVor, R., Kapoor S., and Kline, W., 2002, Use of Radial Forces for Fault Detection in Tapping, International Journal of Machine Tools and Manufacture, 42/4: 479–488.
30. Ko, J., Yun, W.-S., Cho, D.-W., and Ehmann, K.F., 2002, Development of a Virtual Machining System, Part 1: Approximation of the Size Effect for Cutting Force Prediction, International Journal of Machine Tools and Manufacture, 42/15: 1595–1605.
31. Ranganath, S. and Sutherland, J., 2002, An Improved Method for Cutter Runout Modeling in the Peripheral Milling Process, Machining Science and Technology, 6/1: 1–20.
32. Wang, J.-J. and Zheng, C., 2003, Identification of Cutter Offset in End Milling without a Prior Knowledge of Cutting Coefficients, International Journal of Machine Tools and Manufacture, 43/7: 687–697.
33. Lazoglu, I., 2003, Sculpture Surface Machining: A Generalized Model of Ball-End Milling Force System, International Journal of Machine Tools and Manufacture, 43/5: 453–462.
34. Atabey, F., Lazoglu, I., and Altintas, Y., 2003, Mechanics of Boring Processes – Part II. Multi-Insert Boring Heads, International Journal of Machine Tools and Manufacture, 43/5: 477–484.
35. Slavicek, J., 1965, The effect of Irregular Tooth Pitch on Stability of Milling, Proceedings of the 6th Machine Tool Design and Research Conference, Pergamon Press, London, pp. 15–22.
36. Vanherck, P., 1967, Increasing Milling Machine Productivity by Use of Cutter with Non-Constant Edge Pitch, Proceedings of the 8th Machine Tool Design and Research Conference, pp. 947–960.
37. Doolan, P., Phadke, M.S., and Wu, S., 1975, Computer Design of Vibration Free Face Milling Cutters, Journal of Engineering for Industry, 97B/3: 925–930.
38. Doolan, P., Burney, F., and Wu, S., 1976, Computer Design of a Multi-purpose Minimum Vibration Face Milling Cutter, International Journal of Machine Tool Design and Research, 16/3: 187–192.
39. Tlusty, J., Ismail, F., and Zaton W., 1983, Use of Special Milling Cutters Against Chatter, Transactions of the NAMRI/SME, 11: 408–415.
40. Shirase, K. and Altintas, Y., 1995, Cutting Force and Dimensional Surface Error Generation in Peripheral Milling with Variable Pitch Helical End Mills, International Journal of Machine Tools and Manufacture, 36/5: 567–584.
41. Choudhurya, S. and Mathew, J., 1995, Investigations of the Effect of Non-uniform Insert Pitch on Vibration during Face Milling, International Journal of Machine Tools and Manufacture, 35/10: 1435–1444.
42. Altintas, Y, Engin, S., and Budak, E., 1999, Analytical Stability Prediction and Design of Variable Pitch Cutters, Journal of Manufacturing Science and Engineering, 121: 173–178.
43. Budak, E., 2003, An Analytical Design Method for Milling Cutters with Nonconstant Pitch to Increase Stability, Part 1: Theory, Part 2: Application, Journal of Manufacturing Science and Engineering, 123: 29–38.

44. Stone, B., 1970, The Effect on the Chatter Behavior of Cutters With Different Helix Angles on Adjacent Teeth, Proceedings of the 11th International Machine Tool Design and Research Conference, pp. 169–180.
45. Davies, M., Pratt, J., Dutterer, B., and Burns, T., 2000, The Stability of Low Radial Immersion Milling, Annals of the CIRP 49/1: 37–40.
46. Davies, M., Pratt, J., Dutterer, B., and Burns, T., 2002, Stability Prediction for Low Radial Immersion Milling, Journal of Manufacturing Science and Engineering, 124/2: 217–225.
47. Insperger, T., Mann, B., Stépàn, G., and Bayly, P., 2003, Stability of Up-milling and Down-milling, Part 1: Alternative Analytical Methods, International Journal of Machine Tools and Manufacture, 43/1: 25–34.
48. Mann, B., Insperger, T., Bayly, P., and Stépàn, G., 2003, Stability of Up-milling and Down-milling, Part 2: Experimental Verification, International Journal of Machine Tools and Manufacture, 43/1: 35–40.
49. Campomanes, M. and Altintas, Y., 2003, An Improved Time Domain Simulation for Dynamic Milling at Small Radial Immersions, Journal of Manufacturing Science and Engineering, 125/3: 416–422.
50. Merdol, S. and Altintas, Y., 2004, Multi Frequency Solution of Chatter Stability for Low Immersion Milling, Journal of Manufacturing Science and Engineering, 126/3: 459–466.
51. Bayly, P., Halley, J., Mann, B., and Davies, M., 2004, Stability of Interrupted Cutting by Temporal Finite Element Analysis, Journal of Manufacturing Science and Engineering, 125/2: 220–225.
52. Mann, B., Bayly, P, Davies, M., and Halley, J., 2004, Limit Cycles, Bifurcations, and Accuracy of the Milling Process, Journal of Sound and Vibration, 277/1-2: 31–48.
53. Govekar, E., Gradišek, J., Kalveram, M., Insperger, T., Weinert, K., Stépàn, G., and Grabec, I., 2005, On Stability and Dynamics of Milling at Small Radial Immersion, Annals of the CIRP, 54/1: 357–362.
54. Taylor, B. and Kuyatt, C., 1994, Guidelines for Evaluating and Expressing the Uncertainty of NIST Measurement Results, NIST Technical Note 1297 1994 Edition.
55. International Standards Organization (ISO), 1993, Guide to the Expression of Uncertainty in Measurement (Corrected and Reprinted 1995).
56. American National Standards Institute, 1997, ANSI/NCSL Z540-2-1997, US Guide to the Expression of Uncertainty in Measurement.
57. Bevington, P. and Robinson, D., 1992, Data Reduction and Error Analysis for the Physical Sciences, 2nd Edition, WCB/McGraw-Hill, Boston, MA.
58. Altintas, Y. and Budak, E., 1995, Analytical Prediction of Stability Lobes in Milling, Annals of the CIRP, 44/1: 357–362.
59. Duncan, G.S., Kurdi, M., Schmitz, T., and Snyder, J., 2006, Uncertainty Propagation for Selected Analytical Milling Stability Limit Analyses, Transactions of the NAMRI/SME, 34: 17–24.
60. Weisstein, E., "Incommensurate" from MathWorld – A Wolfram Web Resource, http://mathworld.wolfram.com/Incommensurate.html, accessed May, 2008.
61. Weisstein, E., "Bifurcation" from MathWorld – A Wolfram Web Resource, http://mathworld.wolfram.com/Bifurcation.html, accessed May, 2008.

Chapter 7
Tool Point Dynamics Prediction

The secret to creativity is knowing how to hide your sources.

- Albert Einstein

In Chapters 4, 5, and 6 we analyzed several aspects of milling with the ultimate goal of enabling *a priori* stability and surface location error predictions in order to improve productivity. We described frequency-domain solutions that offer closed form expressions for both stability (Chapter 4) and surface location error (Chapter 5). The primary inputs to these analyses are the force model coefficients and tool point frequency response function, or FRF. In this chapter, we apply the receptance coupling technique to prediction of the tool point response in order to complement the frequency-domain process models detailed previously.[1] We also include a brief review of Euler-Bernoulli beam theory and provide closed form solutions for the direct and cross receptances (FRFs) under free-free and clamped-free boundary conditions.

7.1 Motivation

While impact testing (Section 2.6) provides a convenient approach to obtaining the tool-holder-spindle-machine[2] FRF (typically measured at the tool point), it requires a separate set of measurements for each assembly. For example, if there are 25 tools in a machine tool's magazine, then a minimum of 50 measurements are required (one each for the x and y directions, assuming the axial compliance in the z direction is negligible). Further, if the tool insertion length is modified due to new requirements or inadequate control during tooling setup, the measurements must be repeated for that tool-holder combination. This requirement

[1] If the tool point FRF is known, the modal fitting procedure described in Section 2.5 can be applied to identify the modal parameters required for time-domain simulation as well.

[2] The workpiece can also be the source of significant dynamic compliance. However, we will limit our discussions to situations where the workpiece can be assumed to be rigid relative to the tool-holder-spindle-machine assembly.

T.L. Schmitz, K.S. Smith, *Machining Dynamics*, DOI 10.1007/978-0-387-09645-2_7,

for multiple measurements certainly does not preclude the application of the process analyses we've discussed in production environments. However, the required time and cost for impact testing does pose an obstacle to convenient implementation at the shop floor level.

IN A NUTSHELL Although FRF measurements provide us with essential information, they are specific to the setup. The tool-holder-spindle-machine FRF changes strongly with tool length and diameter, for example. Unfortunately, modern CNC systems encourage poor setup repeatability by including tool length offset correction. Although this may enable you to put the tool point at the correct geometric location, it also leads to almost certain variability in day to day machining performance due to the change in assembly dynamics.

While no predictive approach replaces actual data, the requirement for a separate measurement of every conceivable setup can lead to a substantial number of measurements. In this chapter we describe a new method that eliminates the need to measure each tool-holder in a particular spindle. By combining measurements of a simple artifact inserted in the spindle in question with models of tools and holders, we can perform off line FRF predictions for a range of tools.

7.2 Basic Receptance Coupling

Rather than consider the tool-holder-spindle-machine combination as a single assembly, we can view it as being composed of three separate entities, specifically, the tool, the holder, and the spindle-machine. Of these three, the tool and holder are convenient to model because they are not structurally complicated. The spindle-machine, on the other hand, is much more challenging. Spindle dynamics modeling, often completed using finite element analysis, requires detailed knowledge of the mechanical design, bearing stiffness values (which depend on the assembly tolerances), and damping levels[3]. For commercial machining centers, the spindle design is often proprietary or unavailable to the end user, and the bearing stiffness values are difficult to obtain. Also, first principle estimates of the spindle damping remains an active research area. Comparable problems are encountered in modeling the machine dynamics, particularly obtaining the damping values, which often depend on multiple energy dissipation mechanisms and locations. Even more problematic, different lengths and diameters of tools may activate these mechanisms by differing amounts (especially within the spindle).

---

[3] Author T. Schmitz acknowledges conversations with N. Arakere, University of Florida, regarding dynamic modeling of rotating systems.

FOR INSTANCE A short, large diameter tool may be very stiff and emphasize the spindle modes in the assembly FRF. The primary damping source in this case would be the spindle bearings and interface between the spindle and motor rotor. If the tool is long and slender, by contrast, the spindle may appear nearly rigid (but not always![4]). This tool geometry would generally emphasize the damping in the tool-tool holder-spindle interfaces with relatively less energy dissipated by the spindle bearings.

This leads us to a scenario where we can consider modeling those components that lend themselves to this activity (the tool and holder) and measuring the difficult-to-model element (the spindle-machine)[5]. Additionally, rather than describing the modeled and measured parts, or substructures, using modal parameters, for example, it is sensible to develop substructure FRFs since the assembly FRF is a primary input for the analytical milling process analyses. A convenient approach for joining these substructure FRFs to obtain the assembly response is receptance coupling [1, 2], or receptance coupling substructure analysis (RCSA) as referenced in recent literature [3–12]. Prior to detailing RCSA for tool point FRF prediction, let's examine receptance coupling solutions for some simple dynamic systems.

7.2.1 Two Component Rigid Coupling

As shown in Fig. 7.2.1, two components, I and II, are to be rigidly coupled to form assembly III. The coupling coordinates are x_{1a} and x_{1b} for the two

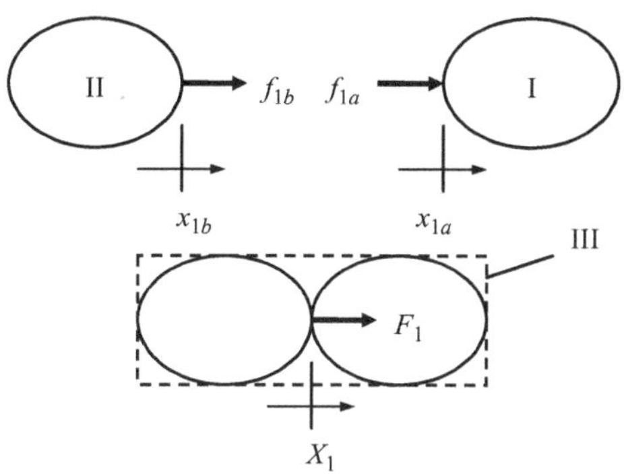

Fig. 7.2.1 Rigid coupling of components I and II to form assembly III. The force F_1 is applied to the assembly in order to determine H_{11}

[4] See Example 7.5.2.

[5] Author T. Schmitz recognizes "hallway" discussions with Dr. J. Pratt, National Institute of Standards and Technology, Gaithersburg, MD, in forming this observation.

substructures I and II, respectively. The corresponding assembly coordinate, X_1, is located at the same physical location as x_{1a} and x_{1b} after they are joined. An attractive aspect of receptance coupling is that the component FRFs are only required at the coupling locations and any point where the assembly response is to be predicted [13–15]. Therefore, the direct (see Section 2.4) assembly response, $H_{11}(\omega) = \frac{X_1}{F_1}$, due to a harmonic force applied at coordinate X_1 can be fully described using the direct component receptances $h_{1a1a}(\omega) = \frac{x_{1a}}{f_{1a}}$ and $h_{1b1b}(\omega) = \frac{x_{1b}}{f_{1b}}$ obtained from harmonic forces applied at x_{1a} and x_{1b}, respectively. Note that we've used upper case variables to designate assembly terms and lower case variables to identify component terms.

To determine the assembly response, we must first state the compatibility condition, $x_{1b} - x_{1a} = 0$, which represents the rigid coupling between component coordinates x_{1a} and x_{1b}. We can therefore write $x_{1b} = x_{1a} = X_1$ due to our decision to locate assembly coordinate X_1 at the (rigid) coupling point. We must also define the equilibrium condition, $f_{1a} + f_{1b} = F_1$, which equates the internal (component) and external (assembly) forces. Let's substitute for the displacements in the compatibility equation.

$$x_{1b} - x_{1a} = 0$$

$$h_{1b1b}f_{1b} - h_{1a1a}f_{1a} = 0$$

We next use the equilibrium condition, rewritten as $f_{1a} = F_1 - f_{1b}$, to eliminate f_{1a}. Rearranging enables us to solve for f_{1b}.

$$h_{1b1b}f_{1b} - h_{1a1a}F_1 + h_{1a1a}f_{1b} = 0$$

$$(h_{1a1a} + h_{1b1b})f_{1b} = h_{1a1a}F_1$$

$$f_{1b} = (h_{1a1a} + h_{1b1b})^{-1}h_{1a1a}F_1$$

Now that we have f_{1b}, we can again use the equilibrium condition to determine f_{1a}.

$$f_{1a} = F_1 - f_{1b}$$

$$f_{1a} = \left(1 - (h_{1a1a} + h_{1b1b})^{-1}h_{1a1a}\right)F_1$$

We solve for H_{11} as shown in Eq. 7.2.1. This equation gives the direct assembly response at the coupling coordinate, X_1, as a function of the component receptances. These frequency dependent, complex valued receptances may have any number of modes. There are no restrictions on the relationship between the number of modes and coordinates as with modal analysis (i.e., we saw in Section 2.4 that the number of modeled modes and coordinates must be equal to obtain square matrices when using modal techniques).

$$H_{11} = \frac{X_1}{F_1} = \frac{x_{1a}}{F_1} = \frac{h_{1a1a}f_{1a}}{F_1} = h_{1a1a} - h_{1a1a}(h_{1a1a} + h_{1b1b})^{-1}h_{1a1a} \qquad (7.2.1)$$

IN A NUTSHELL The spindle-machine FRF, which is difficult to model, is determined by measurement. As we've discussed, the FRF shows the real and imaginary parts of the frequency dependent motion at a selected coordinate in response to a force. The tool and holder, on the other hand, are modeled using beam theory. The measured and modeled FRFs are connected at the appropriate coordinate; this is the essence of receptance coupling. Using this approach, the FRF of the assembled structure can be predicted at any coordinate on the modeled portion of the assembly.

Similarly, we can predict the assembly response at another coordinate, not coincident with the coupling point, by defining the component receptance at the desired location. Consider Fig. 7.2.2, where the direct assembly response at X_1 is again desired, but this location is now at another point on component I. We again assume x_1 and X_1 are collocated before and after coupling. The new coupling coordinates at the rigid coupling point are x_{2a} and x_{2b}. The component receptances corresponding to Fig. 7.2.2 are $h_{11} = \frac{x_1}{f_1}$ and $h_{2a2a} = \frac{x_{2a}}{f_{2a}}$ for I, and $h_{2b2b} = \frac{x_{2b}}{f_{2b}}$ for II. The compatibility condition for the rigid coupling is $x_{2b} - x_{2a} = 0$ and we can therefore write $x_{2a} = x_{2b} = X_2$. Also, $x_1 = X_1$. The equilibrium conditions are $f_{2a} + f_{2b} = 0$ and $f_1 = F_1$.

To determine $H_{11} = \frac{X_1}{F_1}$, we'll first write the component displacements. For I, we now have two forces acting on the body, so the displacements are:

$$x_1 = h_{11}f_1 + h_{12a}f_{2a} \quad \text{and} \quad x_{2a} = h_{2a1}f_1 + h_{2a2a}f_{2a} \qquad (7.2.2)$$

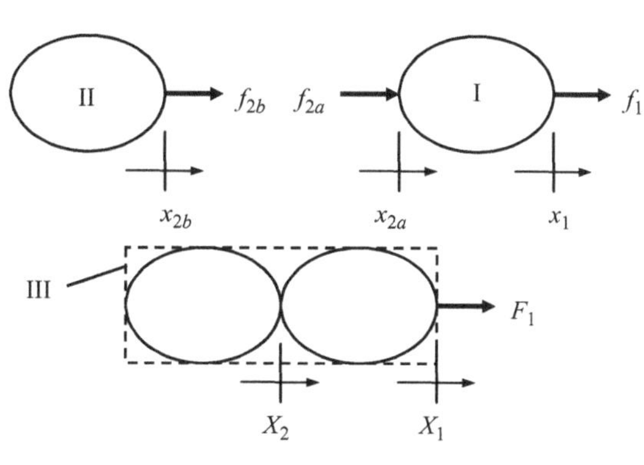

Fig. 7.2.2 Example showing rigid coupling of components I and II to form assembly III. The force F_1 is applied to the assembly in order to determine H_{11} and H_{21}

For II, we have $x_{2b} = h_{2b2b}f_{2b}$. Substitution into the compatibility condition gives:

$$h_{2b2b}f_{2b} - h_{2a1}f_1 - h_{2a2a}f_{2a} = 0. \tag{7.2.3}$$

We apply the equilibrium conditions to replace f_1 and eliminate f_{2a} ($f_{2a} = -f_{2b}$).

$$h_{2b2b}f_{2b} - h_{2a1}F_1 + h_{2a2a}f_{2b} = 0. \tag{7.2.4}$$

This enables us to group terms and solve for f_{2b}. Specifically, we have that $f_{2b} = (h_{2a2a} + h_{2b2b})^{-1}h_{2a1}F_1$. Therefore, we can also write $f_{2a} = -(h_{2a2a} + h_{2b2b})^{-1}h_{2a1}F_1$. Substitution of this force value into the H_{11} expression gives us the desired result; see Eq. 7.2.5. Again, the assembly response is written as a function of the component direct (h_{11}, h_{2a2a}, and h_{2b2b}) and cross (h_{12a} and h_{2a1}) receptances.

$$\begin{aligned} H_{11} &= \frac{X_1}{F_1} = \frac{x_1}{F_1} = \frac{h_{11}f_1 + h_{12a}f_{2a}}{F_1} = \frac{h_{11}f_1 - h_{12a}(h_{2a2a} + h_{2b2b})^{-1}h_{2a1}F_1}{F_1} \\ H_{11} &= \frac{h_{11}F_1 - h_{12a}(h_{2a2a} + h_{2b2b})^{-1}h_{2a1}F_1}{F_1} = h_{11} - h_{12a}(h_{2a2a} + h_{2b2b})^{-1}h_{2a1} \end{aligned} \tag{7.2.5}$$

We can also use f_{2a} to determine the cross receptance H_{21}. See Eq. 7.2.6.

$$\begin{aligned} H_{21} &= \frac{X_2}{F_1} = \frac{x_{2a}}{F_1} = \frac{h_{2a1}f_1 + h_{2a2a}f_{2a}}{F_1} = \frac{h_{2a1}f_1 - h_{2a2a}(h_{2a2a} + h_{2b2b})^{-1}h_{2a1}F_1}{F_1} \\ H_{21} &= \frac{h_{2a1}F_1 - h_{2a2a}(h_{2a2a} + h_{2b2b})^{-1}h_{2a1}F_1}{F_1} = h_{2a1} - h_{2a2a}(h_{2a2a} + h_{2b2b})^{-1}h_{2a1} \end{aligned} \tag{7.2.6}$$

In an analogous way, we can find the direct and cross receptances, H_{22} and H_{12}, respectively, due to a force applied at X_2. See Fig. 7.2.3. The component

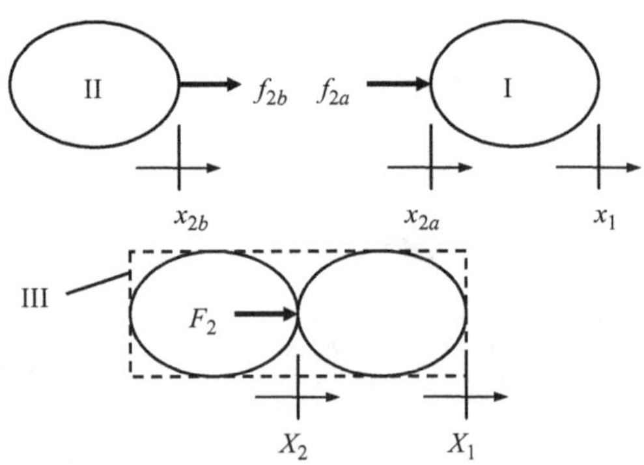

Fig. 7.2.3 Example showing rigid coupling of components I and II to form assembly III. The force F_2 is applied to the assembly in order to determine H_{22} and H_{12}

receptances are again $h_{11} = \frac{x_1}{f_1}$ and $h_{2a2a} = \frac{x_{2a}}{f_{2a}}$ for I, and $h_{2b2b} = \frac{x_{2b}}{f_{2b}}$ for II. The compatibility condition for the rigid coupling remains as $x_{2b} - x_{2a} = 0$. However, the equilibrium condition is $f_{2a} + f_{2b} = F_2$.

To determine $H_{22} = \frac{X_2}{F_2}$, we begin by writing the component displacements. For I, the displacements are:

$$x_1 = h_{12a} f_{2a} \quad \text{and} \quad x_{2a} = h_{2a2a} f_{2a}. \tag{7.2.7}$$

For II, we have $x_{2b} = h_{2b2b} f_{2b}$. Substitution in the compatibility condition gives:

$$h_{2b2b} f_{2b} - h_{2a2a} f_{2a} = 0. \tag{7.2.8}$$

We apply the equilibrium condition, $f_{2a} = F_2 - f_{2b}$, to eliminate f_{2a} in Eq. 7.2.8.

$$h_{2b2b} f_{2b} - h_{2a2a} F_2 + h_{2a2a} f_{2b} = 0 \tag{7.2.9}$$

This enables us to group terms and solve for f_{2b}. We find that $f_{2b} = (h_{2a2a} + h_{2b2b})^{-1} h_{2a2a} F_2$. Again using the equilibrium condition, we can write $f_{2a} = \left(1 - (h_{2a2a} + h_{2b2b})^{-1} h_{2a2a}\right) F_2$. Equation 7.2.10 gives the desired H_{22} expression.

$$\begin{aligned} H_{22} &= \frac{X_2}{F_2} = \frac{x_{2a}}{F_2} = \frac{h_{2a2a} f_{2a}}{F_2} = \frac{h_{2a2a}\left(1 - (h_{2a2a} + h_{2b2b})^{-1} h_{2a2a}\right) F_2}{F_2} \\ H_{22} &= \frac{h_{2a2a} F_2 - h_{2a2a}(h_{2a2a} + h_{2b2b})^{-1} h_{2a2a} F_2}{F_2} = h_{2a2a} - h_{2a2a}(h_{2a2a} + h_{2b2b})^{-1} h_{2a2a} \end{aligned} \tag{7.2.10}$$

We use f_{2a} to find the cross receptance H_{12} as well. See Eq. 7.2.11.

$$\begin{aligned} H_{12} &= \frac{X_1}{F_2} = \frac{x_1}{F_2} = \frac{h_{12a} f_{2a}}{F_2} = \frac{h_{12a}\left(1 - (h_{2a2a} + h_{2b2b})^{-1} h_{2a2a}\right) F_2}{F_2} \\ H_{12} &= h_{12a} - h_{12a}(h_{2a2a} + h_{2b2b})^{-1} h_{2a2a} \end{aligned} \tag{7.2.11}$$

7.2.2 Two Component Flexible Coupling

Let's continue with the system shown in Fig. 7.2.1, but now couple the two components through a linear spring, described by the constant k. This is displayed in Fig. 7.2.4. The component receptances are $h_{1a1a} = \frac{x_{1a}}{f_{1a}}$ and $h_{1b1b} = \frac{x_{1b}}{f_{1b}}$ and the equilibrium condition is $f_{1a} + f_{1b} = F_{1a}$. These are analogous to the rigid coupling case. However, the compatibility condition now becomes:

Fig. 7.2.4 Flexible coupling of components I and II to form assembly III. The force F_{1a} is applied to the assembly in order to determine H_{1a1a} and H_{1b1a}

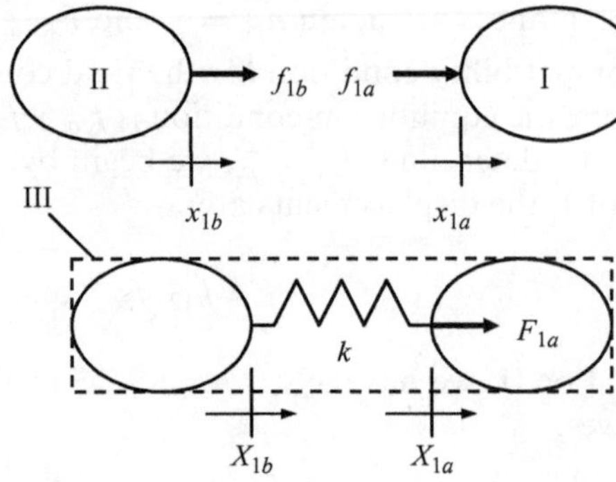

$$k(x_{1b} - x_{1a}) = -f_{1b}. \tag{7.2.12}$$

Because the component and assembly coordinates are coincident, we have that $x_{2a} = X_{2a}$ and $x_{2b} = X_{2b}$. To determine $H_{1a1a} = \frac{X_{1a}}{F_{1a}}$, let's first substitute the component displacements in the compatibility condition. See Eq. 7.2.13.

$$k(h_{1b1b}f_{1b} - h_{1a1a}f_{1a}) = -f_{1b} \tag{7.2.13}$$

Using the equilibrium condition, $f_{1a} = F_{1a} - f_{1b}$, we can eliminate f_{1a} to obtain the equation for f_{1b}.

$$k(h_{1b1b}f_{1b} - h_{1a1a}F_{1a} + h_{1a1a}f_{1b}) = -f_{1b}$$

$$kh_{1b1b}f_{1b} - kh_{1a1a}F_{1a} + kh_{1a1a}f_{1b} = -f_{1b}$$

$$\left(h_{1a1a} + h_{1b1b} + \frac{1}{k}\right)f_{1b} = h_{1a1a}F_{1a}$$

$$f_{1b} = \left(h_{1a1a} + h_{1b1b} + \frac{1}{k}\right)^{-1} h_{1a1a}F_{1a}$$

Using f_{1b} and the equilibrium condition, we find that $f_{1a} = \left(1 - \left(h_{1a1a} + h_{1b1b} + \frac{1}{k}\right)^{-1} h_{1a1a}\right)F_{1a}$. Substitution then yields the direct assembly receptance H_{1a1a} as shown in Eq. 7.2.14. We can see that this equation simplifies to Eq. 7.2.1 as k approaches infinity (rigid connection).

$$H_{1a1a} = \frac{X_{1a}}{F_{1a}} = \frac{x_{1a}}{F_{1a}} = \frac{h_{1a1a}f_{1a}}{F_{1a}} = \frac{h_{1a1a}\left(1 - \left(h_{1a1a} + h_{1b1b} + \frac{1}{k}\right)^{-1} h_{1a1a}\right)F_{1a}}{F_{1a}}$$

$$H_{1a1a} = h_{1a1a} - h_{1a1a}\left(h_{1a1a} + h_{1b1b} + \frac{1}{k}\right)^{-1} h_{1a1a} \tag{7.2.14}$$

The cross receptance due to the force F_{1a} is provided in Eq. 7.2.15.

$$H_{1b1a} = \frac{X_{1b}}{F_{1a}} = \frac{x_{1b}}{F_{1a}} = \frac{h_{1b1b}f_{1b}}{F_{1a}} = \frac{h_{1b1b}\left(h_{1a1a} + h_{1b1b} + \frac{1}{k}\right)^{-1}h_{1a1a}F_{1a}}{F_{1a}}$$
$$H_{1b1a} = h_{1b1b}\left(h_{1a1a} + h_{1b1b} + \frac{1}{k}\right)^{-1}h_{1a1a} \tag{7.2.15}$$

As shown in Fig. 7.2.5, we can alternately apply the assembly force to coordinate X_{1b}. The component receptances and displacements are unchanged, but the equilibrium condition is $f_{1a} + f_{1b} = F_{1b}$. Similarly, we modify the compatibility condition to be:

$$k(x_{1a} - x_{1b}) = -f_{1a}. \tag{7.2.16}$$

Substitution for the component displacements and f_{1b} (from the equilibrium condition) yields the expression for f_{1a}.

$$k(h_{1a1a}f_{1a} - h_{1b1b}F_{1b} + h_{1b1b}f_{1a}) = -f_{1a}$$
$$kh_{1a1a}f_{1a} - kh_{1b1b}F_{1b} + kh_{1b1b}f_{1a} = -f_{1a}$$
$$\left(h_{1a1a} + h_{1b1b} + \frac{1}{k}\right)f_{1a} = h_{1b1b}F_{1b}$$
$$f_{1a} = \left(h_{1a1a} + h_{1b1b} + \frac{1}{k}\right)^{-1}h_{1b1b}F_{1b}$$

Again applying the equilibrium condition, $f_{1b} = F_{1b} - f_{1a}$, we obtain $f_{1b} = \left(1 - \left(h_{1a1a} + h_{1b1b} + \frac{1}{k}\right)^{-1}h_{1b1b}\right)F_{1b}$. Substitution then gives the assembly direct and cross receptances due to F_{1b}.

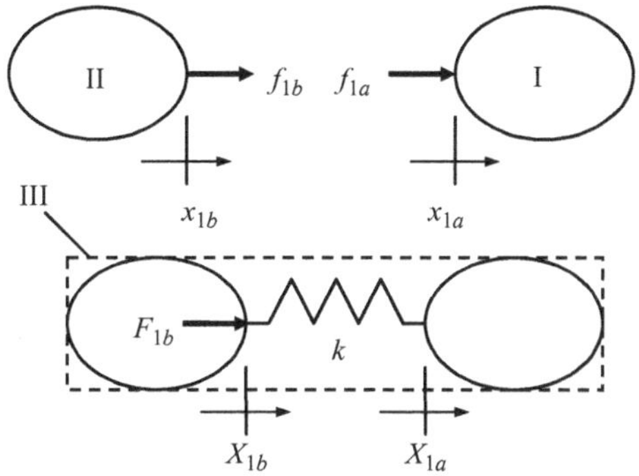

Fig. 7.2.5 Flexible coupling of components I and II to form assembly III. The force F_{1b} is applied to the assembly in order to determine H_{1b1b} and H_{1a1b}

$$H_{1b1b} = \frac{X_{1b}}{F_{1b}} = \frac{x_{1b}}{F_{1b}} = \frac{h_{1b1b} f_{1b}}{F_{1b}} = \frac{h_{1b1b}\left(1 - \left(h_{1a1a} + h_{1b1b} + \frac{1}{k}\right)^{-1} h_{1b1b}\right) F_{1b}}{F_{1b}}$$
$$H_{1b1b} = h_{1b1b} - h_{1b1b}\left(h_{1a1a} + h_{1b1b} + \frac{1}{k}\right)^{-1} h_{1b1b} \quad (7.2.17)$$

$$H_{1a1b} = \frac{X_{1a}}{F_{1b}} = \frac{x_{1a}}{F_{1b}} = \frac{h_{1a1a} f_{1a}}{F_{1b}} = \frac{h_{1a1a}\left(h_{1a1a} + h_{1b1b} + \frac{1}{k}\right)^{-1} h_{1b1b} F_{1b}}{F_{1b}}$$
$$H_{1a1b} = h_{1a1a}\left(h_{1a1a} + h_{1b1b} + \frac{1}{k}\right)^{-1} h_{1b1b} \quad (7.2.18)$$

Similar to the rigid connection example depicted in Fig. 7.2.2, we can again add another coordinate, not located at the coupling location, and apply the external force at that point. See Fig. 7.2.6. The component displacements are again $x_1 = h_{11} f_1 + h_{12a} f_{2a}$ and $x_{2a} = h_{2a1} f_1 + h_{2a2a} f_{2a}$ for substructure I and $x_{2b} = h_{2b2b} f_{2b}$ for substructure II. The equilibrium conditions are $f_{2a} + f_{2b} = 0$ and $f_1 = F_1$. The compatibility condition is:

$$k(x_{2b} - x_{2a}) = -f_{2b}. \quad (7.2.19)$$

As before, the component and assembly coordinates are coincident, so we have that $x_1 = X_1$, $x_{2a} = X_{2a}$, and $x_{2b} = X_{2b}$. To determine $H_{11} = \frac{X_1}{F_1}$, let's first substitute the component displacements in the compatibility condition. See Eq. 7.2.20.

$$k(h_{2b2b} f_{2b} - h_{2a1} f_1 - h_{2a2a} f_{2a}) = -f_{2b} \quad (7.2.20)$$

Using the equilibrium conditions, we can eliminate f_{2a} and replace f_1 with F_1 to obtain the equation for f_{2b}.

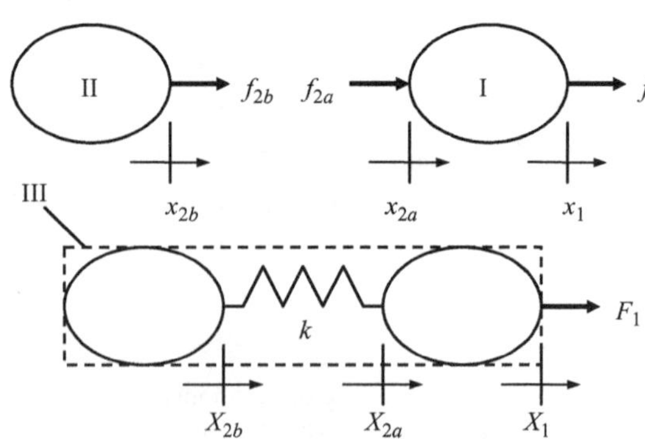

Fig. 7.2.6 Flexible coupling of components I and II to form assembly III. The force F_1 is applied to the assembly in order to determine H_{11}, H_{2a1}, and H_{2b1}

$$k(h_{2b2b}f_{2a} - h_{2a1}F_1 + h_{2a2a}f_{2b}) = -f_{2b}$$

$$kh_{2b2b}f_{2b} - kh_{2a1}F_1 + kh_{2a2a}f_{2b} = -f_{2b}$$

$$\left(h_{2a2a} + h_{2b2b} + \frac{1}{k}\right)f_{2b} = h_{2a1}F_1$$

$$f_{2b} = \left(h_{2a2a} + h_{2b2b} + \frac{1}{k}\right)^{-1} h_{2a1}F_1$$

Applying the equilibrium condition $f_{2a} = -f_{2b}$, we obtain:

$$f_{2a} = -\left(h_{2a2a} + h_{2b2b} + \frac{1}{k}\right)^{-1} h_{2a1}F_1.$$

This enables us to write the direct and cross receptances as shown in Eqs. 7.2.21 and 7.2.22, respectively. We note that these equations simplify to the rigid coupling results provided in Eqs. 7.2.5 and 7.2.6 as k approaches infinity. The assembly cross receptance at coordinate X_{2b} is given by Eq. 7.2.23.

$$H_{11} = \frac{X_1}{F_1} = \frac{x_1}{F_1} = \frac{h_{11}f_1 + h_{12a}f_{2a}}{F_1} = \frac{h_{11}f_1 - h_{12a}\left(h_{2a2a} + h_{2b2b} + \frac{1}{k}\right)^{-1} h_{2a1}F_1}{F_1}$$
$$H_{11} = \frac{h_{11}F_1 - h_{12a}\left(h_{2a2a} + h_{2b2b} + \frac{1}{k}\right)^{-1} h_{2a1}F_1}{F_1} = h_{11} - h_{12a}\left(h_{2a2a} + h_{2b2b} + \frac{1}{k}\right)^{-1} h_{2a1} \quad (7.2.21)$$

$$H_{2a1} = \frac{X_{2a}}{F_1} = \frac{x_{2a}}{F_1} = \frac{h_{2a1}f_1 + h_{2a2a}f_{2a}}{F_1} = \frac{h_{2a1}F_1 - h_{2a2a}\left(h_{2a2a} + h_{2b2b} + \frac{1}{k}\right)^{-1} h_{2a1}F_1}{F_1}$$
$$H_{2a1} = h_{2a1} - h_{2a2a}\left(h_{2a2a} + h_{2b2b} + \frac{1}{k}\right)^{-1} h_{2a1} \quad (7.2.22)$$

$$H_{2b1} = \frac{X_{2b}}{F_1} = \frac{x_{2b}}{F_1} = \frac{h_{2b2b}f_{2b}}{F_1} = \frac{h_{2b2b}\left(h_{2a2a} + h_{2b2b} + \frac{1}{k}\right)^{-1} h_{2a1}F_1}{F_1}$$
$$H_{2b1} = h_{2b2b}\left(h_{2a2a} + h_{2b2b} + \frac{1}{k}\right)^{-1} h_{2a1} \quad (7.2.23)$$

Let's now apply the external force, F_{2a}, to coordinate X_{2a} as shown in Fig. 7.2.7 in order to determine the assembly receptances H_{2a2a}, H_{2b2a}, and H_{12a}. The component displacements are $x_1 = h_{12a}f_{2a}$ and $x_{2a} = h_{2a2a}f_{2a}$ < ?A3B2tlsb for substructure I and $x_{2b} = h_{2b2b}f_{2b}$ for substructure II. The equilibrium condition is $f_{2a} + f_{2b} = F_{2a}$ and the compatibility condition is:

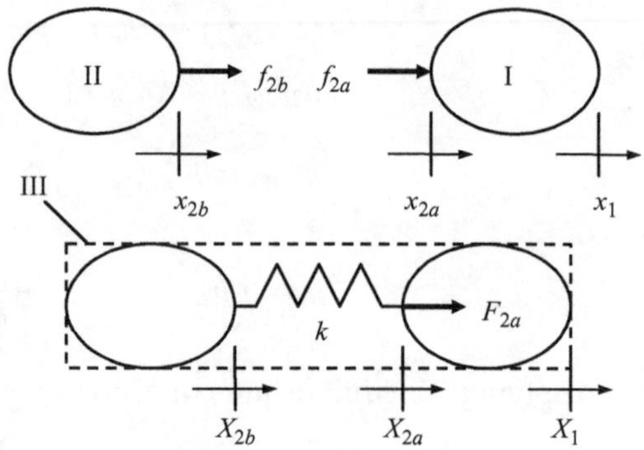

Fig. 7.2.7 Flexible coupling of components I and II to form assembly III. The force F_{2a} is applied to the assembly in order to determine H_{2a2a}, H_{2b2a}, and H_{12a}

$$k(x_{2b} - x_{2a}) = -f_{2b}. \tag{7.2.24}$$

We first determine the force f_{2b} by substituting the component displacements in Eq. 7.2.24 and replacing f_{2a} with $F_{2a} - f_{2b}$.

$$k(h_{2b2b}f_{2b} - h_{2a2a}F_{2a} + h_{2a2a}f_{2b}) = -f_{2b}$$

$$kh_{2b2b}f_{2b} - kh_{2a2a}F_{2a} + kh_{2a2a}f_{2b} = -f_{2b}$$

$$\left(h_{2a2a} + h_{2b2b} + \frac{1}{k}\right)f_{2b} = h_{2a2a}F_{2a}$$

$$f_{2b} = \left(h_{2a2a} + h_{2b2b} + \frac{1}{k}\right)^{-1} h_{2a2a}F_{2a}$$

Again using the equilibrium condition we find the equation for f_{2a}.

$$f_{2a} = F_{2a} - f_{2b} = \left(1 - \left(h_{2a2a} + h_{2b2b} + \frac{1}{k}\right)^{-1} h_{2a2a}\right)F_{2a}$$

The direct and cross receptances for this situation (depicted in Fig. 7.2.7) are provided in Eqs. 7.2.25 through 7.2.27.

$$H_{2a2a} = \frac{X_{2a}}{F_{2a}} = \frac{x_{2a}}{F_{2a}} = \frac{h_{2a2a}f_{2a}}{F_{2a}} = \frac{h_{2a2a}\left(1 - \left(h_{2a2a} + h_{2b2b} + \frac{1}{k}\right)^{-1} h_{2a2a}\right)F_{2a}}{F_{2a}}$$
$$H_{2a2a} = h_{2a2a} - h_{2a2a}\left(h_{2a2a} + h_{2b2b} + \frac{1}{k}\right)^{-1} h_{2a2c} \tag{7.2.25}$$

$$H_{2b2a} = \frac{X_{2b}}{F_{2a}} = \frac{x_{2b}}{F_{2a}} = \frac{h_{2b2b}f_{2b}}{F_{2b}} = \frac{h_{2b2b}\left(h_{2a2a} + h_{2b2b} + \frac{1}{k}\right)^{-1} h_{2a2a}F_{2a}}{F_{2a}}$$
$$H_{2b2a} = h_{2b2b}\left(h_{2a2a} + h_{2b2b} + \frac{1}{k}\right)^{-1} h_{2a2a} \tag{7.2.26}$$

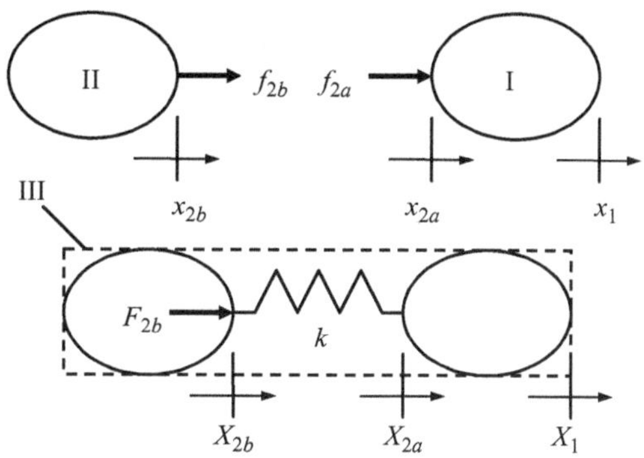

Fig. 7.2.8 Flexible coupling of components I and II to form assembly III. The force F_{2b} is applied to the assembly in order to determine H_{2b2b}, H_{2a2b}, and H_{12b}

$$H_{12a} = \frac{X_1}{F_{2a}} = \frac{x_1}{F_{2a}} = \frac{h_{12a}f_{2a}}{F_{2a}} = \frac{h_{12a}\left(1 - \left(h_{2a2a} + h_{2b2b} + \frac{1}{k}\right)^{-1} h_{2a2a}\right)F_{2a}}{F_{2a}} \tag{7.2.27}$$
$$H_{12a} = h_{12a} - h_{12a}\left(h_{2a2a} + h_{2b2b} + \frac{1}{k}\right)^{-1} h_{2a2a}$$

Our final scenario for the two component flexible coupling is shown in Fig. 7.2.8. Here, we apply the external force F_{2b} to coordinate X_{2b} to obtain the direct and cross assembly receptances H_{2b2b}, H_{2a2b}, and H_{12b}. The component displacements are the same as the previous case: $x_1 = h_{12a}f_{2a}$ and $x_{2a} = h_{2a2a}f_{2a}$ for substructure I and $x_{2b} = h_{2b2b}f_{2b}$ for substructure II. However, the equilibrium condition is modified to be $f_{2a} + f_{2b} = F_{2b}$ and the compatibility condition is rewritten as:

$$k(x_{2a} - x_{2b}) = -f_{2a}. \tag{7.2.28}$$

We find f_{2a} by substituting the component displacements in Eq. 7.2.28 and replacing f_{2b} with $F_{2a} - f_{2a}$.

$$k(h_{2a2a}f_{2a} - h_{2b2b}F_{2b} + h_{2b2b}f_{2a}) = -f_{2a}$$
$$kh_{2a2a}f_{2a} - kh_{2b2b}F_{2b} + kh_{2b2b}f_{2a} = -f_{2a}$$
$$\left(h_{2a2a} + h_{2b2b} + \frac{1}{k}\right)f_{2a} = h_{2b2b}F_{2b}$$
$$f_{2a} = \left(h_{2a2a} + h_{2b2b} + \frac{1}{k}\right)^{-1} h_{2b2b}F_{2b}$$

Again using the equilibrium condition we find the equation for f_{2b}.

$$f_{2b} = F_{2b} - f_{2a} = \left(1 - \left(h_{2a2a} + h_{2b2b} + \frac{1}{k}\right)^{-1} h_{2b2b}\right) F_{2b}$$

The direct and cross receptances for the case shown in Fig. 7.2.8 are given in Eqs. 7.2.29 through 7.2.31.

$$H_{2b2b} = \frac{X_{2b}}{F_{2b}} = \frac{x_{2b}}{F_{2b}} = \frac{h_{2b2b} f_{2b}}{F_{2b}} = \frac{h_{2b2b}\left(1 - \left(h_{2a2a} + h_{2b2b} + \frac{1}{k}\right)^{-1} h_{2b2b}\right) F_{2b}}{F_{2b}}$$
$$H_{2b2b} = h_{2b2b} - h_{2b2b}\left(h_{2a2a} + h_{2b2b} + \frac{1}{k}\right)^{-1} h_{2b2b} \tag{7.2.29}$$

$$H_{2a2b} = \frac{X_{2a}}{F_{2b}} = \frac{x_{2a}}{F_{2b}} = \frac{h_{2a2a} f_{2a}}{F_{2b}} = \frac{h_{2a2a}\left(h_{2a2a} + h_{2b2b} + \frac{1}{k}\right)^{-1} h_{2b2b} F_{2b}}{F_{2b}}$$
$$H_{2a2b} = h_{2a2a}\left(h_{2a2a} + h_{2b2b} + \frac{1}{k}\right)^{-1} h_{2b2b} \tag{7.2.30}$$

$$H_{12b} = \frac{X_1}{F_{2b}} = \frac{x_1}{F_{2b}} = \frac{h_{12a} f_{2a}}{F_{2a}} = \frac{h_{12a}\left(h_{2a2a} + h_{2b2b} + \frac{1}{k}\right)^{-1} h_{2b2b} F_{2b}}{F_{2b}}$$
$$H_{12b} = h_{12a}\left(h_{2a2a} + h_{2b2b} + \frac{1}{k}\right)^{-1} h_{2b2b} \tag{7.2.31}$$

7.2.3 Two Component Flexible, Damped Coupling[6]

As we discussed in Section 2.1, damping is always present in mechanical systems. Therefore, as a final step in our receptance coupling of bodies I and II to form assembly III, we can expand the model in Fig. 7.2.6 to include viscous damping at the coupling interface. See Fig. 7.2.9.

IN A NUTSHELL We recognize that the interface between the tool and holder, as with any common interface, is not rigid. Small relative motions between these bodies can produce damping.

---

[6] Author T. Schmitz gratefully acknowledges collaboration with Dr. T. Burns, National Institute of Standards and Technology, Gaithersburg, MD, in developing the damping analysis.

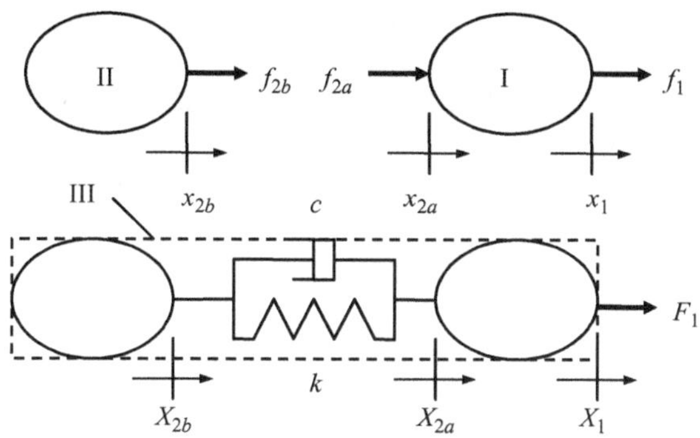

Fig. 7.2.9 Viscously damped, flexible coupling of components I and II to form assembly III. As with the two component flexible coupling case, the force F_1 is applied to the assembly in order to determine H_{11}, H_{2a1}, and H_{2b1}

The expressions for the component displacements and equilibrium conditions remain unchanged relative to the flexible coupling derivation when we add damping. However, the compatibility condition is now:

$$k(x_{2b} - x_{2a}) + i\omega c(x_{2b} - x_{2a}) = -f_{2b}, \tag{7.2.32}$$

where we have assumed harmonic motion so that the velocity dependent damping forces can be express in the form $i\omega cx$. Equation 7.2.32 can be rewritten as:

$$(k + i\omega c)(x_{2b} - x_{2a}) = -f_{2b}. \tag{7.2.33}$$

If we substitute the complex, frequency dependent variable k' for $(k + i\omega c)$, then we see that the compatibility equation takes the same form as shown in Eq. 7.2.19 and we can simply replace k in Eq. 7.2.21 with k' to obtain Eq. 7.2.34 [5]. This defines the direct FRF at coordinate X_1 on assembly III in Fig. 7.2.9. The same substitution can be made in the other assembly receptances derived for the two component flexible coupling in order to obtain the two component flexible, damped coupling results.

$$H_{11} = h_{11} - h_{12a}\left(h_{2a2a} + h_{2b2b} + \frac{1}{k'}\right)^{-1} h_{2a1} \tag{7.2.34}$$

Before proceeding with a numerical example, we present Table 7.2.1 which summarizes the receptance coupling equations developed in the previous paragraphs.

Table 7.2.1 Direct and cross receptances for two component coupling. The connection type (labeled C-type) is R, rigid, or F, flexible. The receptance type (labeled R-type) is D, direct, or C, cross. The corresponding figure and equation numbers are also included

C-type	Substructure coordinates I	II	R-type	Receptances	Fig.	Eq.
R	x_{1a}	x_{1b}	D	$H_{11} = h_{1a1a} - h_{1a1a}(h_{1a1a} + h_{1b1b})^{-1} h_{1a1a}$	7.2.1	7.2.1
R	x_1, x_{2a}	x_{2b}	D	$H_{11} = h_{11} - h_{12a}(h_{2a2a} + h_{2b2b})^{-1} h_{2a1}$	7.2.2	7.2.5
			C	$H_{21} = h_{2a1} - h_{2a2a}(h_{2a2a} + h_{2b2b})^{-1} h_{2a1}$		7.2.6
			D	$H_{22} = h_{2a2a} - h_{2a2a}(h_{2a2a} + h_{2b2b})^{-1} h_{2a2a}$	7.2.3	7.2.10
			C	$H_{12} = h_{12a} - h_{12a}(h_{2a2a} + h_{2b2b})^{-1} h_{2a2a}$		7.2.11
F	x_{1a}	x_{1b}	D	$H_{1a1a} = h_{1a1a} - h_{1a1a}\left(h_{1a1a} + h_{1b1b} + \frac{1}{k}\right)^{-1} h_{1a1a}$	7.2.4	7.2.14
			C	$H_{1b1a} = h_{1b1b}\left(h_{1a1a} + h_{1b1b} + \frac{1}{k}\right)^{-1} h_{1a1a}$		7.2.15
			D	$H_{1b1b} = h_{1b1b} - h_{1b1b}\left(h_{1a1a} + h_{1b1b} + \frac{1}{k}\right)^{-1} h_{1b1b}$	7.2.5	7.2.17
			C	$H_{1a1b} = h_{1a1a}\left(h_{1a1a} + h_{1b1b} + \frac{1}{k}\right)^{-1} h_{1b1b}$		7.2.18
F	x_1, x_{2a}	x_{2b}	D	$H_{11} = h_{11} - h_{12a}\left(h_{2a2a} + h_{2b2b} + \frac{1}{k}\right)^{-1} h_{2a1}$	7.2.6	7.2.21
			C	$H_{2a1} = h_{2a1} - h_{2a2a}\left(h_{2a2a} + h_{2b2b} + \frac{1}{k}\right)^{-1} h_{2a1}$		7.2.22
			C	$H_{2b1} = h_{2b2b}\left(h_{2a2a} + h_{2b2b} + \frac{1}{k}\right)^{-1} h_{2a1}$		7.2.23
			D	$H_{2a2a} = h_{2a2a} - h_{2a2a}\left(h_{2a2a} + h_{2b2b} + \frac{1}{k}\right)^{-1} h_{2a2a}$	7.2.7	7.2.25
			C	$H_{2b2a} = h_{2b2b}\left(h_{2a2a} + h_{2b2b} + \frac{1}{k}\right)^{-1} h_{2a2a}$		7.2.26
			C	$H_{12a} = h_{12a} - h_{12a}\left(h_{2a2a} + h_{2b2b} + \frac{1}{k}\right)^{-1} h_{2a2a}$		7.2.27
			D	$H_{2b2b} = h_{2b2b} - h_{2b2b}\left(h_{2a2a} + h_{2b2b} + \frac{1}{k}\right)^{-1} h_{2b2b}$	7.2.8	7.2.29
			C	$H_{2a2b} = h_{2a2a}\left(h_{2a2a} + h_{2b2b} + \frac{1}{k}\right)^{-1} h_{2b2b}$		7.2.30
			C	$H_{12b} = h_{12a}\left(h_{2a2a} + h_{2b2b} + \frac{1}{k}\right)^{-1} h_{2b2b}$		7.2.31

IN A NUTSHELL As always, improved accuracy of the model and prediction comes at the expense of increased model complexity and/or measurement requirements. Prediction of the natural frequency of the assembly, for example, does not place the same demands on the model as prediction of the dynamic stiffness (including damping). Accurate frequency response function prediction and, by extension, the corresponding stability lobe diagram requires more care.

Example 7.2.1: Comparison of assembly modeling techniques Let's now complete an example where we compare receptance coupling to the methods we discussed in Sections 2.3 and 2.4: modal analysis and complex matrix inversion. As shown in Fig. 7.2.10, two single degree of freedom spring-mass-damper systems, I and II, are to be connected using the linear spring element, k_c, to form the new two degree of freedom assembly, III [16]. The assembled system's equations of motion are determined as shown in Sections 2.3 and 2.4. The matrix representation of these equations, after substituting the assumed harmonic form of the solution, is provided in Eq. 7.2.35. This equation takes the same form as Eq. 2.4.2, $(s^2[M] + s[C] + [K])\{X\}e^{st} = \{F\}e^{st}$, where we've substituted the

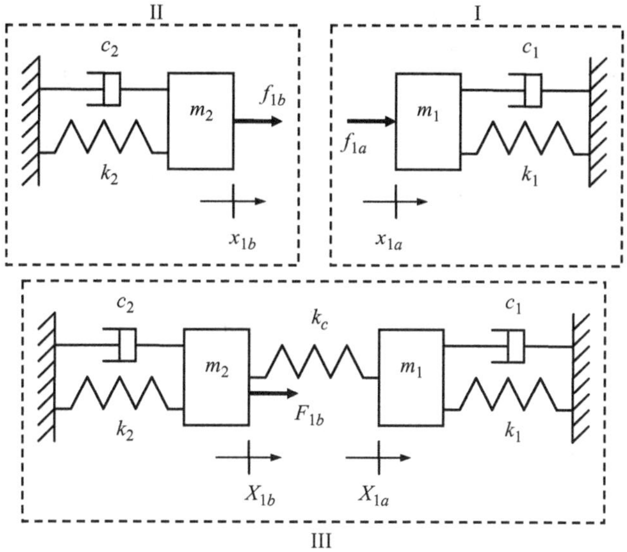

Fig. 7.2.10 Flexible coupling of spring-mass-damper systems I and II to form the two degree of freedom assembly III

Laplace variable s for the product $i\omega$ and $[M]$, $[C]$, and $[K]$ are the lumped parameter mass, damping, and stiffness matrices in local coordinates, respectively.

$$\left(s^2\begin{bmatrix} m_1 & 0 \\ 0 & m_2 \end{bmatrix} + s\begin{bmatrix} c_1 & 0 \\ 0 & c_2 \end{bmatrix} + \begin{bmatrix} k_1+k_c & -k_c \\ -k_c & k_2+k_c \end{bmatrix}\right)\begin{Bmatrix} X_{1a} \\ X_{1b} \end{Bmatrix} = \begin{Bmatrix} 0 \\ F_{1b} \end{Bmatrix}$$

$$\begin{bmatrix} m_1s^2+c_1s+(k_1+k_c) & -k_c \\ -k_c & m_2s^2+c_2s+(k_2+k_c) \end{bmatrix}\begin{Bmatrix} X_{1a} \\ X_{1b} \end{Bmatrix} = \begin{Bmatrix} 0 \\ F_{1b} \end{Bmatrix} \tag{7.2.35}$$

7.2.4 Modal Analysis

We can use the equations of motion shown in Eq. 7.2.35 to find the modal solution for the assembled system. If we assume that proportional damping exists (i.e., $[C] = \alpha[M] + \beta[K]$, where α and β are real numbers), damping can be neglected in the modal solution. Note that this solution is also independent of the external force, F_{1b}. We write the characteristic equation for this system as shown in Eq. 7.2.36. The quadratic roots of this 4th order equation, s_1^2 and s_2^2, give the two eigenvalues ($s_1^2 = -\omega_{n1}^2$ and $s_2^2 = -\omega_{n2}^2$, where $\omega_{n1} < \omega_{n2}$) for the two degree of freedom system.

$$\begin{aligned}&\left(m_1 s^2+(k_1+k_c)\right)\left(m_2 s^2+(k_2+k_c)\right)-k_c^2=0\\&m_1 m_2 s^4+(m_1(k_2+k_c)+m_2(k_1+k_c))s^2+(k_1+k_c)(k_2+k_c)-k_c^2=0\end{aligned} \tag{7.2.36}$$

Substitution of these eigenvalues, normalized to the coordinate of interest (coordinate X_{1b} in this case), into either of the original equations of motion, again neglecting damping and the external force, yields the eigenvectors (mode shapes). Selecting the top equation from Eq. 7.2.35, for example, gives:

$$\frac{X_{1a}}{X_{1b}}=\frac{k_c}{m_1 s^2+(k_1+k_c)}. \tag{7.2.37}$$

The mass, damping, and stiffness matrices are diagonalized using the modal matrix (composed of columns of the eigenvectors), P, defined in Eq. 7.2.38. Specifically, we have that: $[M_q]=[P]^T[M][P]=\begin{bmatrix} m_{q1} & 0\\ 0 & m_{q2}\end{bmatrix}$, $[C_q]=[P]^T[C][P]=\begin{bmatrix} c_{q1} & 0\\ 0 & c_{q2}\end{bmatrix}$, and $[K_q]=[P]^T[K][P]=\begin{bmatrix} k_{q1} & 0\\ 0 & k_{q2}\end{bmatrix}$. Based on these modal values, we calculate the associated damping ratios, $\zeta_{q1,2}=\frac{c_{q1,2}}{2\sqrt{k_{q1,2}m_{q1,2}}}$. The modal solution for the direct FRF at coordinate X_{1b} of the assembled system is then expressed as shown in Eq. 7.2.39, where $r_{1,2}=\frac{\omega}{\omega_{n1,2}}$.

$$P=\begin{bmatrix} \frac{X_{1a}}{X_{1b}}\left(s_1^2\right) & \frac{X_{1a}}{X_{1b}}\left(s_2^2\right)\\ 1 & 1\end{bmatrix} \tag{7.2.38}$$

$$H_{1b1b}=\frac{X_{1b}}{F_{1b}}=\frac{1}{k_{q1}}\left(\frac{(1-r_1^2)-i(2\zeta_{q1}r_1)}{(1-r_1^2)^2+(2\zeta_{q1}r_1)^2}\right)+\frac{1}{k_{q2}}\left(\frac{(1-r_2^2)-i(2\zeta_{q2}r_2)}{(1-r_2^2)^2+(2\zeta_{q2}r_2)^2}\right) \tag{7.2.39}$$

7.2.5 Complex Matrix Inversion[7]

Equation 7.2.35 can be compactly written as $[A]\{X\}=\{F\}$. As shown in Section 2.4, complex matrix inversion is carried out using $\{X\}\{F\}^{-1}=[A]^{-1}$ to determine the assembly direct and cross FRFs. The inverted $[A]$ matrix for this two degree of freedom example is:

---

[7] As discussed in Section 2.4, complex matrix inversion, rather than modal analysis, is applied when the damping may not be proportional.

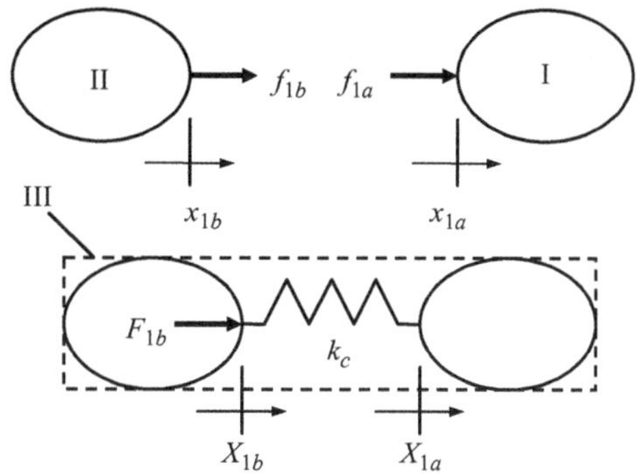

Fig. 7.2.11 Receptance coupling representation of joining spring-mass-damper systems I and II to form the two degree of freedom assembly III

$$[A]^{-1}=\frac{\begin{bmatrix} a_{22} & -a_{12} \\ -a_{21} & a_{11} \end{bmatrix}}{a_{11}\cdot a_{22}-a_{12}\cdot a_{21}}=\frac{\begin{bmatrix} -\omega^2 m_2+i\omega c_2+(k_2+k_c) & k_c \\ k_c & -\omega^2 m_1+i\omega c_1+(k_1+k_c) \end{bmatrix}}{(-\omega^2 m_1+i\omega c_1+(k_1+k_c))(-\omega^2 m_2+i\omega c_2+(k_2+k_c))-k_c^2},$$

where we've replaced s with $i\omega$ relative to Eq. 7.2.35. The individual terms in the inverted $[A]$ matrix are:

$$[A]^{-1}=\begin{bmatrix} \frac{X_{1a}}{F_{1a}} & \frac{X_{1a}}{F_{1b}} \\ \frac{X_{1b}}{F_{1a}} & \frac{X_{1b}}{F_{1b}} \end{bmatrix}=\begin{bmatrix} H_{1a1a} & H_{1a1b} \\ H_{1b1a} & H_{1b1b} \end{bmatrix}. \tag{7.2.40}$$

7.2.6 Receptance Coupling

This case is the same as the two component flexible coupling example shown in Fig. 7.2.5. Replacing k with k_c in Eq. 7.2.17, we obtain Eq. 7.2.41. See Fig. 7.2.11.

$$\begin{aligned} \frac{X_{1b}}{F_{1b}}&=\frac{x_{1b}}{F_{1b}}=\frac{h_{1b1b}f_{1b}}{F_{1b}}=\frac{h_{1b1b}\left(1-\left(h_{1a1a}+h_{1b1b}+\frac{1}{k_c}\right)^{-1}h_{1b1b}\right)F_{1b}}{F_{1b}} \\ \frac{X_{1b}}{F_{1b}}&=H_{1b1b}=h_{1b1b}-h_{1b1b}\left(h_{1a1a}+h_{1b1b}+\frac{1}{k_c}\right)^{-1}h_{1b1b} \end{aligned} \tag{7.2.41}$$

To compare the three methods, we select the mass, damping, and stiffness values shown in Table 7.2.2 for the model in Fig. 7.2.10. We note that proportional

Table 7.2.2 Mass, damping, and stiffness values for Ex. 7.2.1

Parameter	Value
m_1	3 kg
c_1	200 N-s/m
k_1	2×10^6 N/m
m_2	2 kg
c_2	100 N-s/m
k_2	1×10^6 N/m
k_c	5×10^5 N/m

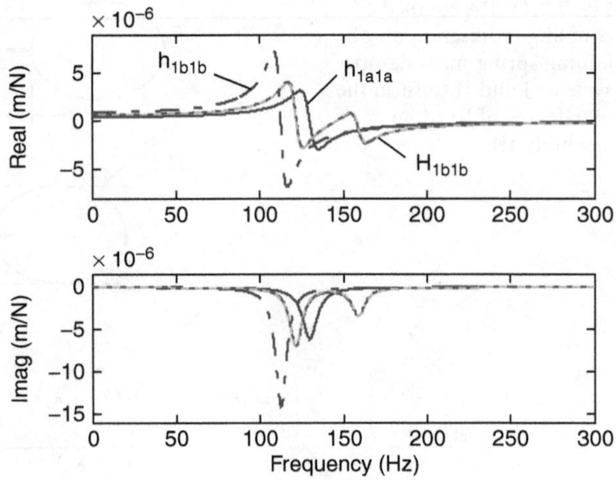

Fig. 7.2.12 Comparison of three methods for H_{1b1b} calculation. It is seen that the modal analysis, complex matrix inversion, and receptance coupling methods nominally agree (superimposed solid lines). The component receptances, h_{1a1a} and h_{1b1b}, are also shown

damping exists ($\alpha = 0$ and $\beta = 1 \times 10^{-4}$) for the selected system, so the modal approach may be applied. The MATLAB® program used to produce Fig. 7.2.12, which displays both the component receptances and the assembly receptance computed using the three methods, is provided on the companion CD as p_7_2_1_1.m. The frequency dependent differences between the complex matrix inversion result, which was obtained through vector manipulations only by calculating the H_{1b1b} result directly:

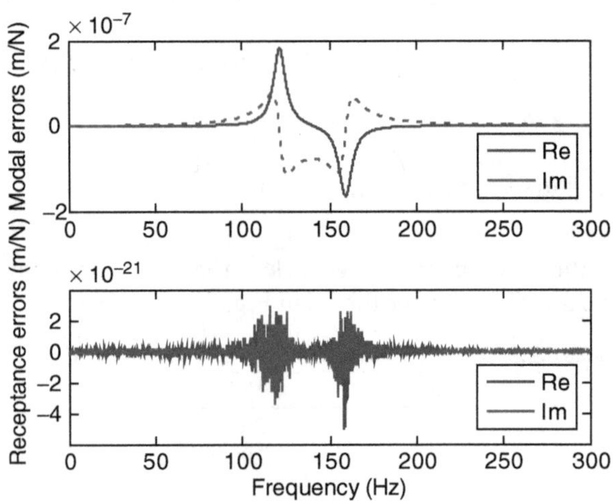

Fig. 7.2.13 Real and imaginary parts of difference between complex matrix inversion and modal analysis (top) and real and imaginary parts of difference between complex matrix inversion and receptance coupling (bottom). Receptance coupling agrees much more closely

$$H_{1b1b} = \frac{-\omega^2 m_1 + i\omega c_1 + (k_1 + k_c)}{(-\omega^2 m_1 + i\omega c_1 + (k_1 + k_c))(-\omega^2 m_2 + i\omega c_2 + (k_2 + k_c)) - k_c^2},$$

and the modal and receptance coupling method results are shown in Fig. 7.2.13. It is seen that the errors introduced by the modal method (top) are approximately 4×10^{13} times greater than the errors associated with the receptance technique (bottom). The differences between the three techniques are introduced by numerical round off errors in the mathematical manipulations. However, the improved numerical accuracy obtained with receptance coupling (vector manipulations) over modal coupling (matrix manipulations) is another benefit of the receptance coupling approach.

7.3 Advanced Receptance Coupling

The primary difference between the simple examples we've considered so far and tool-holder-spindle-machine modeling is that we now need to consider not only lateral displacements (x_i/X_i) and forces (f_j/F_j), but also rotations about lines perpendicular to the beam axis (θ_i/Θ_i) and bending couples (m_j/M_j). [8,9] To begin this discussion, let's consider the solid cylinder-prismatic cantilever beam assembly shown in Fig. 7.3.1. To determine the assembly dynamics, all four bending receptances must be included in the component descriptions (i.e., displacement-to-force, h_{ij}, displacement-to-couple, l_{ij}, rotation-to-force, n_{ij}, and rotation-to-couple, p_{ij}).

Let's summarize the steps required to predict the Fig. 7.3.1 assembly receptances.

1. Define the components and coordinates for the model. In this simple example, we can select two components: a prismatic beam with fixed-free (or cantilever) boundary conditions and a cylinder with free-free (or unsupported) boundary conditions; see Fig. 7.3.2.

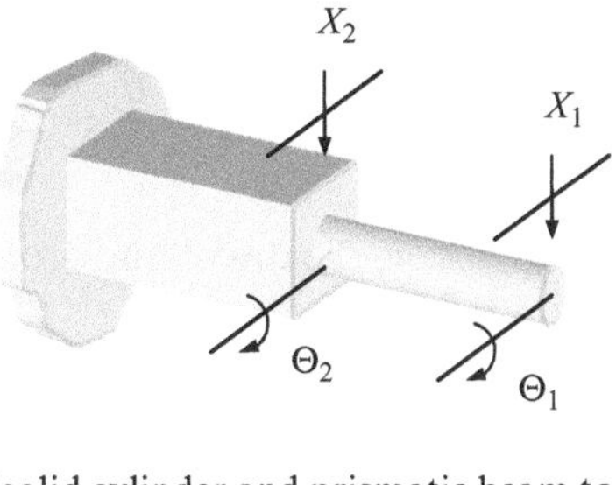

Fig. 7.3.1 Rigid coupling of solid cylinder and prismatic beam to form cantilevered assembly

[8] We will not consider axial or torsional vibrations in this analysis.

[9] Author T. Schmitz acknowledges significant collaboration with Dr. M. Davies, University of North Carolina at Charlotte, in the early application of receptance coupling to tool point FRF prediction.

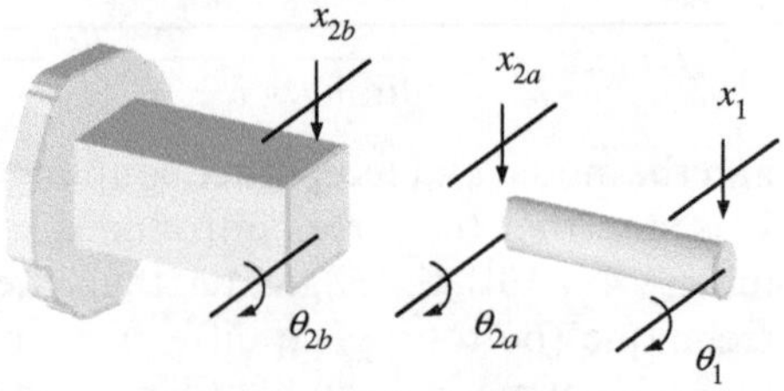

Fig. 7.3.2 Solid cylinder and prismatic beam components used to form cantilevered assembly

2. Determine the component receptances. We can use either measurements or models. For the models, an elegant choice is the closed form receptances presented by Bishop and Johnson [1] for flexural vibrations of uniform Euler-Bernoulli beams with free, fixed, sliding, and pinned boundary conditions. Of course, the Timoshenko beam model or other approaches may also be applied. See Section 7.4. For measurements, we can follow the procedures outlined in Section 2.6.
3. Based on the selected model from step 1, express the assembly receptances as a function of the component receptances. As demonstrated in Section 7.2, we determine the assembly receptances using the component displacements/rotations, equilibrium conditions, and compatibility conditions.

We begin the analysis of the system shown in Figs. 7.3.1 and 7.3.2 by writing the component receptances. Note that we have placed coordinates at the prediction location (1), which would represent the tool point for a tool-holder-spindle assembly, and coupling locations (2*a* and 2*b*). For the cylinder, we have the following direct receptances at the coordinate 1 end:

$$h_{11} = \frac{x_1}{f_1} \quad l_{11} = \frac{x_1}{m_1} \quad n_{11} = \frac{\theta_1}{f_1} \quad p_{11} = \frac{\theta_1}{m_1}. \tag{7.3.1}$$

The corresponding cross receptances at the same location are:

$$h_{12a} = \frac{x_1}{f_{2a}} \quad l_{12a} = \frac{x_1}{m_{2a}} \quad n_{12a} = \frac{\theta_1}{f_{2a}} \quad p_{12a} = \frac{\theta_1}{m_{2a}}. \tag{7.3.2}$$

At coordinate 2*a* on the cylinder, the direct and cross receptances are written as shown in Eqs. 7.3.3 and 7.3.4, respectively.

$$h_{2a2a} = \frac{x_{2a}}{f_{2a}} \quad l_{2a2a} = \frac{x_{2a}}{m_{2a}} \quad n_{2a2a} = \frac{\theta_{2a}}{f_{2a}} \quad p_{2a2a} = \frac{\theta_{2a}}{m_{2a}} \tag{7.3.3}$$

$$h_{2a1} = \frac{x_{2a}}{f_1} \quad l_{2a1} = \frac{x_{2a}}{m_1} \quad n_{2a1} = \frac{\theta_{2a}}{f_1} \quad p_{2a1} = \frac{\theta_{2a}}{m_1} \tag{7.3.4}$$

Similarly, for the prismatic beam, the direct receptances at the coupling location 2*b* are described by Eq. 7.3.5.

$$h_{2b2b} = \frac{x_{2b}}{f_{2b}} \quad l_{2b2b} = \frac{x_{2b}}{m_{2b}} \quad n_{2b2b} = \frac{\theta_{2b}}{f_{2b}} \quad p_{2b2b} = \frac{\theta_{2b}}{m_{2b}} \tag{7.3.5}$$

To simplify notation, the component receptances can be compactly represented in matrix form as shown in Eqs. 7.3.6 through 7.3.9 for the cylinder and Eq. 7.3.10 for the prismatic beam:

$$\begin{Bmatrix} x_1 \\ \theta_1 \end{Bmatrix} = \begin{bmatrix} h_{11} & l_{11} \\ n_{11} & p_{11} \end{bmatrix} \begin{Bmatrix} f_1 \\ m_1 \end{Bmatrix} \text{ or } \{u_1\} = [R_{11}]\{q_1\}, \tag{7.3.6}$$

$$\begin{Bmatrix} x_{2a} \\ \theta_{2a} \end{Bmatrix} = \begin{bmatrix} h_{2a2a} & l_{2a2a} \\ n_{2a2a} & p_{2a2a} \end{bmatrix} \begin{Bmatrix} f_{2a} \\ m_{2a} \end{Bmatrix} \text{ or } \{u_{2a}\} = [R_{2a2a}]\{q_{2a}\}, \tag{7.3.7}$$

$$\begin{Bmatrix} x_1 \\ \theta_1 \end{Bmatrix} = \begin{bmatrix} h_{12a} & l_{12a} \\ n_{12a} & p_{12a} \end{bmatrix} \begin{Bmatrix} f_{2a} \\ m_{2a} \end{Bmatrix} \text{ or } \{u_1\} = [R_{12a}]\{q_{2a}\}, \tag{7.3.8}$$

$$\begin{Bmatrix} x_{2a} \\ \theta_{2a} \end{Bmatrix} = \begin{bmatrix} h_{2a1} & l_{2a1} \\ n_{2a1} & p_{2a1} \end{bmatrix} \begin{Bmatrix} f_1 \\ m_1 \end{Bmatrix} \text{ or } \{u_{2a}\} = [R_{2a1}]\{q_1\}, \text{ and} \tag{7.3.9}$$

$$\begin{Bmatrix} x_{2b} \\ \theta_{2b} \end{Bmatrix} = \begin{bmatrix} h_{2b2b} & l_{2b2b} \\ n_{2b2b} & p_{2b2b} \end{bmatrix} \begin{Bmatrix} f_{2b} \\ m_{2b} \end{Bmatrix} \text{ or } \{u_{2b}\} = [R_{2b2b}]\{q_{2b}\}, \tag{7.3.10}$$

where R_{ij} is the generalized receptance matrix that describes both translational and rotational component behavior [6, 10, 17] and u_i and q_j are the corresponding generalized displacement/rotation and force/couple vectors. To visualize R_{ij}, we can think of each frequency dependent 2×2 R_{ij} matrix as a page in a book with every page representing a different frequency value. Flipping through the book from front to back scans the frequency values from low to high through the modeled or measured bandwidth. Naturally, all receptances in the coupling analysis must be based on the same frequency vector (resolution and range).

We write the component receptances, using the new notation, as: $u_1 = R_{11}q_1 + R_{12a}q_{2a}$ and $u_{2a} = R_{2a1}q_1 + R_{2a2a}q_{2a}$ for the cylinder, and $u_{2b} = R_{2b2b}q_{2b}$ for the prismatic beam. If we apply a rigid connection between the two components, the compatibility condition is: $u_{2b} - u_{2a} = 0$. Additionally, if we again specify that the component and assembly coordinates are at the same physical locations, then we have that $u_1 = U_1$ and $u_{2a} = u_{2b} = U_2$ (due to the rigid coupling).

We can write the assembly receptances as shown in Eq. 7.3.11, which again implements the generalized notation:

$$\begin{Bmatrix} U_1 \\ U_2 \end{Bmatrix} = \begin{bmatrix} G_{11} & G_{12} \\ G_{21} & G_{22} \end{bmatrix} \begin{Bmatrix} Q_1 \\ Q_2 \end{Bmatrix}, \tag{7.3.11}$$

where $U_i = \begin{Bmatrix} X_i \\ \Theta_i \end{Bmatrix}$, $G_{ij} = \begin{bmatrix} H_{ij} & L_{ij} \\ N_{ij} & P_{ij} \end{bmatrix}$, and $Q_j = \begin{Bmatrix} F_j \\ M_j \end{Bmatrix}$. To determine the assembly receptance at the free end of the cylinder, G_{11}, we apply Q_1 to coordinate U_1 as shown in Fig. 7.3.3, where the generalized U_i and u_i vectors

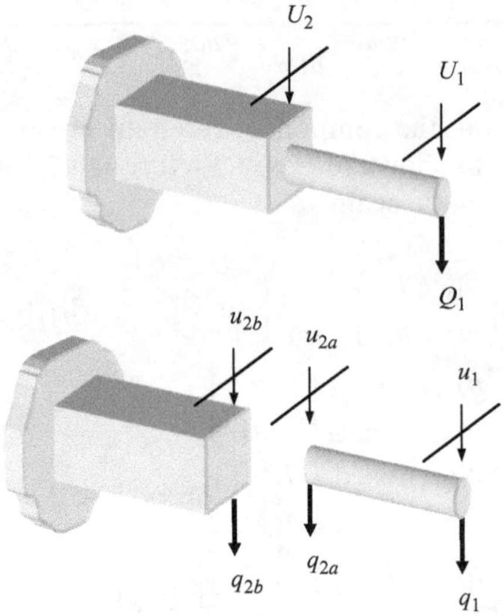

Fig. 7.3.3 Receptance coupling model for determining G_{11} and G_{21}. Rigid coupling is assumed

are shown schematically as "displacements", although we recognize that they describe both lateral translation and rotation. The associated equilibrium conditions are: $q_{2a} + q_{2b} = 0$ and $q_1 = Q_1$. By substituting the component displacements/rotations and equilibrium conditions into the compatibility condition, we obtain the expression for q_{2b} shown in Eq. 7.3.12. The component force q_{2a} is then determined from the equilibrium condition $q_{2a} = -q_{2b}$. The expression for G_{11} is given by Eq. 7.3.13. We find the corresponding cross receptance matrix, G_{21}, in a similar manner; see Eq. 7.3.14. Note that G_{11} and G_{21} comprise the first column of the receptance matrix in Eq. 7.3.11.

$$\begin{aligned} & u_{2b} - u_{2a} = 0 \\ & R_{2b2b}q_{2b} - R_{2a1}q_1 - R_{2a2a}q_{2a} = 0 \\ & (R_{2a2a} + R_{2b2b})q_{2b} - R_{2a1}Q_1 = 0 \\ & q_{2b} = (R_{2a2a} + R_{2b2b})^{-1}R_{2a1}Q_1 \end{aligned} \tag{7.3.12}$$

$$\begin{aligned} G_{11} &= \frac{U_1}{Q_1} = \frac{u_1}{Q_1} = \frac{R_{11}q_1 + R_{12a}q_{2a}}{Q_1} = \frac{R_{11}Q_1 - R_{12a}(R_{2a2a} + R_{2b2b})^{-1}R_{2a1}Q_1}{Q_1} \\ G_{11} &= R_{11} - R_{12a}(R_{2a2a} + R_{2b2b})^{-1}R_{2a1} = \begin{bmatrix} H_{11} & L_{11} \\ N_{11} & P_{11} \end{bmatrix} \end{aligned} \tag{7.3.13}$$

$$\begin{aligned} G_{21} &= \frac{U_2}{Q_1} = \frac{u_{2a}}{Q_1} = \frac{R_{2a1}q_1 + R_{2a2a}q_{2a}}{Q_1} = \frac{R_{2a1}Q_1 - R_{2a2a}(R_{2a2a} + R_{2b2b})^{-1}R_{2a1}Q_1}{Q_1} \\ G_{21} &= R_{2a1} - R_{2a2a}(R_{2a2a} + R_{2b2b})^{-1}R_{2a1} = \begin{bmatrix} H_{21} & L_{21} \\ N_{21} & P_{21} \end{bmatrix} \end{aligned} \tag{7.3.14}$$

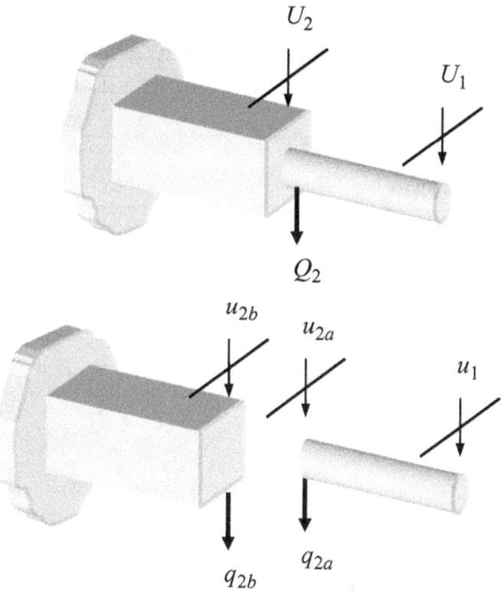

Fig. 7.3.4 Receptance coupling model for determining G_{22} and G_{12}. Rigid coupling is assumed

To find the receptances in the second column of Eq. 7.3.11, we apply Q_2 at U_2, as shown in Fig. 7.3.4. The component receptances are: $u_1 = R_{12a}q_{2a}$ and $u_{2a} = R_{2a2a}q_{2a}$ for the cylinder, and $u_{2b} = R_{2b2b}q_{2b}$ for the prismatic beam. For the assumed rigid connection the compatibility condition is again: $u_{2b} - u_{2a} = 0$. The equilibrium condition is: $q_{2a} + q_{2b} = Q_2$. By substituting the component displacements/rotations and equilibrium condition into the compatibility condition, we obtain the expression for q_{2b} shown in Eq. 7.3.15. The component force q_{2a} is then determined from the equilibrium condition $q_{2a} = Q_2 - q_{2b}$. The expression for G_{22} is provided by Eq. 7.3.16. We find the corresponding cross receptance matrix, G_{12}, in a similar manner as shown in Eq. 7.3.17.

$$\begin{aligned}
&u_{2b} - u_{2a} = 0\\
&R_{2b2b}q_{2b} - R_{2a2a}q_{2a} = 0\\
&R_{2b2b}q_{2b} - R_{2a2a}Q_2 + R_{2a2a}q_{2b} = 0\\
&(R_{2a2a} + R_{2b2b})q_{2b} - R_{2a2a}Q_2 = 0\\
&q_{2b} = (R_{2a2a} + R_{2b2b})^{-1}R_{2a2a}Q_2
\end{aligned} \tag{7.3.15}$$

$$\begin{aligned}
G_{22} &= \frac{U_2}{Q_2} = \frac{u_{2a}}{Q_2} = \frac{R_{2a2a}q_{2a}}{Q_2} = \frac{R_{2a2a}\left(1-(R_{2a2a}+R_{2b2b})^{-1}R_{2a2a}\right)Q_2}{Q_2}\\
G_{22} &= R_{2a2a} - R_{2a2a}(R_{2a2a} + R_{2b2b})^{-1}R_{2a2a} = \begin{bmatrix} H_{22} & L_{22} \\ N_{22} & P_{22} \end{bmatrix}
\end{aligned} \tag{7.3.16}$$

Table 7.3.1 Direct and cross receptances for generalized two component coupling. The connection type (labeled C-type) is R, rigid. The receptance type (labeled R-type) is D, direct, or C, cross. The figure and equation numbers are also included. Similarities to the corresponding entries in Table 7.2.1 are evident

C-type	Substructure coordinates I	II	Receptances R-type		Fig.	Eq.
R	u_1, u_{2a}	u_{2b}	D	$G_{11} = R_{11} - R_{12a}(R_{2a2a} + R_{2b2b})^{-1}R_{2a1}$	7.3.3	7.3.13
			C	$G_{21} = R_{2a1} - R_{2a2a}(R_{2a2a} + R_{2b2b})^{-1}R_{2a1}$		7.3.14
			D	$G_{22} = R_{2a2a} - R_{2a2a}(R_{2a2a} + R_{2b2b})^{-1}R_{2a2a}$	7.3.4	7.3.16
			C	$G_{12} = R_{12a} - R_{12a}(R_{2a2a} + R_{2b2b})^{-1}R_{2a2a}$		7.3.17

$$G_{12} = \frac{U_1}{Q_2} = \frac{u_1}{Q_2} = \frac{R_{12a}q_{2a}}{Q_2} = \frac{R_{12a}\left(1-(R_{2a2a}+R_{2b2b})^{-1}R_{2a2a}\right)Q_2}{Q_2}$$
$$G_{12} = R_{12a} - R_{12a}(R_{2a2a} + R_{2b2b})^{-1}R_{2a2a} = \begin{bmatrix} H_{12} & L_{12} \\ N_{12} & P_{12} \end{bmatrix} \tag{7.3.17}$$

We see that the procedure to model the systems with both displacement and rotations is analogous to the examples provided in Section 7.2. Let's again summarize the receptance terms in tabular form; see Table 7.3.1. Due to the clear similarities to Table 7.2.1, we will not derive the receptances for the other two component coupling cases. The only consideration is that for non-rigid coupling, we replace the scalar stiffness term, $\frac{1}{k}$, from the displacement-to-force analyses with the matrix expression $\left[\tilde{k}\right]^{-1}$, where:

$$\left[\tilde{k}\right] = \begin{bmatrix} k_{xf} & k_{\theta f} \\ k_{xm} & k_{\theta m} \end{bmatrix}.$$

The subscripts for the stiffness matrix entries indicate their function. For example, $k_{\theta f}$ represents resistance to rotation due to an applied force. As shown in Eq. 7.2.34, these four real valued stiffness terms are augmented by the corresponding damping expressions if viscous damping is included at the coupling location [11]. The new complex, frequency dependent stiffness matrix is:

$$\left[\tilde{k}'\right] = \begin{bmatrix} k_{xf} + i\omega c_{xf} & k_{\theta f} + i\omega c_{\theta f} \\ k_{xm} + i\omega c_{xm} & k_{\theta m} + i\omega c_{\theta m} \end{bmatrix}.$$

7.4 Beam Receptances

To describe the lateral vibration and associated rotation at the ends of uniform beams, we may apply Euler-Bernoulli beam theory. Many sources are available for full equation development, such as [1, 18], and we refer the reader to these texts for a detailed analysis. However, we'll review some of the basic concepts before presenting the closed form equations developed by Bishop and Johnson [1].

For a uniform elastic beam subject to lateral vibrations, y, we can write the differential equation:

$$\frac{\partial^2 y}{\partial t^2} + \frac{EI}{\rho A}\frac{\partial^4 y}{\partial x^4} = 0 \tag{7.4.1}$$

to describe its lateral motion as a function of time, t, and position along the beam, x, where E is the beam's elastic modulus (N/m^2), I is the 2nd moment of area/area moment of inertia (m^4), ρ is the density (kg/m^3), and A is the cross sectional area (m^2). This equation assumes that the axis of the undeflected beam lies along the x direction and an infinitesimal slice of the deflected beam is bounded by plane faces. If we assume a harmonic disturbance, we can use the trial function $y = Y(x)\sin(\omega t)$, where ω is the frequency, to eliminate the time dependence. See Eq. 7.4.2, where $\lambda^4 = \omega^2 \frac{\rho A}{EI}$.

$$\frac{\partial^4 Y}{\partial x^4} - \lambda^4 Y = 0 \tag{7.4.2}$$

The general solution to Eq. 7.4.2 is given in Eq. 7.4.3. In this equation, A, B, C, and D are constants that are determined from the boundary conditions. See Table 7.4.1.

$$Y = A\cos(\lambda x) + B\sin(\lambda x) + C\cosh(\lambda x) + D\sinh(\lambda x) \tag{7.4.3}$$

Example 7.4.1: Tip receptances for clamped-free beam In this example, we'll use Eq. 7.4.3 and Table 7.4.1 to determine the receptances at the free end of a clamped-free beam. Let's first consider a harmonic force applied in the positive y direction at the free end of the uniform beam of length L as shown in Fig. 7.4.1 and find $H_{11} = \frac{Y_1}{F_1}$ and $N_{11} = \frac{\Theta_1}{F_1}$. From Table 7.4.1, the boundary conditions at coordinate 2 (clamped), where $x = 0$, are $y = 0$ and $\frac{\partial y}{\partial x} = 0$. At coordinate 1

Table 7.4.1 Boundary conditions for Euler-Bernoulli beam receptance calculations [1]

End description	Boundary conditions
Clamped	$y = 0, \frac{\partial y}{\partial x} = 0$
Free	$\frac{\partial^2 y}{\partial x^2} = 0, \frac{\partial^3 y}{\partial x^3} = 0$
Pinned	$y = 0, \frac{\partial^2 y}{\partial x^2} = 0$
Sliding	$\frac{\partial y}{\partial x} = 0, \frac{\partial^3 y}{\partial x^3} = 0$
Harmonic force $F\sin(\omega t)$	$\frac{\partial^2 y}{\partial x^2} = 0, \frac{\partial^3 y}{\partial x^3} = -\frac{F}{EI}\sin(\omega t)$
Harmonic bending couple $M\sin(\omega t)$	$\frac{\partial^2 y}{\partial x^2} = \frac{M}{EI}\sin(\omega t), \frac{\partial^3 y}{\partial x^3} = 0$

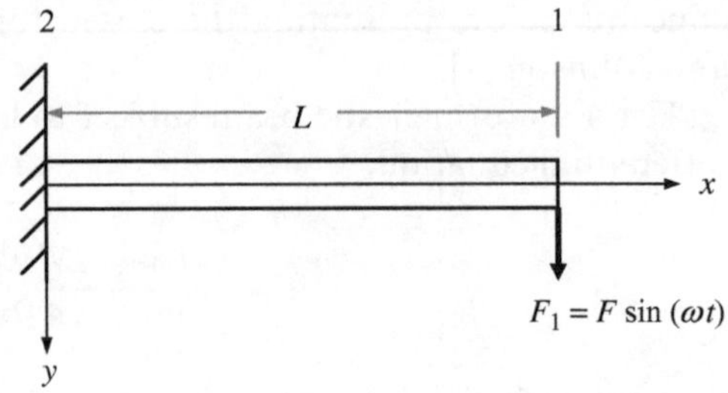

Fig. 7.4.1 Clamped-free uniform beam with a harmonic force applied at the free end

(free), where $x = L$, the boundary conditions are $\frac{\partial^2 y}{\partial x^2} = 0$ and $\frac{\partial^3 y}{\partial x^3} = -\frac{F}{EI}\sin(\omega t)$. Using these four end conditions let's determine the four coefficients A, B, C, and D in Eq. 7.4.3.

At $x = 0$, we have that $y = 0$ and $\frac{\partial y}{\partial x} = 0$. Therefore, we can use Eq. 7.4.3 directly and substitute $x = 0$ to obtain Eq. 7.4.4.

$$y|_{x=0} = Y|_{x=0} = A + C = 0 \tag{7.4.4}$$

This gives $A = -C$. Using the slope boundary condition at $x = 0$, we calculate the first derivative of Eq. 7.4.3 with respect to y and evaluate it at $x = 0$ to get Eq. 7.4.5, which yields the relationship $B = -D$.

$$\left.\frac{\partial y}{\partial x}\right|_{x=0} = \left.\frac{\partial Y}{\partial x}\right|_{x=0} = \lambda(B + D) = 0 \tag{7.4.5}$$

At $x = L$, we first apply $\frac{\partial^2 y}{\partial x^2} = 0$. See Eq. 7.4.6.

$$\left.\frac{\partial^2 y}{\partial x^2}\right|_{x=L} = \left.\frac{\partial^2 Y}{\partial x^2}\right|_{x=L} = \lambda^2(-A\cos(\lambda L) - B\sin(\lambda L) + C\cosh(\lambda L) + D\sinh(\lambda L)) = 0 \tag{7.4.6}$$

Substituting for A and B in Eq. 7.4.6, we obtain:

$$C(\cos(\lambda L) + \cosh(\lambda L)) + D(\sin(\lambda L) + \sinh(\lambda L)) = 0. \tag{7.4.7}$$

We next use $\frac{\partial^3 y}{\partial x^3} = -\frac{F}{EI}\sin(\omega t)$ at $x = L$ and again substitute for A and B to find:

$$C(-\sin(\lambda L) + \sinh(\lambda L)) + D(\cos(\lambda L) + \cosh(\lambda L)) = -\frac{F}{\lambda^3 EI}\sin(\omega t). \tag{7.4.8}$$

Let's now rewrite Eqs. 7.4.7 and 7.4.8 in matrix form and solve for the coefficients C and D. See Eq. 7.4.9.

$$\begin{bmatrix} \cos(\lambda L)+\cosh(\lambda L) & \sin(\lambda L)+\sinh(\lambda L) \\ -\sin(\lambda L)+\sinh(\lambda L) & \cos(\lambda L)+\cosh(\lambda L) \end{bmatrix} \begin{Bmatrix} C \\ D \end{Bmatrix} = \begin{Bmatrix} 0 \\ -\frac{F}{\lambda^3 EI} \end{Bmatrix} \sin(\omega t) \quad (7.4.9)$$

We can use Cramer's rule [19] to determine C and D. Writing Eq. 7.4.9 in the generic form $[A]\{x\}=\{B\}$, or $\begin{bmatrix} a_{11} & a_{12} \\ a_{21} & a_{22} \end{bmatrix}\begin{Bmatrix} x_1 \\ x_2 \end{Bmatrix}=\begin{Bmatrix} b_1 \\ b_2 \end{Bmatrix}$, we find x_1 by Eq. 7.4.10 and x_2 by 7.4.11.

$$x_1 = \frac{\begin{vmatrix} b_1 & a_{12} \\ b_2 & a_{22} \end{vmatrix}}{\begin{vmatrix} a_{11} & a_{12} \\ a_{21} & a_{22} \end{vmatrix}} = \frac{b_1 a_{22} - b_2 a_{12}}{a_{11}a_{22} - a_{21}a_{12}} \quad (7.4.10)$$

$$x_2 = \frac{\begin{vmatrix} a_{11} & b_1 \\ a_{21} & b_2 \end{vmatrix}}{\begin{vmatrix} a_{11} & a_{12} \\ a_{21} & a_{22} \end{vmatrix}} = \frac{a_{11} b_2 - a_{21} b_1}{a_{11}a_{22} - a_{21}a_{12}} \quad (7.4.11)$$

Substitution in Eq. 7.4.10 gives C, while Eq. 7.4.11 is used to find D. See Eqs. 7.4.12 and 7.4.13.

$$C = \frac{\begin{vmatrix} 0 & \sin(\lambda L)+\sinh(\lambda L) \\ -\frac{F}{\lambda^3 EI}\sin(\omega t) & \cos(\lambda L)+\cosh(\lambda L) \end{vmatrix}}{\begin{vmatrix} \cos(\lambda L)+\cosh(\lambda L) & \sin(\lambda L)+\sinh(\lambda L) \\ -\sin(\lambda L)+\sinh(\lambda L) & \cos(\lambda L)+\cosh(\lambda L) \end{vmatrix}} =$$

$$C = \frac{\frac{F}{\lambda^3 EI}(\sin(\lambda L)+\sinh(\lambda L))}{(\cos(\lambda L)+\cosh(\lambda L))^2 - (-\sin(\lambda L)+\sinh(\lambda L))(\sin(\lambda L)+\sinh(\lambda L))}\sin(\omega t) \quad (7.4.12)$$

$$C = \frac{F(\sin(\lambda L)+\sinh(\lambda L))}{2\lambda^3 EI(1+\cos(\lambda L)\cosh(\lambda L))}\sin(\omega t)$$

$$D = \frac{\begin{vmatrix} \cos(\lambda L)+\cosh(\lambda L) & 0 \\ -\sin(\lambda L)+\sinh(\lambda L) & -\frac{F}{\lambda^3 EI}\sin(\omega t) \end{vmatrix}}{\begin{vmatrix} \cos(\lambda L)+\cosh(\lambda L) & \sin(\lambda L)+\sinh(\lambda L) \\ -\sin(\lambda L)+\sinh(\lambda L) & \cos(\lambda L)+\cosh(\lambda L) \end{vmatrix}} =$$

$$D = \frac{-\frac{F}{\lambda^3 EI}(\cos(\lambda L)+\cosh(\lambda L))}{(\cos(\lambda L)+\cosh(\lambda L))^2 - (-\sin(\lambda L)+\sinh(\lambda L))(\sin(\lambda L)+\sinh(\lambda L))}\sin(\omega t) \quad (7.4.13)$$

$$D = -\frac{F(\cos(\lambda L)+\cosh(\lambda L))}{2\lambda^3 EI(1+\cos(\lambda L)\cosh(\lambda L))}\sin(\omega t)$$

We determine Y_1 by substituting the equations for C and D, together with the relationships $A = -C$ and $B = -D$, in Eq. 7.4.3. Because $x = L$ at coordinate 1, we also substitute L for x. The result is provided in Eq. 7.4.14.

$$
\begin{aligned}
Y_1 &= A\cos(\lambda L) + B\sin(\lambda L) + C\cosh(\lambda L) + D\sinh(\lambda L) \\
Y_1 &= -C\cos(\lambda L) - D\sin(\lambda L) + C\cosh(\lambda L) + D\sinh(\lambda L) \\
Y_1 &= C(-\cos(\lambda L) + \cosh(\lambda L)) + D(-\sin(\lambda L) + \sinh(\lambda L)) \\
Y_1 &= \begin{array}{l} \dfrac{F(\sin(\lambda L) + \sinh(\lambda L))}{2\lambda^3 EI(1 + \cos(\lambda L)\cosh(\lambda L))}\sin(\omega t)(-\cos(\lambda L) + \cosh(\lambda L)) - \\ \dfrac{F(\cos(\lambda L) + \cosh(\lambda L))}{2\lambda^3 EI(1 + \cos(\lambda L)\cosh(\lambda L))}\sin(\omega t)(-\sin(\lambda L) + \sinh(\lambda L)) \end{array} \\
Y_1 &= -\left(\begin{array}{l} \dfrac{(\sin(\lambda L) + \sinh(\lambda L))(\cos(\lambda L) - \cosh(\lambda L))}{2\lambda^3 EI(1 + \cos(\lambda L)\cosh(\lambda L))} - \\ \dfrac{(\cos(\lambda L) + \cosh(\lambda L))(\sin(\lambda L) - \sinh(\lambda L))}{2\lambda^3 EI(1 + \cos(\lambda L)\cosh(\lambda L))} \end{array} \right) F\sin(\omega t)
\end{aligned}
\tag{7.4.14}
$$

Finally, the displacement-to-force tip receptance at the free end of the beam is written as shown in Eq. 7.4.15.

$$
\begin{aligned}
H_{11} &= \frac{Y_1}{F_1} = \frac{-\left(\begin{array}{l} \dfrac{(\sin(\lambda L) + \sinh(\lambda L))(\cos(\lambda L) - \cosh(\lambda L))}{2\lambda^3 EI(1 + \cos(\lambda L)\cosh(\lambda L))} - \\ \dfrac{(\cos(\lambda L) + \cosh(\lambda L))(\sin(\lambda L) - \sinh(\lambda L))}{2\lambda^3 EI(1 + \cos(\lambda L)\cosh(\lambda L))} \end{array} \right) F\sin(\omega t)}{F\sin(\omega t)} \\
H_{11} &= -\left(\begin{array}{l} \dfrac{(\sin(\lambda L) + \sinh(\lambda L))(\cos(\lambda L) - \cosh(\lambda L))}{2\lambda^3 EI(1 + \cos(\lambda L)\cosh(\lambda L))} - \\ \dfrac{(\cos(\lambda L) + \cosh(\lambda L))(\sin(\lambda L) - \sinh(\lambda L))}{2\lambda^3 EI(1 + \cos(\lambda L)\cosh(\lambda L))} \end{array} \right) \\
H_{11} &= \frac{\sin(\lambda L)\cosh(\lambda L) - \cos(\lambda L)\sinh(\lambda L)}{\lambda^3 EI(1 + \cos(\lambda L)\cosh(\lambda L))}
\end{aligned}
\tag{7.4.15}
$$

To determine the rotation-to-force tip receptance, we return to Eq. 7.4.3 and substitute for A, B, C, and D.

$$
Y = (C(-\cos(\lambda x) + \cosh(\lambda x)) + D(-\sin(\lambda x) + \sinh(\lambda x))) \tag{7.4.16}
$$

We then obtain rotation by differentiating Y with respect to x, $\Theta = \frac{dY}{dx}$, and evaluate this expression at $x = L$. Finally, we divide this result by F_1 to find $N_{11} = \frac{\Theta_1}{F_1}$. See Eq. 7.4.17.

$$
N_{11} = \frac{\sin(\lambda L)\sinh(\lambda L)}{\lambda^2 EI(1 + \cos(\lambda L)\cosh(\lambda L))} \tag{7.4.17}
$$

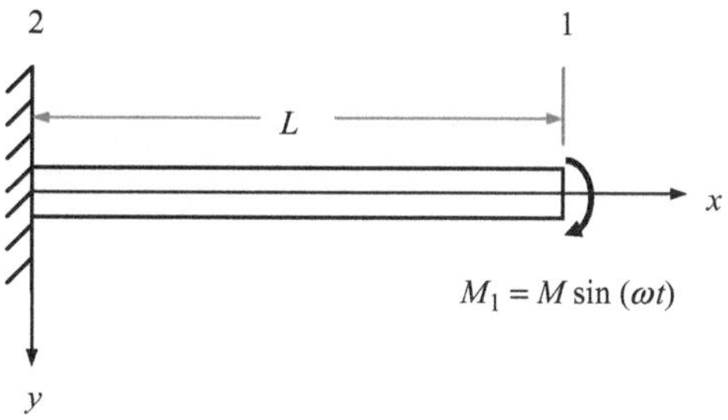

Fig. 7.4.2 Clamped-free uniform beam with a harmonic bending couple applied at the free end

The remaining tip receptances are $L_{11} = \frac{Y_1}{M_1}$ and $P_{11} = \frac{\Theta_1}{M_1}$. We find these terms by applying the harmonic bending couple $M_1 = M\sin(\omega t)$ at coordinate 1 as shown in Fig. 7.4.2. The boundary conditions at coordinate 2 ($x = 0$) are $y = 0$ and $\frac{\partial y}{\partial x} = 0$. The boundary conditions at coordinate 1 ($x = L$) are $\frac{\partial^2 y}{\partial x^2} = \frac{M}{EI}\sin(\omega t)$ and $\frac{\partial^3 y}{\partial x^3} = 0$. We find the coefficients A, B, C, and D from Eq. 7.4.3 in the same manner as described in the previous paragraphs.

At $x = 0$, the situation is identical to the force application case shown in Fig. 7.4.1 so we obtain $A = -C$ and $B = -D$. At $x = L$, we first use $\frac{\partial^2 y}{\partial x^2} = \frac{M}{\lambda^2 EI}$ as demonstrated in Eq. 7.4.18.

$$\left.\frac{\partial^2 y}{\partial x^2}\right|_{x=L} = \left.\frac{\partial^2 Y}{\partial x^2}\right|_{x=L} = \lambda^2 \begin{pmatrix} -A\cos(\lambda L) - B\sin(\lambda L) + \\ C\cosh(\lambda L) + D\sinh(\lambda L) \end{pmatrix} = \frac{M}{\lambda^2 EI}\sin(\omega t) \quad (7.4.18)$$

Substitution for A and B in Eq. 7.4.18 gives:

$$C(\cos(\lambda L) + \cosh(\lambda L)) + D(\sin(\lambda L) + \sinh(\lambda L)) = \frac{M}{\lambda^2 EI}\sin(\omega t). \quad (7.4.19)$$

We next apply $\frac{\partial^3 y}{\partial x^3} = 0$ (at $x = L$) and substitute for A and B to get:

$$C(-\sin(\lambda L) + \sinh(\lambda L)) + D(\cos(\lambda L) + \cosh(\lambda L)) = 0. \quad (7.4.20)$$

Expressing Eqs. 7.4.19 and 7.4.20 in matrix form yields:

$$\begin{bmatrix} \cos(\lambda L) + \cosh(\lambda L) & \sin(\lambda L) + \sinh(\lambda L) \\ -\sin(\lambda L) + \sinh(\lambda L) & \cos(\lambda L) + \cosh(\lambda L) \end{bmatrix} \begin{Bmatrix} C \\ D \end{Bmatrix} = \begin{Bmatrix} \frac{M}{\lambda^2 EI} \\ 0 \end{Bmatrix} \sin(\omega t). \quad (7.4.21)$$

Again applying Cramer's rule, we obtain equations for C and D.

$$C = \frac{M(\cos(\lambda L) + \cosh(\lambda L))}{2\lambda^2 EI(1 + \cos(\lambda L)\cosh(\lambda L))}\sin(\omega t) \tag{7.4.22}$$

$$D = -\frac{M(-\sin(\lambda L) + \sinh(\lambda L))}{2\lambda^2 EI(1 + \cos(\lambda L)\cosh(\lambda L))}\sin(\omega t) \tag{7.4.23}$$

We find Y_1 by substituting Eqs. 7.4.22 and 7.4.23, together with the relationships $A = -C$ and $B = -D$, in Eq. 7.4.3. We also set $x = L$. The result is given in Eq. 7.4.24.

$$Y_1 = -\left(\begin{array}{c}\frac{(\cos(\lambda L) + \cosh(\lambda L))(\cos(\lambda L) - \cosh(\lambda L))}{2\lambda^2 EI(1 + \cos(\lambda L)\cosh(\lambda L))} + \\ \frac{(\sin(\lambda L) - \sinh(\lambda L))(\sin(\lambda L) - \sinh(\lambda L))}{2\lambda^2 EI(1 + \cos(\lambda L)\cosh(\lambda L))}\end{array}\right) M\sin(\omega t) \tag{7.4.24}$$

We obtain the displacement-to-couple tip receptance at the free end of the beam by dividing Eq. 7.4.24 by M_1. A comparison of Eqs. 7.4.25 and 7.4.17 shows us that the displacement-to-couple and rotation-to-force receptances are identical.

$$L_{11} = \frac{Y_1}{M_1} = \frac{-\left(\begin{array}{c}\frac{(\cos(\lambda L) + \cosh(\lambda L))(\cos(\lambda L) - \cosh(\lambda L))}{2\lambda^2 EI(1 + \cos(\lambda L)\cosh(\lambda L))} + \\ \frac{(\sin(\lambda L) - \sinh(\lambda L))(\sin(\lambda L) - \sinh(\lambda L))}{2\lambda^2 EI(1 + \cos(\lambda L)\cosh(\lambda L))}\end{array}\right) M\sin(\omega t)}{M\sin(\omega t)}$$

$$L_{11} = -\left(\begin{array}{c}\frac{(\cos(\lambda L) + \cosh(\lambda L))(\cos(\lambda L) - \cosh(\lambda L))}{2\lambda^2 EI(1 + \cos(\lambda L)\cosh(\lambda L))} + \\ \frac{(\sin(\lambda L) - \sinh(\lambda L))(\sin(\lambda L) - \sinh(\lambda L))}{2\lambda^2 EI(1 + \cos(\lambda L)\cosh(\lambda L))}\end{array}\right) \tag{7.4.25}$$

$$L_{11} = \frac{\sin(\lambda L)\sinh(\lambda L)}{\lambda^2 EI(1 + \cos(\lambda L)\cosh(\lambda L))}$$

To determine the rotation-to-couple tip receptance, we return to Eq. 7.4.3 and substitute for A, B, C, and D (according to Eqs. 7.4.22 and 7.4.23).

$$Y = (C(-\cos(\lambda x) + \cosh(\lambda x)) + D(-\sin(\lambda x) + \sinh(\lambda x))) \tag{7.4.26}$$

We then find $P_{11} = \frac{\Theta_1}{M_1}$ by: 1) differentiating Y with respect to x to obtain rotation $\Theta = \frac{dY}{dx}$; 2) evaluating this expression at $x = L$; and 3) dividing this result by M_1. See Eq. 7.4.27.

$$P_{11} = \frac{\sin(\lambda L)\cosh(\lambda L) + \cos(\lambda L)\sinh(\lambda L)}{\lambda EI(1 + \cos(\lambda L)\cosh(\lambda L))} \tag{7.4.27}$$

This process can be repeated for any of the boundary conditions shown in Table 7.4.1. The tip receptance results for clamped-free and free-free conditions

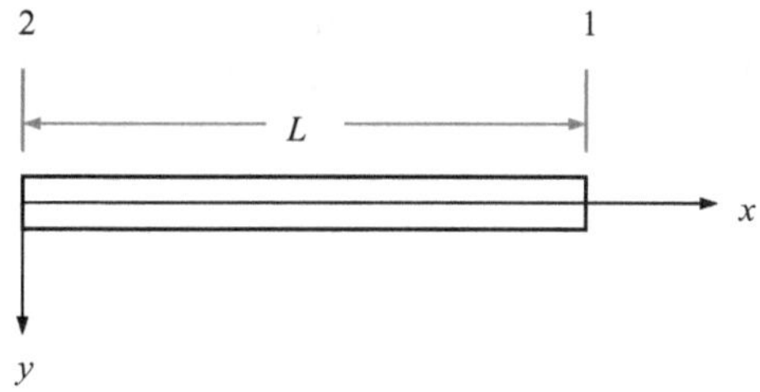

Fig. 7.4.3 Free-free uniform beam

Table 7.4.2 Euler-Bernoulli beam tip receptances for clamped-free and free-free boundary conditions [1]. Coordinates 1 and 2 are defined in Figs. 7.4.1 through 7.4.3. Also, $\lambda^4 = \omega^2 \frac{\rho A}{EI}$ and L is the beam length

H_{22}	N_{22} L_{22}	H_{12} H_{21}	N_{12} L_{21}	P_{22}	L_{12} N_{21}	P_{12} P_{21}	H_{11}	N_{11} L_{11}	P_{11}
Free-free									
$\frac{-c_1}{\lambda^3 c_7}$	$\frac{-c_2}{\lambda^2 c_7}$	$\frac{c_3}{\lambda^3 c_7}$	$\frac{c_4}{\lambda^2 c_7}$	$\frac{c_5}{\lambda c_7}$	$\frac{-c_4}{\lambda^2 c_7}$	$\frac{c_6}{\lambda c_7}$	$\frac{-c_1}{\lambda^3 c_7}$	$\frac{c_2}{\lambda^2 c_7}$	$\frac{c_5}{\lambda c_7}$
Clamped-free									
–	–	–	–	–	–	–	$\frac{-c_1}{\lambda^3 c_8}$	$\frac{c_2}{\lambda^2 c_8}$	$\frac{c_5}{\lambda c_8}$

Terms c_1 through c_8

$c_1 = \cos(\lambda L)\sinh(\lambda L) - \sin(\lambda L)\cosh(\lambda L)$

$c_2 = \sin(\lambda L)\sinh(\lambda L)$

$c_3 = \sin(\lambda L) - \sinh(\lambda L)$

$c_4 = \cos(\lambda L) - \cosh(\lambda L)$

$c_5 = \cos(\lambda L)\sinh(\lambda L) + \sin(\lambda L)\cosh(\lambda L)$

$c_6 = \sin(\lambda L) + \sinh(\lambda L)$

$c_7 = EI(\cos(\lambda L)\cosh(\lambda L) - 1)$

$c_8 = EI(\cos(\lambda L)\cosh(\lambda L) + 1)$

(see Fig. 7.4.3) are summarized in Table 7.4.2, where both direct and cross receptances are included for the free-free beam. No cross receptances are shown for the clamped-free beam because the response at the tip is zero for any excitation at the clamped end and the response is always zero at the clamped end. We should note that the free-free receptances also include the two (zero frequency) rigid body modes for the uniform beam. The first displacement-to-force rigid body mode, for example, represents the motion of the unsupported beam (imagine the beam floating in space) when a force is applied at its mass center and rigid body translation occurs. The second rigid body mode corresponds to a force applied at any other location, which causes rigid body rotation about the mass center.

An important omission for the receptance expressions provided in Table 7.4.2 is damping. Because we are considering solid beams under lateral vibration, the energy dissipation occurs within the beam only; there are no joints to introduce damping. As discussed in Section 2.1, internal damping is typically classified as solid, or structural, damping [20]. It is conveniently included in our harmonic vibration analysis by replacing EI in λ, c_7, and c_8 with the complex stiffness term $EI(1 + i\eta)$, where η is the solid damping factor. Representative values for selected engineering materials are provided in [21]. For tool-holder modeling, typical values for steel and sintered carbide components are in the 0.001 to 0.002 range.

While the closed form Euler-Bernoulli beam tip receptances provided in Table 7.4.2 are convenient to apply, accurate solutions are obtained only for beams which exhibit small cross sectional area to length ratios (i.e., long slender beams). An alternative for beams that do not meet this criterion is the Timoshenko beam model [22]. The corresponding differential equation is given by:

$$\left(\frac{\partial^2 y}{\partial t^2}+\frac{EI}{\rho A}\frac{\partial^4 y}{\partial x^4}\right)+\left(\frac{\rho I}{\hat{k}AG}\frac{\partial^4 y}{\partial t^4}+\frac{EI}{\hat{k}AG}\frac{\partial^4 y}{\partial x^2 \partial t^2}\right)-\left(\frac{I}{A}\frac{\partial^4 y}{\partial x^2 \partial t^2}\right)=0, \quad (7.4.28)$$

where $\hat{k}$ is a shape factor that depends on the beam cross section [23] and G is the shear modulus. Equation 7.4.28 is grouped into three sections (i.e., three parenthetical expressions). We see that the first section matches the Euler-Bernoulli beam equation provided in Eq. 7.4.1. The second and third sections account for shear deformations and rotary inertia, respectively. While these additional terms improve the model accuracy (particularly at higher frequencies), the tradeoff is that a closed form solution is unavailable. Finite element calculations may be applied, but at the expense of computation time. A description of the Timoshenko free-free beam receptances obtained from finite element calculations is given in [24].

7.5 Assembly Receptance Predictions

In Sections 7.3 and 7.4, we provided the building blocks for assembly receptance predictions. In this section, we'll detail coupling examples to demonstrate their implementation.

Example 7.5.1: Free-free beam coupled to rigid support As a test of the receptance coupling procedure, let's couple a free-free beam to a rigid support (i.e., a wall) to verify that it matches the clamped-free beam response we derived in Section 7.4. As described in Section 7.3, we have three primary tasks to complete in order to predict the assembly response. First, we must define the components and coordinates for the model. Here we have two components: a uniform beam with free-free boundary conditions and a rigid support (which exhibits zero receptances); see Fig. 7.5.1. Second, we need to determine the component receptances. We will apply the closed form receptances provided in Table 7.4.2. Third, based on the selected model, we express the assembly receptances as a function of the component receptances as shown in Table 7.3.1.

Let's define the free-free beam to be a solid steel cylinder with a diameter of 10 mm and a length of 125 mm. The elastic modulus is 200 GPa and the density is 7800 kg/m^3. We'll select the solid damping factor to be 0.01 for plotting purposes, but in practice a value near 0.001 would be more realistic. The free-free cylinder's direct and cross receptance equations are given in Table 7.4.2, while the wall receptances are zero. To calculate λ, we need the frequency vector ω (rad/s), cross sectional area, A, and second moment of area, I. We'll use a frequency range of 5000 Hz with a resolution of 0.1 Hz. The variables A and I are defined in Eqs. 7.5.1

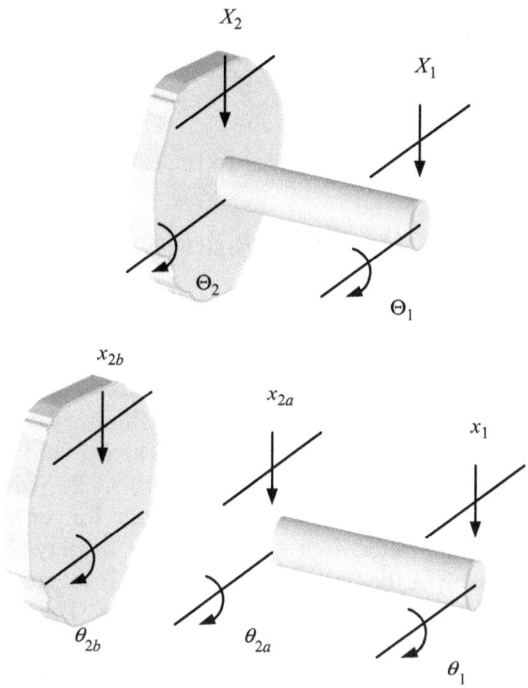

Fig. 7.5.1 Rigid coupling of the free-free cylinder to a wall

and 7.5.2 for the cylinder, where *d* is the cylinder diameter. The displacement-to-force free-free receptance for the cylinder, h_{11}, is shown in Fig. 7.5.2. We see a first bending natural frequency of 2884.9 Hz. The rigid body behavior is exhibited as the dramatic change in the real part as the frequency approaches zero.

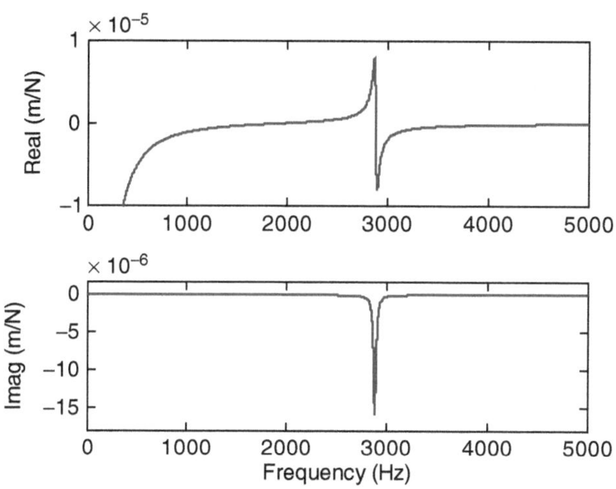

Fig. 7.5.2 Free-free receptance, h_{11}, for 10 mm diameter by 125 mm long steel cylinder

$$A = \frac{\pi d^2}{4} \quad (7.5.1)$$

$$I = \frac{\pi d^4}{64} \quad (7.5.2)$$

To rigidly couple the free-free cylinder to the wall, we apply Eq. 7.3.13:

$$G_{11} = \begin{bmatrix} H_{11} & L_{11} \\ N_{11} & P_{11} \end{bmatrix} = R_{11} - R_{12a}(R_{2a2a} + R_{2b2b})^{-1} R_{2a1},$$

where the generalized receptance matrices R_{11}, R_{12a}, R_{2a2a}, and R_{2a1} correspond to the cylinder and R_{2b2b} characterizes the wall response. The MATLAB® program p_7_5_1_1.m is used to complete the receptance coupling procedure. The results are displayed in Figs. 7.5.3 and 7.5.4. Figure 7.5.3 shows the H_{11} response from the $G_{11}(1,1)$ position (solid line). The dotted line in the figure is the clamped-free response, $H_{11} = \frac{-c_1}{\lambda^3 c_8}$, from Table 7.4.2. We see that the two curves are identical and the rigid body behavior is no longer present due to the coupling conditions. A limited frequency range is displayed in Fig. 7.5.3 to enable close comparison of the first bending mode. However, all bending modes are included in the Euler-Bernoulli beam receptances. The frequency range is increased in Fig. 7.5.4 to show the first two assembly bending modes. The vertical axis (response magnitude) is logarithmic in this plot because the second mode magnitude is much smaller than the first. Again, we observe exact agreement between the receptance coupling result (solid) and clamped-free receptance (dotted). An interesting aspect of Fig. 7.5.4 is that, in addition to the two resonant peaks at 453.4 Hz and 2841.4 Hz, we also see an "anti-resonance" at

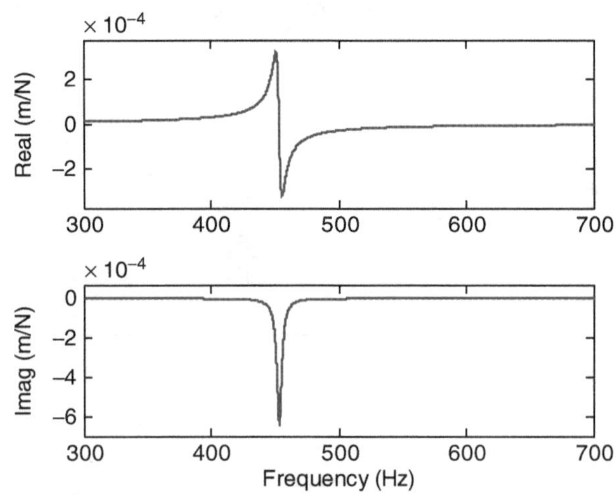

Fig. 7.5.3 Comparison of H_{11} receptance coupling result (solid line) and clamped-free response (dotted) for 10 mm diameter by 125 mm long steel cylinder

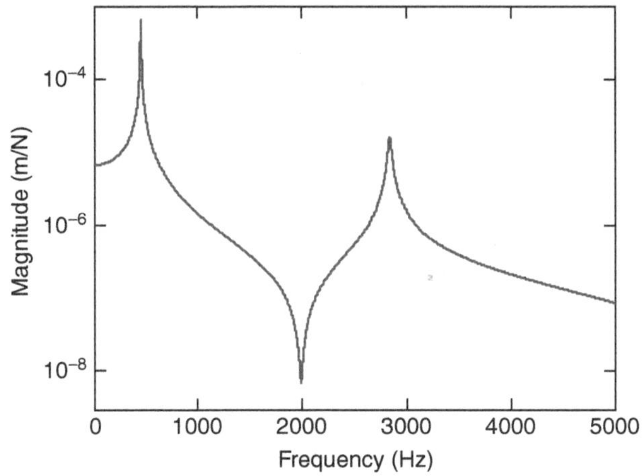

Fig. 7.5.4 Semi-logarithmic plot showing the first two bending modes for H_{11} tip receptances obtained from: 1) rigid coupling of free-free beam to wall (solid line); and 2) clamped-free response (dotted)

1988.1 Hz. At this frequency, the response is very small, even for large input force magnitudes.

The rotation-to-couple tip receptance determined from the rigid free-free beam coupling to the wall is also calculated in p_7_5_1_1.m. This $G_{11}(2,2)$ entry is shown in Fig. 7.5.5 (solid line). The clamped-free response (dotted line), $P_{11} = \frac{c_5}{\lambda c_8}$, again agrees with the receptance coupling result. We also see that the first mode natural frequency matches the H_{11} result

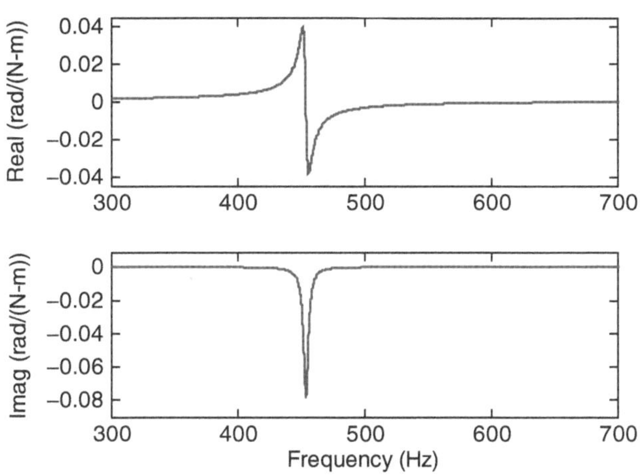

Fig. 7.5.5 Comparison of P_{11} receptance coupling result (solid line) and clamped-free response (dotted) for 10 mm diameter by 125 mm long steel cylinder

(453.4 Hz), but the magnitude is quite different; note the new units of rad/(N-m).

We've already noted that the assembly cross receptances, G_{12} and G_{21}, and the direct receptances at the clamped end, G_{22}, are zero. We can verify this by direct application of Eqs. 7.3.14, 7.3.16, and 7.3.17. For the clamped end direct receptance, Eq. 7.3.16 simplifies as shown in Eq. 7.5.3.

$$G_{22} = R_{2a2a} - R_{2a2a}(R_{2a2a} + R_{2b2b})^{-1} R_{2a2a}$$
$$G_{22} = R_{2a2a} - R_{2a2a}\left(R_{2a2a} + \begin{bmatrix} 0 & 0 \\ 0 & 0 \end{bmatrix}\right)^{-1} R_{2a2a} \quad (7.5.3)$$
$$G_{22} = R_{2a2a} - R_{2a2a}(R_{2a2a})^{-1} R_{2a2a} = R_{2a2a} - R_{2a2a} = 0$$

Similar results are obtained for the cross receptances in Eqs. 7.3.14 and 7.3.17 when substituting $R_{2b2b} = \begin{bmatrix} 0 & 0 \\ 0 & 0 \end{bmatrix}$.

Example 7.5.2: Free-free beam coupled to clamped-free beam Let's now consider the case depicted in Fig. 7.3.3. A 10 mm diameter by 100 mm long steel cylinder (free-free boundary conditions) is to be rigidly coupled to a clamped-free 50 mm by 50 mm by 200 mm long steel prismatic beam. The steel elastic modulus, density, and solid damping factor are 200 GPa, 7800 kg/m^3, and 0.01, respectively. (Again, we selected the solid damping value to be artificially high for display purposes.) The analysis is the same as Ex. 7.5.1 except that the R_{2b2b} receptances are no longer zero. They are now defined as shown in Table 7.4.2, $H_{11} = \frac{-c_1}{\lambda^3 c_8}$, $L_{11} = N_{11} = \frac{c_2}{\lambda^2 c_8}$, and $P_{11} = \frac{c_5}{\lambda c_8}$. We'll again use a frequency range of 5000 Hz with a resolution of 0.1 Hz to calculate λ. The variables A and I are defined in Eqs. 7.5.4 and 7.5.5 for the square prismatic beam, where s is the side length of 50 mm. The displacement-to-force free-free receptance for the cylinder, h_{11}, is shown in Fig. 7.5.6 (solid line). The clamped-free square beam tip receptance, h_{2b2b}, is also displayed (dotted line). We see a first bending natural frequency of 4507.6 Hz for the free-free beam. The clamped-free beam has a first bending frequency of 1022.5 Hz.

$$A = s^2 \quad (7.5.4)$$

$$I = \frac{s^4}{12} \quad (7.5.5)$$

The application of Eq. 7.3.13 to this scenario using program p_7_5_2_1.m gives Fig. 7.5.7, which shows H_{11} for the assembly. We see two modes within the 5000 Hz frequency range: one at 1045.2 Hz, near the original clamped-free response, and a second more flexible mode at 680.9 Hz due to the now coupled cylinder. Because the prismatic beam is much stiffer than the cylinder, it appears to serve as a nearly rigid support for the cylinder. This may lead us to believe that approximating the assembly as a cylinder clamped to a wall is adequate. However, let's investigate

Fig. 7.5.6 Free-free receptance, h_{11}, for 10 mm diameter by 100 mm long steel cylinder (solid line) and clamped-free receptance, h_{2b2b}, for 50 mm square by 200 mm long steel prismatic beam (dotted line)

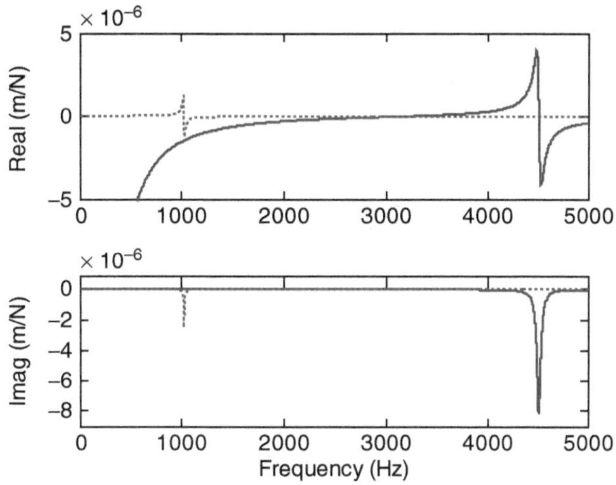

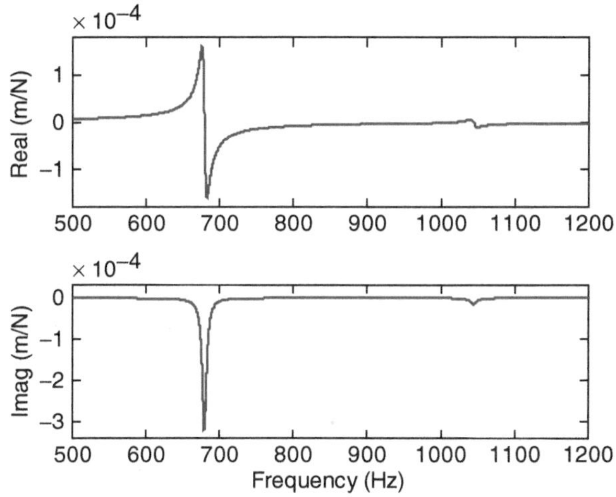

Fig. 7.5.7 Assembly displacement-to-force tip receptance H_{11} for the rigidly coupled cylinder and prismatic beam shown in Fig. 7.3.3

what happens if we modify the prismatic beam to reduce its first bending frequency to a value near the clamped-free cylinder's first bending frequency.

Figure 7.5.8 displays h_{11} for the free-free cylinder (solid line), as well as h_{2b2b} for a longer clamped-free prismatic beam (dotted line). The cylinder's first bending natural frequency remains at 4507.6 Hz for the free-free boundary conditions. However, the first bending frequency for the extended clamped-free beam is reduced to 654.4 Hz. Figure 7.5.9 shows H_{11} for the cylinder rigidly coupled to a 50 mm square by 250 mm long prismatic beam. The response is now quite different than the assembly receptance shown in Fig. 7.5.7 for the 200 mm long prismatic beam. Even though the cylinder is coupled to a more flexible base (i.e., a longer clamped-free

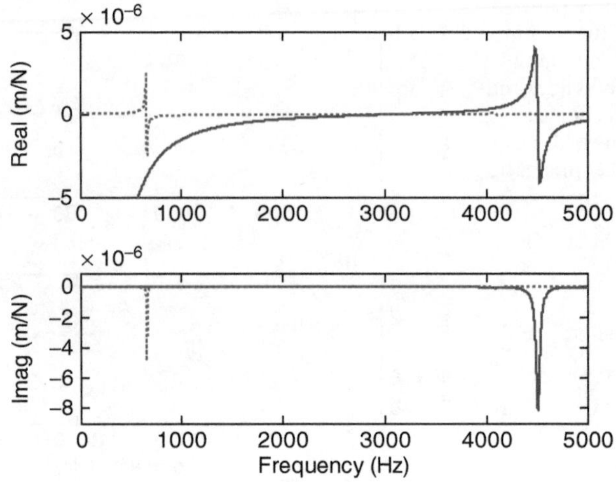

Fig. 7.5.8 Free-free receptance, h_{11}, for 10 mm diameter by 100 mm long steel cylinder (solid line) and clamped-free receptance, h_{2b2b}, for 50 mm square by 250 mm long steel prismatic beam (dotted line)

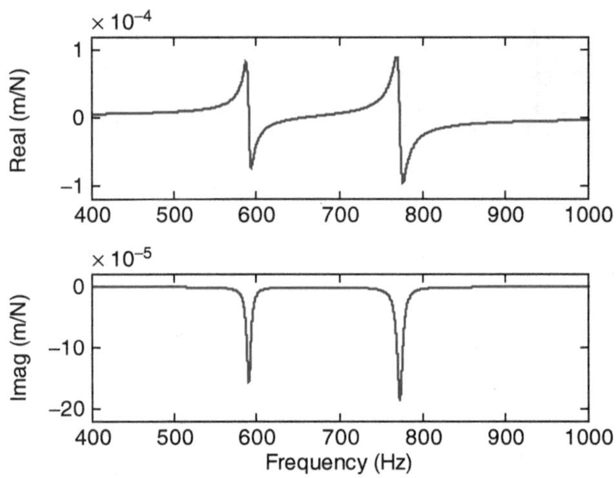

Fig. 7.5.9 The displacement-to-force tip receptance, H_{11}, for rigid coupling of the 10 mm diameter by 100 mm long cylinder to the 50 mm square by 250 mm long prismatic beam is displayed

beam), the assembly response has a smaller peak magnitude. The minimum imaginary value for the new assembly is -1.865×10^{-4} m/N, while the corresponding value for the shorter (and stiffer) prismatic beam assembly is -3.222×10^{-4} m/N; this represents a 42% compliance[10] reduction. The compliance reduction, or

[10] Compliance is the inverse of stiffness.

equivalently the stiffness increase, is due to interaction between the two beams in a manner analogous to the well known dynamic absorber. When the clamped-free prismatic beam's natural frequency is near the coupled cylinder's natural frequency, some energy is able to "pass through" the cylinder and excite the stiffer base. The result is that the energy is more equally partitioned between the two modes and the assembly response appears stiffer [8]. An electrical equivalent is the impedance matching strategy used at cable connections. For example, it is common to use 50 Ω terminations at all connections to encourage signal transmission and avoid reflection.

One application of this phenomenon is to select tool lengths that encourage the interaction between the clamped-free tool's first bending frequency and one of the spindle natural frequencies (in bending). For slender tools, as the tool overhang length from the holder face is adjusted, its natural frequency can be modified to match a spindle natural frequency. This technique, referred to as "tool length tuning" or simply "tool tuning" [8, 25, 26], can lead to improved dynamic stiffness (i.e., a smaller magnitude for the tool point FRF) and increased allowable axial depths of cut. As we saw with the prismatic beam-cylinder coupling, the surprising outcome is that increasing the tool length can reduce the assembly compliance in some instances. We should also note that forcing this interaction and encouraging the two mode response shown in Fig. 7.5.9 will generally lead to competing lobes (Section 3.4) in the corresponding stability lobe diagram.

7.6 Tool-Holder-Spindle-Machine Receptance Predictions[11]

As we noted in Section 7.2, we can consider the tool-holder-spindle-machine combination as being composed of three parts: the tool, holder, and spindle-machine [7]. This enables us to use the closed form Euler-Bernoulli beam receptances contained in Table 7.4.2 to describe the tool and holder dynamics; finite element based Timoshenko beam receptances may also be applied, of course. Due to modeling challenges, we measure the spindle-machine response using impact testing techniques. Given the component responses, we then couple their individual FRFs to obtain the assembly response via the receptance coupling substructure analysis (RCSA) approach [7]. Let's now discuss this procedure in more detail.

Figure 7.6.1 depicts our model composed of the three individual components: I – tool, II – holder, and III – spindle-machine[12]. So far we have not discussed a three component model. However, by sequentially coupling the components we can limit the analysis to the two component case. For example, we can first couple the free-free holder and tool to form the substructure I-II identified in Fig. 7.6.2. To carry out this step, we begin by defining the components and coordinates as

[11] Author T. Schmitz recognizes the significant contributions of G.S. Duncan, Valparaiso University, to this section.

[12] US Patent application 20070088456.

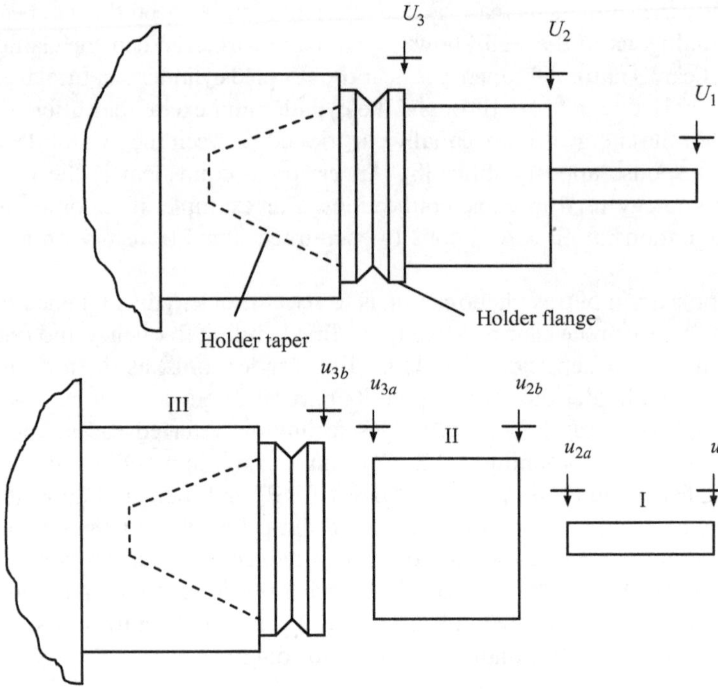

Fig. 7.6.1 Three component receptance coupling model of tool (I), holder (II), and spindle-machine (III)

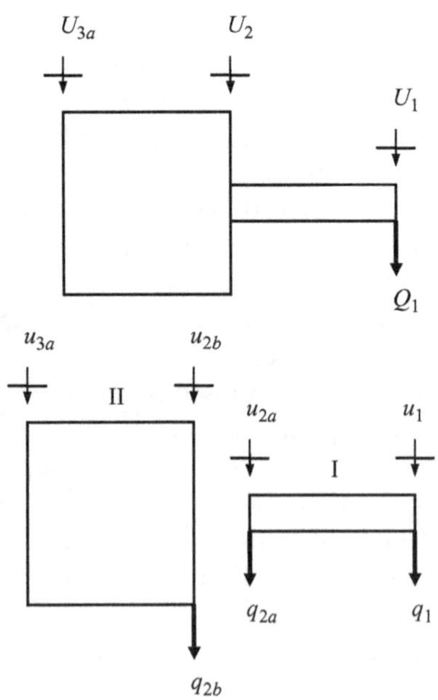

Fig. 7.6.2 (Sub)assembly I-II composed of tool (I) and holder (II). The generalized force Q_1 is applied to U_1 to determine G_{11} and G_{3a1}

displayed in Fig. 7.6.2. We then calculate the component receptances. If we apply the Euler-Bernoulli receptances described by Bishop and Johnson [1], the appropriate equations are provided in Table 7.4.2. For I, we have the free-free receptances: $h_{11} = \frac{-c_1}{\lambda^3 c_7}$, $l_{11} = n_{11} = \frac{c_2}{\lambda^2 c_7}$, $p_{11} = \frac{c_5}{\lambda c_7}$, $h_{2a2a} = \frac{-c_1}{\lambda^3 c_7}$, $l_{2a2a} = n_{2a2a} = \frac{-c_2}{\lambda^2 c_7}$, $p_{2a2a} = \frac{c_5}{\lambda c_7}$, $h_{12a} = h_{2a1} = \frac{c_3}{\lambda^3 c_7}$, $l_{12a} = n_{2a1} = \frac{-c_4}{\lambda^2 c_7}$, $l_{2a1} = n_{12a} = \frac{c_4}{\lambda^2 c_7}$, and $p_{12a} = p_{2a1} = \frac{c_6}{\lambda c_7}$, where $\lambda^4 = \omega^2 \frac{\rho A}{EI(1+i\eta)}$, and L, ρ, A, E, I, and η depend on the tool geometry and material properties. For II, we simply replace coordinate 1 with $2b$ and coordinate $2a$ with $3a$ in the previous equations. Additionally, we must use the holder geometry and material properties to define λ, L, ρ, A, E, I, and η. We will assume a rigid coupling between these two components and follow the approaches described previously to determine the I-II (sub)assembly tip receptances: (direct) G_{11} and G_{3a3a}; and (cross) G_{13a} and G_{3a1}.

To find G_{11} and G_{3a1}, we apply Q_1 to coordinate U_1 as shown in Fig. 7.6.2. The components' displacements/rotations are: $u_1 = R_{11}q_1 + R_{12a}q_{2a}$, $u_{2a} = R_{2a1}q_1 + R_{2a2a}q_{2a}$, $u_{2b} = R_{2b2b}q_{2b}$, and $u_{3a} = R_{3a2b}q_{2b}$. The equilibrium conditions are: $q_{2a} + q_{2b} = 0$ and $q_1 = Q_1$. We substitute the component displacements/rotations and equilibrium conditions into the compatibility condition, $u_{2b} - u_{2a} = 0$, to obtain the expression for q_{2b} shown in Eq. 7.6.1. The component force q_{2a} is then determined from the equilibrium condition $q_{2a} = -q_{2b}$. The expression for G_{11} is provided in Eq. 7.6.2. The cross receptance matrix G_{3a1} is shown in Eq. 7.6.3.

$$\begin{aligned} &u_{2b} - u_{2a} = 0 \\ &R_{2b2b}q_{2b} - R_{2a1}q_1 - R_{2a2a}q_{2a} = 0 \\ &(R_{2a2a} + R_{2b2b})q_{2b} - R_{2a1}Q_1 = 0 \\ &q_{2b} = (R_{2a2a} + R_{2b2b})^{-1}R_{2a1}Q_1 \end{aligned} \tag{7.6.1}$$

$$\begin{aligned} G_{11} &= \frac{U_1}{Q_1} = \frac{u_1}{Q_1} = \frac{R_{11}q_1 + R_{12a}q_{2a}}{Q_1} = \frac{R_{11}Q_1 - R_{12a}(R_{2a2a}+R_{2b2b})^{-1}R_{2a1}Q_1}{Q_1} \\ G_{11} &= R_{11} - R_{12a}(R_{2a2a} + R_{2b2b})^{-1}R_{2a1} = \begin{bmatrix} H_{11} & L_{11} \\ N_{11} & P_{11} \end{bmatrix} \end{aligned} \tag{7.6.2}$$

$$\begin{aligned} G_{3a1} &= \frac{U_{3a}}{Q_1} = \frac{u_{3a}}{Q_1} = \frac{R_{3a2b}q_{2b}}{Q_1} = \frac{R_{3a2b}(R_{2a2a}+R_{2b2b})^{-1}R_{2a1}Q_1}{Q_1} \\ G_{3a1} &= R_{3a2b}(R_{2a2a} + R_{2b2b})^{-1}R_{2a1} = \begin{bmatrix} H_{3a1} & L_{3a1} \\ N_{3a1} & P_{3a1} \end{bmatrix} \end{aligned} \tag{7.6.3}$$

We find the remaining tip receptances G_{3a3a} and G_{13a} by applying Q_{3a} to coordinate U_{3a} as shown in Fig. 7.6.3. The components displacements/rotations are: $u_1 = R_{12a}q_{2a}$, $u_{2a} = R_{2a2a}q_{2a}$, $u_{2b} = R_{2b2b}q_{2b} + R_{2b3a}q_{3a}$, and $u_{3a} = R_{3a2b}q_{2b} + R_{3a3a}q_{3a}$. The equilibrium conditions are: $q_{2a} + q_{2b} = 0$ and $q_{3a} = Q_{3a}$. In the same manner as before, we substitute the component displacements/rotations and equilibrium conditions into the compatibility condition, $u_{2a} - u_{2b} = 0$, to determine q_{2a}; see Eq. 7.6.4. The component force q_{2b} is found

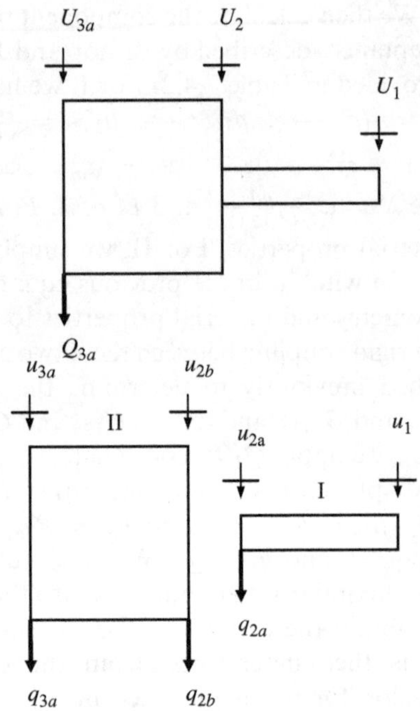

Fig. 7.6.3 (Sub)assembly I-II composed of tool (I) and holder (II). The generalized force Q_{3a} is applied to U_{3a} to determine G_{3a3a} and G_{13a}

from $q_{2b} = -q_{2a}$. The equation for the direct receptance G_{3a3a} is shown in Eq. 7.6.5, while the cross receptance G_{13a} is provided in Eq. 7.6.6.

$$\begin{aligned} & u_{2a} - u_{2b} = 0 \\ & R_{2a2a}q_{2a} - R_{2b2b}q_{2b} - R_{2b3a}q_{3a} = 0 \\ & (R_{2a2a} + R_{2b2b})q_{2a} - R_{2b3a}Q_{3a} = 0 \\ & q_{2a} = (R_{2a2a} + R_{2b2b})^{-1}R_{2b3a}Q_{3a} \end{aligned} \tag{7.6.4}$$

$$\begin{aligned} G_{3a3a} &= \frac{U_{3a}}{Q_{3a}} = \frac{u_{3a}}{Q_{3a}} = \frac{R_{3a3a}q_{3a} + R_{3a2b}q_{2b}}{Q_{3a}} = \frac{R_{3a3a}Q_{3a} - R_{3a2b}(R_{2a2a} + R_{2b2b})^{-1}R_{2b3a}Q_{3a}}{Q_{3a}} \\ G_{3a3a} &= R_{3a3a} - R_{3a2b}(R_{2a2a} + R_{2b2b})^{-1}R_{2b3a} = \begin{bmatrix} H_{3a3a} & L_{3a3a} \\ N_{3a3a} & P_{3a3a} \end{bmatrix} \end{aligned} \tag{7.6.5}$$

$$\begin{aligned} G_{13a} &= \frac{U_1}{Q_{3a}} = \frac{u_1}{Q_{3a}} = \frac{R_{12a}q_{2a}}{Q_{3a}} = \frac{R_{12a}(R_{2a2a} + R_{2b2b})^{-1}R_{2b3a}Q_{3a}}{Q_{3a}} \\ G_{13a} &= R_{12a}(R_{2a2a} + R_{2b2b})^{-1}R_{2b3a} = \begin{bmatrix} H_{13a} & L_{13a} \\ N_{13a} & P_{13a} \end{bmatrix} \end{aligned} \tag{7.6.6}$$

Now that we've coupled components I and II to form the (sub)assembly I-II, we can rigidly couple this result to the spindle-machine. See Fig. 7.6.4, which is

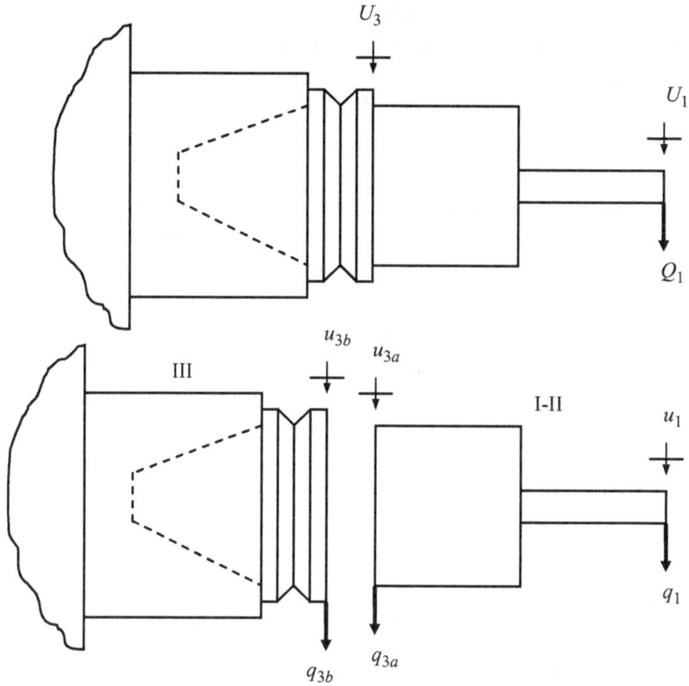

Fig. 7.6.4 The I-II (sub)assembly is rigidly coupled to the spindle-machine (III) to determine the tool point receptances, G_{11}

very similar to Fig. 7.3.3. To apply Eq. 7.3.13, we must make minor modifications to the subscripts to match our new coordinates. Relative to the original equation, we see that we must replace coordinate $2a$ with $3a$ and coordinate $2b$ with $3b$. Therefore, the tool point receptances are defined by:

$$G_{11} = R_{11} - R_{13a}(R_{3a3a} + R_{3b3b})^{-1}R_{3a1}, \tag{7.6.7}$$

where the R_{ij} matrices are the (sub)assembly matrices from the I-II coupling result. Therefore, we have: $R_{11} = G_{11}$ from Eq. 7.6.2, $R_{3a1} = G_{3a1}$ from Eq. 7.6.3, $R_{3a3a} = G_{3a3a}$ from Eq. 7.6.5, and $R_{13a} = G_{13a}$ from Eq. 7.6.6. We still have one matrix in Eq. 7.6.7 that is unknown, however. We do yet not know the spindle-machine receptances R_{3b3b}.

7.6.1 Spindle-Machine Receptances

Before discussing R_{3b3b} in more detail, let's take another look at the model in Fig. 7.6.1. We notice that the section of the tool holder beyond the flange has been artificially separated from the rest of the holder, which includes the portion that is inserted and clamped in the spindle. There are two reasons for this approach. First, for a given spindle, all holders inserted in that spindle will typically have the same flange geometry. In general, it is only the portion of the holder beyond the flange that varies from one holder to the next; this is necessary for automatic tool changes. Second, the portion of the spindle that we need to excite and measure is the spindle shaft itself. Unfortunately, typical spindle designs do not give us access to the spindle shaft. It is located within the spindle housing and interfaces with the holder-spindle coupling mechanism (such as HSK, CAT, or other). Therefore, it is convenient to consider the holder flange and taper to be part of the spindle itself. This also locates potential flexibility in the holder-spindle coupling within the measured spindle receptances so that they do not have to be separately modeled.

Given our discussion of impact testing in Section 2.6, we see that the direct FRF h_{3b3b} would be straightforward to obtain. We'd simply need to modify a holder to remove the portion beyond the flange, place this artifact in the spindle, and then excite spindle-machine at the free end of the holder (on the flange), while measuring the response at the same location. However, as we've discussed, we also require the direct displacement-to-couple, rotation-to-force, and rotation-to-couple FRFs to fully populate the R_{3b3b} matrix. These additional receptances are not so easy to obtain experimentally. Exciting the system with an impulsive couple is particularly challenging. Therefore, we can consider a different approach.

Rather than using a measurement artifact that includes only the flange and taper, let's select an artifact that incorporates some length beyond the flange as shown in Fig. 7.6.5. If we can determine the assembly matrix $G_{22} = \begin{bmatrix} H_{22} & L_{22} \\ N_{22} & P_{22} \end{bmatrix}$ experimentally, then we can use this information, together with a model of the portion of the artifact beyond the flange, to determine R_{3b3b}. By replacing coordinate 1 with 2, coordinate $2a$ with $3a$, and coordinate $2b$ with $3b$ in Eq. 7.3.13, we obtain the free end response for the artifact-spindle-machine assembly:

$$G_{22} = R_{22} - R_{23a}(R_{3a3a} + R_{3b3b})^{-1} R_{3a2}. \tag{7.6.8}$$

We can rearrange Eq. 7.6.8 to isolate R_{3b3b}. See Eq. 7.6.9, where the R_{ij} matrices are obtained from a model of the free-free portion of the artifact beyond the flange (component II in Fig. 7.6.5) and G_{22} is determined from measurements. We can describe this decomposition process of identifying the substructure receptances, R_{3b3b}, from the measured assembly receptances, G_{22}, and modeled substructure receptances, R_{3a2}, R_{22}, R_{23a} , and R_{3a3a}, as "inverse RCSA".

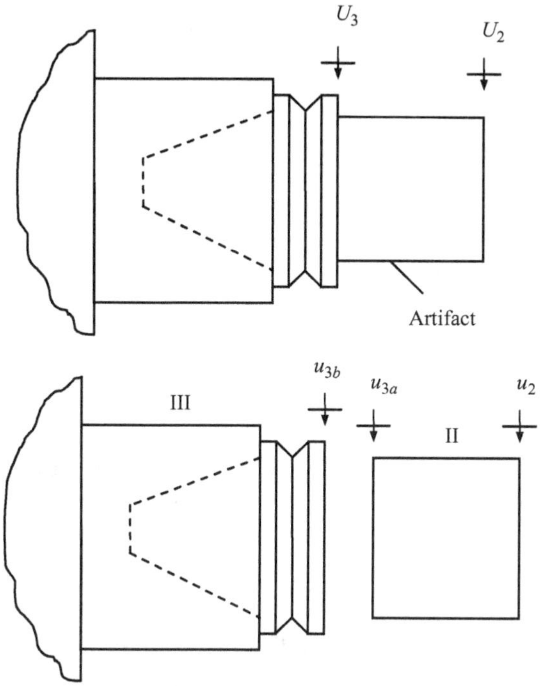

Fig. 7.6.5 Artifact model for G_{22} measurement; these receptances are subsequently used to determine R_{3b3b} by inverse RCSA (Eq. 7.6.9)

$$
\begin{aligned}
G_{22} - R_{22} &= -R_{23a}(R_{3a3a} + R_{3b3b})^{-1}R_{3a2} \\
R_{23a}^{-1}(R_{22} - G_{22})R_{3a2}^{-1} &= (R_{3a3a} + R_{3b3b})^{-1} \\
R_{3a2}(R_{22} - G_{22})^{-1}R_{23a} &= R_{3a3a} + R_{3b3b} \\
R_{3b3b} &= R_{3a2}(R_{22} - G_{22})^{-1}R_{23a} - R_{3a3a}
\end{aligned}
\quad (7.6.9)
$$

Our only remaining task is to define the G_{22} receptances. The displacement-to-force term $H_{22} = \frac{X_2}{F_2}$ is straightforward to obtain. We simply excite the assembly at coordinate 2, typically via an impact hammer, and record the response at the same location using, for example, an accelerometer, laser vibrometer, or capacitance probe. To find the rotation-to-force receptance $N_{22} = \frac{\Theta_2}{F_2}$, we can implement a first order finite difference approach [27]. By measuring both the direct FRF H_{22} and cross FRF $H_{2a2} = \frac{X_{2a}}{F_2}$, we can compute N_{22} according to Eq. 7.6.10. The displacement-to-force cross FRF H_{2a2} is obtained by exciting the assembly at U_2 and measuring the response at coordinate U_{2a}, located a distance S from the artifact's free end, as shown in Fig. 7.6.6.

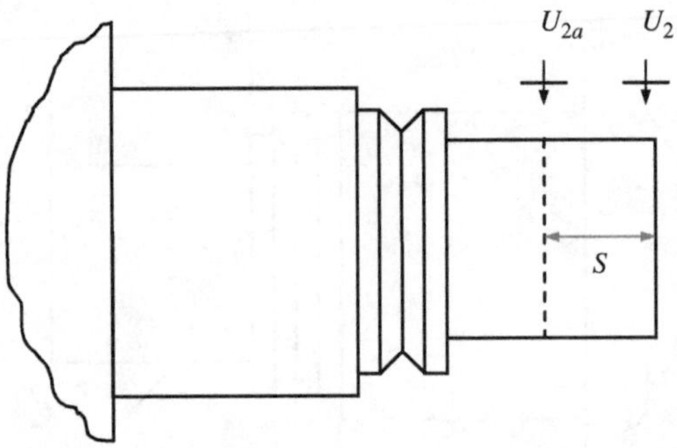

Fig. 7.6.6 Locations for direct and cross artifact-spindle-machine assembly measurements

Equivalently, we could measure H_{22a}, where the linear transducer is placed at U_2 and the force is applied at U_{2a}.

$$N_{22} = \frac{H_{22} - H_{2a2}}{S} = \frac{H_{22} - H_{22a}}{S} \tag{7.6.10}$$

We can assume reciprocity to establish the equality: $L_{22} = N_{22}$. (We see this same behavior for the free-free beam receptances in Table 7.4.2.) We cannot rely on reciprocity or finite difference computations to find P_{22}. However, using the other three artifact-spindle-machine receptances, we can synthesize P_{22} [28]. See Eq. 7.6.11. Given P_{22}, we have now fully populated G_{22} and we can use Eq. 7.6.9 to obtain R_{3b3b}.

$$P_{22} = \frac{\Theta_2}{M_2} = \frac{F_2}{X_2}\frac{X_2}{M_2}\frac{\Theta_2}{F_2} = \frac{1}{H_{22}}L_{22}N_{22} = \frac{N_{22}^2}{H_{22}} \tag{7.6.11}$$

7.6.2 Summary

Let's conclude the chapter by summarizing the RCSA steps and discussing implementation considerations. Given the three component model shown in Fig. 7.6.1, our initial task is to identify the substructure receptances. We find the tool (I) and holder (II) free-free receptances from models. These may be based on Euler-Bernoulli beam theory or finite element computations, for example. We use artifact measurements and inverse RCSA to determine the spindle-machine (III) receptances. Once we have the substructure dynamics defined, we sequentially couple the three component receptance matrices to obtain the tool point FRF, H_{11}, that is required for the stability and surface location error analyses described in Chapters 4, 5, and 6. A benefit of this technique is that

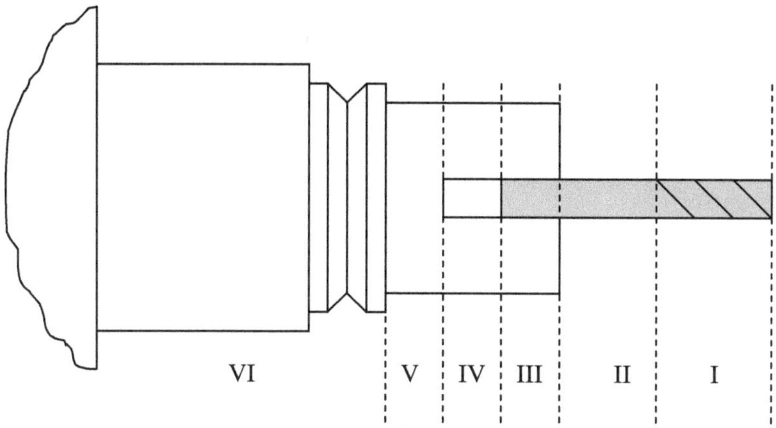

Fig. 7.6.7 Component definitions to account for changes in cross sectional dimensions and the portion of the tool inserted in the holder

once the spindle-machine receptances have been identified, any tool-holder combination can be coupled to the selected spindle without the need for additional measurements.

A final point for consideration is the tool and holder models. One issue is that a portion of the tool is inserted in the holder; see component III in Fig. 7.6.7. This gives the potential for cross sections with different material properties between a steel holder and carbide tool, for example. Equivalent structural rigidity, EI_{eq}, and mass per unit length, ρA_{eq}, values can be calculated as shown in Eqs. 7.6.12 and 7.6.13, where the h and t subscripts indicate the holder and tool, respectively. These may then be substituted for the EI and ρA products in the Euler-Bernoulli beam receptance equations defined in Table 7.4.2.

$$EI_{eq} = E_h I_h + E_t I_t \tag{7.6.12}$$

$$\rho A_{eq} = \rho_h A_h + \rho_t A_t \tag{7.6.13}$$

A second issue for the Euler-Bernoulli receptances is that constant cross sectional dimensions are required. Changes in cross section can be accommodated by defining a new component for each constant cross section portion. Components I through V are defined in Fig. 7.6.7 to represent the required tool and holder substructures. Here we see that a separate component, I, was defined for the fluted portion of the tool. This is necessary because the actual 2nd moment of area, I, and area, A, for the helical flutes differ from the cylindrical shank I and A values. One approach is to define an equivalent diameter that is then used in the I and A calculations. See [29, 30], for example.

IN A NUTSHELL If measurement of the tool tip FRF is convenient for the machine tool user, then the information required for pre-process predictions of machining performance can be directly obtained. However, if FRF measurements are difficult for the user to complete, then receptance coupling offers an alternative.

While receptance coupling requires accurate measurements of the spindle-machine substructure, these measurements certainly fall within the domain of expert consultants. If the spindle-machine response is archived and the computations are embedded in software, then the power of a machine tool user to predict the assembly FRF for a new unmeasured tool is formidable. The predicted FRF can be used to compute stability lobes and surface location error and enable the end user significant advantage because he/she will know which cuts are acceptable and which are not for the selected tool-holder-spindle-machine combination.

The previous scenario highlights the thrust of this book. It is possible to use the techniques described here to predict machining performance with sufficient accuracy that the trial and error process development approach so prevalent in machining operations today can be rendered obsolete.

Exercises

1. Determine the direct frequency response function, $\frac{X_2}{F_2}$, for the two degree of freedom system shown in Fig. e.7.1 using receptance coupling. Express your final result as a function of m, c, k, and the excitation frequency, ω. You may assume a harmonic forcing function, F_2, is applied to coordinate X_2.

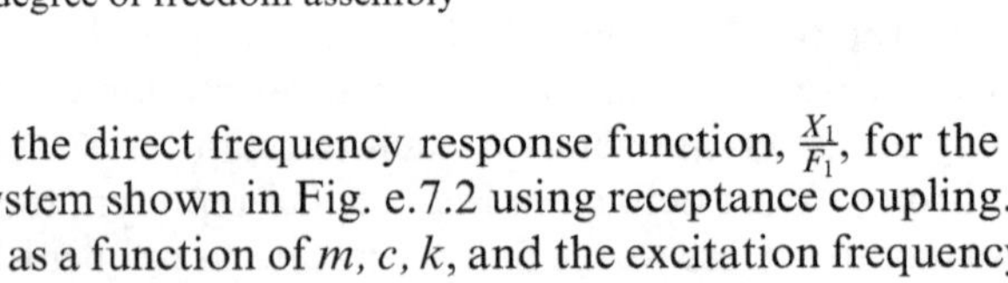

Fig. e.7.1 Two degree of freedom assembly

2. Determine the direct frequency response function, $\frac{X_1}{F_1}$, for the two degree of freedom system shown in Fig. e.7.2 using receptance coupling. Express your final result as a function of m, c, k, and the excitation frequency, ω. You may assume a harmonic forcing function, F_1, is applied to coordinate X_1.
3. Plot the displacement-to-force tip receptance for a sintered carbide cylinder with free-free boundary conditions. The beam is described by the following parameters: 19 mm diameter, 150 mm length, 550 GPa elastic modulus,

and 15000 kg/m^3 density. Assume a solid damping factor of 0.002. Select a frequency range that encompasses the first three bending modes and display your results as magnitude (m/N) vs. frequency (Hz) in a semilog format.

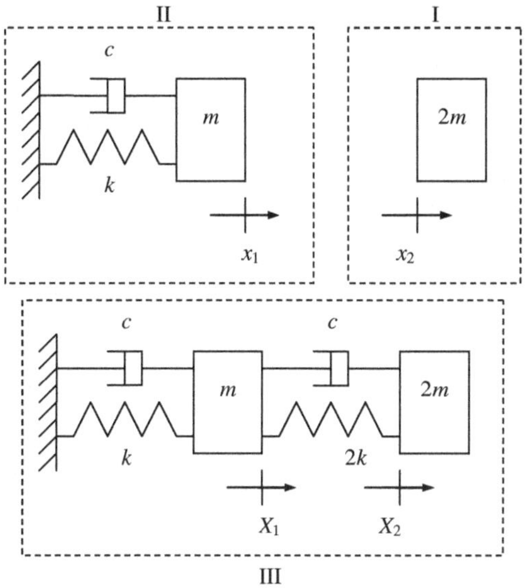

Fig. e.7.2 Flexible damped coupling of mass (I) to spring-mass-damper (II) to form the two degree of freedom assembly (III)

4. Use receptance coupling to rigidly join two free-free beams and find the free-free assembly's displacement-to-force tip receptance. Both steel cylinders are described by the following parameters: 12.7 mm diameter, 100 mm length, 200 GPa elastic modulus, and 7800 kg/m^3 density. Assume a solid damping factor of 0.0015. Once you have determined the assembly response, verify your result against the displacement-to-force tip receptance for a 12.7 mm diameter, 200 mm long free-free steel cylinder with the same material properties. Select a frequency range that encompasses the first three bending modes and display your results as the real (m/N) and imaginary (m/N) parts vs. frequency (Hz) using a linear scale.

References

1. Bishop, R. and Johnson, D., 1960, The Mechanics of Vibration, Cambridge University Press, Cambridge.
2. Jetmundsen, B., Bielawa, R., and Flannelly, W., 1988, Generalized Frequency-Domain Substructure Synthesis, Journal of the American Helicopter Society, 33: 55–64.
3. Schmitz, T. and Donaldson, R., 2000, Predicting High-Speed Machining Dynamics by Substructure Analysis, Annals of the CIRP, 49/1: 303–308.

4. Schmitz, T., Davies, M., and Kennedy, M., 2001, Tool Point Frequency Response Prediction for High-Speed Machining by RCSA, Journal of Manufacturing Science and Engineering, 123: 700–707.
5. Schmitz, T., Davies, M., Medicus, K., and Snyder, J., 2001, Improving High-Speed Machining Material Removal Rates by Rapid Dynamic Analysis, Annals of the CIRP, 50/1: 263–268.
6. Burns, T., Schmitz, T., 2004, Receptance Coupling Study of Tool-Length Dependent Dynamic Absorber Effect, Proceedings of American Society of Mechanical Engineers International Mechanical Engineering Congress and Exposition, IMECE2004-60081, Anaheim, CA.
7. Schmitz, T., and Duncan, G.S., 2005, Three-Component Receptance Coupling Substructure Analysis for Tool Point Dynamics Prediction, Journal of Manufacturing Science and Engineering, 127/4: 781–790.
8. Duncan, G.S., Tummond, M., and Schmitz, T., 2005, An Investigation of the Dynamic Absorber Effect in High-Speed Machining, International Journal of Machine Tools and Manufacture, 45: 497–507.
9. Cheng, C.-H., Schmitz, T., Arakere, N., and Duncan, G.S., 2005, An Approach for Micro End mill Frequency Response Predictions, Proceedings of American Society of Mechanical Engineers International Mechanical Engineering Congress and Exposition, IMECE2005-81215, Orlando, FL.
10. Burns, T. and Schmitz, T., 2005, A Study of Linear Joint and Tool Models in Spindle-Holder-Tool Receptance Coupling, Proceedings of 2005 American Society of Mechanical Engineers International Design Engineering Technical Conferences and Computers and Information in Engineering Conference, DETC2005-85275, Long Beach, CA.
11. Schmitz, T., Powell, K., Won, D., Duncan, G.S., Sawyer, W.G., Ziegert, J., 2007, Shrink Fit Tool Holder Connection Stiffness/Damping Modeling for Frequency Response Prediction in Milling, International Journal of Machine Tools and Manufacture, 47/9: 1368–1380.
12. Cheng, C.-H., Duncan, G.S., and Schmitz, T., 2007, Rotating Tool Point Frequency Response Prediction using RCSA, Machining Science and Technology, 11/3: 433–446.
13. Ewins, D., 1986, Analysis of Modified or Coupled Structures using FRF Properties, Report No. 86002, Imperial College London, Dynamics Section, Mechanical Engineering, London, England.
14. Ferreira, J. and Ewins, D., 1995, Nonlinear Receptance Coupling Approach Based on Describing Functions, Proceedings of the 14th International Modal Analysis Conference, Dearborn, MI, 1034–1040.
15. Lui, W. and Ewins, D., 2002, Substructure Synthesis via Elastic Media, Journal of Sound and Vibration, 257/2: 361-379.
16. Schmitz, T., Davies, M., and Kennedy, M., 2000, High-Speed Machining Frequency Response Prediction for Process Optimization, Proceedings of the 2nd International Seminar on Improving Machine Tool Performance, La Baule, France.
17. Park, S., Altintas, Y., and Movahhedy, M., 2003, Receptance Coupling for End Mills, International Journal of Machine Tools and Manufacture, 43: 889–896.
18. Thomson, W. and Dahleh, M., 1998, Theory of Vibration with Application, 5th Ed., Prentice Hall, Upper Saddle River, NJ, Section 9.5.
19. Chapra, S. and Canale, R., 1985, Numerical Methods for Engineers with Personal Computer Applications, McGraw-Hill, Inc., New York, NY, Section 7.1.
20. Thomson, W. and Dahleh, M., 1998, Theory of Vibration with Application, 5th Ed., Prentice Hall, Upper Saddle River, NJ, Section 3.9.
21. Beards, C., 1996, Structural Vibration: Analysis and Damping, Arnold, London, Section 2.2.5.
22. Weaver, W., Jr., Timoshenko, S., and Young, D., 1990, Vibration Problems in Engineering, 5th Ed., John Wiley and Sons, New York, NY, Section 5.12.

23. Hutchinson, J., 2001, Shear Coefficients for Timoshenko Beam Theory, Journal of Applied Mechanics, 68: 87–92.
24. Schmitz, T., and Duncan, G.S., 2005, Three-Component Receptance Coupling Substructure Analysis for Tool Point Dynamics Prediction, Journal of Manufacturing Science and Engineering, 127/4: 781-790, Appendix A: Beam Receptance Modeling.
25. Tlusty, J., Smith, S., and Winfough, W.R., 1996, Techniques for the Use of Long Slender End Mills in High-speed Milling, Annals of the CIRP, 45/1: 393–396.
26. Davies, M., Dutterer, B., Pratt, J., and Schaut, A., 1998, On the Dynamics of High-Speed Milling with Long, Slender Endmills, Annals of the CIRP, 47/1: 55–60.
27. Sattinger, S., 1980, A Method for Experimentally Determining Rotational Mobilities of Structures, Shock and Vibration Bulletin, 50: 17–27.
28. Ewins, D., 2000, Modal Testing: Theory, Practice and Application, 2nd Ed., Research Studies Press, Philadelphia, PA.
29. Kops, L. and Vo, D., 1990, Determination of the Equivalent Diameter of an End Mill Based on its Compliance, Annals of the CIRP, 39/1: 93–96.
30. Kivanc, E. and Budak, E., 2004, Structural Modeling of End Mills for Form Error and Stability Analysis, International Journal of Machine Tools and Manufacture, 44: 1151–1161.

Appendix A

Orthogonality of Eigenvectors

As described in Chapter 2, the orthogonality of eigenvectors with respect to the system mass and stiffness matrices is the basis for modal analysis. In general, we can say that two vectors are perpendicular if their scalar, or dot, product is zero. Consider the two vectors:

$$[U] = \begin{Bmatrix} u_{11} \\ u_{21} \end{Bmatrix} \quad \text{and} \quad [V] = \begin{Bmatrix} v_{11} \\ v_{21} \end{Bmatrix}. \tag{A1}$$

Their dot product is:

$$[U] \bullet [V] = [U]^T[V] = \{\, u_{11} \quad u_{21} \,\}\begin{Bmatrix} v_{11} \\ v_{21} \end{Bmatrix} = u_{11} \cdot v_{11} + u_{21} \cdot v_{21}. \tag{A2}$$

This product is zero if the vectors are perpendicular. Orthogonality can be considered a generalization of the concept of perpendicularity.

From Chapter 2, we have seen that we can write the matrix form of the system equations of motion $\left([M]s^2 + [K]\right)\{X\}e^{st} = \{0\}$ if we assume harmonic vibration. We used the characteristic equation, $\left|[M]s^2 + [K]\right| = 0$, to find the eigenvalues, s_1^2 and s_2^2. We then substituted the eigenvalues into either of the linearly dependent equations of motion to find the eigenvectors, or mode shapes. Using ${s_1}^2 = -{\omega_{n1}}^2$, we can write:

$$\left(-[M]{\omega_{n1}}^2 + [K]\right)\{\psi_1\} = \{0\}, \tag{A3}$$

where ψ_1 is the corresponding mode shape. Equation A3 can be expanded to:

$$-{\omega_{n1}}^2[M]\{\psi_1\} + [K]\{\psi_1\} = \{0\}. \tag{A4}$$

Premultiplying Eq. A4 by the transpose of the second mode shape ψ_2, which corresponds to vibration at ω_{n2}, yields:

$$-\omega_{n1}^2\{\psi_2\}^T[M]\{\psi_1\} + \{\psi_2\}^T[K]\{\psi_1\} = 0. \tag{A5}$$

Performing the transpose operation on Eq. A5 gives:

$$-\omega_{n1}^2\{\psi_1\}^T[M]\{\psi_2\} + \{\psi_1\}^T[K]\{\psi_2\} = 0, \tag{A6}$$

where the transpose properties $([A][B])^T = [B]^T[A]^T$ and $\left([A]^T\right)^T = [A]$ (using matrices of appropriate dimensions) have been applied.

Completing the same operations using $s_2^2 = -\omega_{n2}^2$ gives:

$$-\omega_{n2}^2\{\psi_1\}^T[M]\{\psi_2\} + \{\psi_1\}^T[K]\{\psi_2\} = 0. \tag{A7}$$

Taking the difference of Eqs. A6 and A7 yields:

$$\left(\omega_{n2}^2 - \omega_{n1}^2\right)\{\psi_1\}^T[M]\{\psi_2\} = 0. \tag{A8}$$

Provided $\omega_{n2}^2 \neq \omega_{n1}^2$, then $\{\psi_1\}^T[M]\{\psi_2\} = 0$. Substituting this result into either Eq. A6 or Eq. A7 gives $\{\psi_1\}^T[K]\{\psi_2\} = 0$. Collecting these results, we obtain the orthogonality conditions shown in Eqs. A9 through A12.

$$\begin{aligned} \{\psi_1\}^T[M]\{\psi_2\} &= 0 \\ \{\psi_2\}^T[M]\{\psi_1\} &= 0 \end{aligned} \tag{A9}$$

$$\begin{aligned} \{\psi_1\}^T[M]\{\psi_1\} &= m_{q1} \\ \{\psi_2\}^T[M]\{\psi_2\} &= m_{q2} \end{aligned} \text{(These products are not necessarily zero.)} \tag{A10}$$

$$\begin{aligned} \{\psi_1\}^T[K]\{\psi_2\} &= 0 \\ \{\psi_2\}^T[K]\{\psi_1\} &= 0 \end{aligned} \tag{A11}$$

$$\begin{aligned} \{\psi_1\}^T[K]\{\psi_1\} &= k_{q1} \\ \{\psi_2\}^T[K]\{\psi_2\} &= k_{q2} \end{aligned} \text{(These products are not necessarily zero.)} \tag{A12}$$

Using the modal matrix, $[P] = [\psi_1, \psi_2]$, and the orthogonality conditions we obtain the diagonalized modal mass and stiffness matrices:

$$[P]^T[M][P] = \begin{bmatrix} \{\psi_1\}^T[M]\{\psi_1\} & \{\psi_1\}^T[M]\{\psi_2\} \\ \{\psi_2\}^T[M]\{\psi_1\} & \{\psi_2\}^T[M]\{\psi_2\} \end{bmatrix} = \begin{bmatrix} m_{q1} & 0 \\ 0 & m_{q2} \end{bmatrix} = \left[M_q\right] \text{and} \tag{A13}$$

$$[P]^T[K][P] = \begin{bmatrix} \{\psi_1\}^T[K]\{\psi_1\} & \{\psi_1\}^T[K]\{\psi_2\} \\ \{\psi_2\}^T[K]\{\psi_1\} & \{\psi_2\}^T[K]\{\psi_2\} \end{bmatrix} = \begin{bmatrix} k_{q1} & 0 \\ 0 & k_{q2} \end{bmatrix} = [K_q]. \quad \text{(A14)}$$

These diagonal modal mass and stiffness matrices uncouple the equations of motion and enable the solution of independent single degree of freedom systems in modal coordinates. The individual modal contributions can then be transformed back into local (physical) coordinates as discussed in Chapter 2.

Appendix B

Reformulation of Fourier Series Eigenvalue Problem

In Section 4.3, we detailed the truncated Fourier series approach to obtaining the analytical stability limit for milling [1]. The dynamic milling equation (Eq. 4.3.27) was presented as:

$$\begin{pmatrix} F_x \\ F_y \end{pmatrix} e^{i\omega_c t} = \frac{1}{2} bK_t[A_0]\left(1 - e^{i\omega_c \tau}\right)\begin{bmatrix} FRF_{xx} & 0 \\ 0 & FRF_{yy} \end{bmatrix}\begin{pmatrix} F_x \\ F_y \end{pmatrix} e^{i\omega_c t}, \quad \text{(B1)}$$

which is true if $\begin{pmatrix} F_x \\ F_y \end{pmatrix} e^{i\omega_c t} = \begin{bmatrix} 1 & 0 \\ 0 & 1 \end{bmatrix}\begin{pmatrix} F_x \\ F_y \end{pmatrix} e^{i\omega_c t} = [I]\begin{pmatrix} F_x \\ F_y \end{pmatrix} e^{i\omega_c t}$. This enables us to write $\frac{1}{2} bK_t[A_0]\left(1 - e^{i\omega_c \tau}\right)\begin{bmatrix} FRF_{xx} & 0 \\ 0 & FRF_{yy} \end{bmatrix} = [I]$ or $\frac{N_t}{4\pi} bK_t\left(1 - e^{-i\omega_c \tau}\right)[FRF_{or}] = [I]$. The eigenvalue problem was then posed in [1] as:

$$\det([I] + \Lambda[FRF_{or}]) = 0, \quad \text{(B2)}$$

where $\Lambda = -\frac{N_t}{4\pi} bK_t\left(1 - e^{-i\omega_c \tau}\right)$ gives the eigenvalues. To apply the MATLAB® `eig` function, we must restate the eigenvalue problem as $\det([FRF_{or}] - \lambda[I]) = 0$. The new complex eigenvalues are therefore $\lambda = \lambda_{\mathrm{Re}} + i\lambda_{\mathrm{Im}} = \frac{4\pi}{N_t} \frac{1}{bK_t(1 - e^{-i\omega_c \tau})}$. We determine the corresponding stability limit by solving this expression for b, rationalizing the result, and then substituting for $e^{-i\omega_c \tau}$ using the Euler identity $e^{-i\omega_c \tau} = \cos(\omega_c \tau) - i\sin(\omega_c \tau)$.

$$\begin{aligned} \tilde{b}_{\mathrm{lim}} &= \frac{4\pi}{N_t K_t(\lambda_{\mathrm{Re}} + i\lambda_{\mathrm{Im}})(1 - e^{-i\omega_c \tau})} \\ \tilde{b}_{\mathrm{lim}} &= \frac{4\pi(\lambda_{\mathrm{Re}} - i\lambda_{\mathrm{Im}})}{N_t K_t(\lambda_{\mathrm{Re}} + i\lambda_{\mathrm{Im}})(\lambda_{\mathrm{Re}} - i\lambda_{\mathrm{Im}})(1 - e^{-i\omega_c \tau})} \\ \tilde{b}_{\mathrm{lim}} &= \frac{4\pi(\lambda_{\mathrm{Re}} - i\lambda_{\mathrm{Im}})}{N_t K_t\left(\lambda_{\mathrm{Re}}^2 + \lambda_{\mathrm{Im}}^2\right)(1 - \cos(\omega_c \tau) + i\sin(\omega_c \tau))} \end{aligned} \quad \text{(B3)}$$

After the Euler identity substitution, we again rationalize to obtain Eq. B4.

$$\tilde{b}_{\text{lim}} = \frac{4\pi}{N_t K_t \left(\lambda_{\text{Re}}{}^2 + \lambda_{\text{Im}}^2\right)} \frac{(\lambda_{\text{Re}} - i\lambda_{\text{Im}})(1 + \cos(\omega_c \tau) - i\sin(\omega_c \tau))}{(1 - \cos(\omega_c \tau) + i\sin(\omega_c \tau))(1 - \cos(\omega_c \tau) - i\sin(\omega_c \tau))}$$

$$\tilde{b}_{\text{lim}} = \frac{4\pi}{N_t K_t \left(\lambda_{\text{Re}}{}^2 + \lambda_{\text{Im}}^2\right)} \frac{(\lambda_{\text{Re}} - i\lambda_{\text{Im}})(1 + \cos(\omega_c \tau) - i\sin(\omega_c \tau))}{(2 - 2\cos(\omega_c \tau))} \tag{B4}$$

$$\tilde{b}_{\text{lim}} = \frac{2\pi}{N_t K_t \left(\lambda^2 + \lambda_{\text{Im}}{}^2\right)} \frac{\begin{pmatrix} (\lambda_{\text{Re}}(1 - \cos(\omega_c \tau)) + \lambda_{\text{Im}} \sin(\omega_c \tau)) + \\ i(\lambda_{\text{Im}}(1 - \cos(\omega_c \tau)) - \lambda_{\text{Re}} \sin(\omega_c \tau)) \end{pmatrix}}{(1 - \cos(\omega_c \tau))}$$

Because $\tilde{b}_{\text{lim}}$ must be real valued, the imaginary part from the parenthetical portion of the numerator in the final line in Eq. B4 must be equal to zero:

$$\lambda_{\text{Im}}(1 - \cos(\omega_c \tau)) - \lambda_{\text{Re}} \sin(\omega_c \tau) = 0. \tag{B5}$$

This gives $\frac{\lambda_{\text{Im}}}{\lambda_{\text{Re}}} = \frac{\sin(\omega_c \tau)}{1 - \cos(\omega_c \tau)} = \tilde{\kappa}$. Substitution in Eq. B4 yields Eq. B6.

$$\tilde{b}_{\text{lim}} = \frac{2\pi}{N_t K_t \left(\lambda_{\text{Re}}{}^2 + \lambda_{\text{Im}}{}^2\right)} \left(\frac{\lambda_{\text{Re}}(1 - \cos(\omega_c \tau))}{(1 - \cos(\omega_c \tau))} + \frac{\lambda_{\text{Im}} \sin(\omega_c \tau)}{(1 - \cos(\omega_c \tau))} \right)$$

$$\tilde{b}_{\text{lim}} = \frac{2\pi}{N_t K_t \left(\lambda_{\text{Re}}{}^2 + \lambda_{\text{Im}}{}^2\right)} (\lambda_{\text{Re}} + \lambda_{\text{Im}} \tilde{\kappa}) \tag{B6}$$

$$\tilde{b}_{\text{lim}} = \frac{2\pi}{N_t K_t \left(\lambda_{\text{Re}}{}^2 + \lambda_{\text{Im}}{}^2\right)} \lambda_{\text{Re}} \left(1 + \frac{\lambda_{\text{Im}}}{\lambda_{\text{Re}}} \tilde{\kappa}\right) = \frac{2\pi}{N_t K_t \left(\lambda_{\text{Re}}{}^2 + \lambda_{\text{Im}}{}^2\right)} \lambda_{\text{Re}} \left(1 + \tilde{\kappa}^2\right)$$

Appendix B references

1. Altintas, Y. and Budak, E., 1995, Analytical Prediction of Stability Lobes in Milling, Annals of the CIRP, 44/1: 357-362.

Appendix C

Fourier Force Series Coefficients

The Fourier coefficients, a_n and b_n, for the y-direction force series:

$$F_y(\phi) = \sum_{j=1}^{A}\sum_{i=1}^{N_t}\left(a_0 + \sum_{n=1}^{\infty}\left(a_n\cos(n\phi_i) + b_n\sin(n\phi_i)\right)\right),$$

where $\phi_i = \omega t + \frac{2\pi}{N_t}(i-1) - \chi(j-1)$, are provided here [1]. The terms in Eqs. C1 through C6 were determined using Eqs. 5.2.5 and 5.2.6. The integration limits for down milling (φ_1 to π) are shown. For up milling, the limits are modified to be zero to φ_1.

$$a_1 = -\frac{bN_t}{\pi}\left[\begin{array}{c} k_tc(-\frac{1}{4}\sin\phi + \frac{1}{12}\sin 3\phi) + k_nc(-\frac{1}{4}\cos\phi - \frac{1}{12}\cos 3\phi) + \\ k_{te}(\frac{1}{4}\cos 2\phi) + k_{ne}(\frac{1}{2}\phi + \frac{1}{4}\sin 2\phi) \end{array}\right]_{\phi_1}^{\pi} \tag{C1}$$

$$a_2 = -\frac{bN_t}{\pi}\left[\begin{array}{c} k_tc(\frac{1}{4}\phi - \frac{1}{4}\sin 2\phi + \frac{1}{16}\sin 4\phi) + k_nc(-\frac{1}{16}\cos 4\phi) + \\ k_{te}(-\frac{1}{2}\cos\phi + \frac{1}{6}\cos 3\phi) + k_{ne}(\frac{1}{2}\sin\phi + \frac{1}{6}\sin 3\phi) \end{array}\right]_{\phi_1}^{\pi} \tag{C2}$$

$$a_n = -\frac{bN_t}{\pi}\left[\begin{array}{c} k_tc\left(-\frac{1}{2n}\sin n\phi + \frac{1}{4(n-2)}\sin(n-2)\phi + \frac{1}{4(n+2)}\sin(n+2)\phi\right) + \\ k_nc\left(\frac{1}{4(n-2)}\cos(n-2)\phi - \frac{1}{4(n+2)}\cos(n+2)\phi\right) + \\ k_{te}\left(-\frac{1}{2(n-1)}\cos(n-1)\phi + \frac{1}{2(n+1)}\cos(n+1)\phi\right) + \\ k_{ne}\left(\frac{1}{2(n-1)}\sin(n-1)\phi + \frac{1}{2(n+1)}\sin(n+1)\phi\right) \end{array}\right]_{\phi_1}^{\pi},$$

$n = 3, 4, \ldots$

(C3)

$$b_1 = -\frac{bN_t}{\pi}\left[\begin{array}{c} k_tc(\frac{3}{4}\cos\phi - \frac{1}{12}\cos 3\phi) + k_nc(\frac{1}{4}\sin\phi - \frac{1}{12}\sin 3\phi) + \\ k_{te}(-\frac{1}{2}\phi + \frac{1}{4}\sin 2\phi) + k_{ne}(-\frac{1}{4}\cos 2\phi) \end{array}\right]_{\phi_1}^{\pi} \tag{C4}$$

$$b_2 = -\frac{bN_t}{\pi}\left[\begin{array}{c} k_t c\left(\frac{1}{4}\cos 2\phi - \frac{1}{16}\cos 4\phi\right) + k_n c\left(\frac{1}{4}\phi - \frac{1}{16}\sin 4\phi\right) + \\ k_{te}\left(-\frac{1}{2}\sin\phi + \frac{1}{6}\sin 3\phi\right) + k_{ne}\left(-\frac{1}{2}\cos\phi - \frac{1}{6}\cos 3\phi\right) \end{array}\right]_{\phi_1}^{\pi} \quad \text{(C5)}$$

$$b_n = -\frac{bN_t}{\pi}\left[\begin{array}{c} k_t c\left(\frac{1}{2n}\cos n\phi - \frac{1}{4(n-2)}\cos(n-2)\phi - \frac{1}{4(n+2)}\cos(n+2)\phi\right) + \\ k_n c\left(\frac{1}{4(n-2)}\sin(n-2)\phi - \frac{1}{4(n+2)}\sin(n+2)\phi\right) + \\ k_{te} c\left(-\frac{1}{2(n-1)}\sin(n-1)\phi + \frac{1}{2(n+1)}\sin(n+1)\phi\right) + \\ k_{ne} c\left(-\frac{1}{2(n-1)}\cos(n-1)\phi - \frac{1}{2(n+1)}\cos(n+1)\phi\right) \end{array}\right]_{\phi_1}^{\pi},$$

$$n = 3, 4, \ldots \quad \text{(C6)}$$

Appendix C references

1. Schmitz, T. and Mann, B., 2006, Closed Form Solutions for Surface Location Error in Milling, International Journal of Machine Tools and Manufacture, 46: 1369-1377.

Index

Springer

Kolibri 360 GmbH
Ferdinand-Jühlke-Straße 7
99095 Erfurt, Deutschland
produktsicherheit@kolibri360.de